L'Organisation de Coopération et de Développement Économiques (OCDE), qui a été instituée par une Convention signée le 14 décembre 1960, à Paris, a pour objectif de promouvoir des politiques visant :
- à réaliser la plus forte expansion possible de l'économie et de l'emploi et une progression du niveau de vie dans les pays Membres, tout en maintenant la stabilité financière, et contribuer ainsi au développement de l'économie mondiale ;
- à contribuer à une saine expansion économique dans les pays Membres, ainsi que non membres, en voie de développement économique ;
- à contribuer à l'expansion du commerce mondial sur une base multilatérale et non discriminatoire, conformément aux obligations internationales.

Les Membres de l'OCDE sont : la République Fédérale d'Allemagne, l'Australie, l'Autriche, la Belgique, le Canada, le Danemark, l'Espagne, les États-Unis, la Finlande, la France, la Grèce, l'Irlande, l'Islande, l'Italie, le Japon, le Luxembourg, la Norvège, la Nouvelle-Zélande, les Pays-Bas, le Portugal, le Royaume-Uni, la Suède, la Suisse et la Turquie.

L'Agence de l'OCDE pour l'Énergie Nucléaire (AEN) a été créée le 20 avril 1972, en remplacement de l'Agence Européenne pour l'Énergie Nucléaire de l'OCDE (ENEA) lors de l'adhésion du Japon à titre de Membre de plein exercice.

L'AEN groupe désormais tous les pays Membres européens de l'OCDE ainsi que l'Australie, le Canada, les États-Unis et le Japon. La Commission des Communautés Européennes participe à ses travaux.

L'AEN a pour principaux objectifs de promouvoir, entre les gouvernements qui en sont Membres, la coopération dans le domaine de la sécurité et de la réglementation nucléaires, ainsi que l'évaluation de la contribution de l'énergie nucléaire au progrès économique.

Pour atteindre ces objectifs, l'AEN :
- *encourage l'harmonisation des politiques et pratiques réglementaires dans le domaine nucléaire, en ce qui concerne notamment la sûreté des installations nucléaires, la protection de l'homme contre les radiations ionisantes et la préservation de l'environnement, la gestion des déchets radioactifs, ainsi que la responsabilité civile et les assurances en matière nucléaire ;*
- *examine régulièrement les aspects économiques et techniques de la croissance de l'énergie nucléaire et du cycle du combustible nucléaire, et évalue la demande et les capacités disponibles pour les différentes phases du cycle du combustible nucléaire, ainsi que le rôle que l'énergie nucléaire jouera dans l'avenir pour satisfaire la demande énergétique totale ;*
- *développe les échanges d'informations scientifiques et techniques concernant l'énergie nucléaire, notamment par l'intermédiaire de services communs ;*
- *met sur pied des programmes internationaux de recherche et développement, ainsi que des activités organisées et gérées en commun par les pays de l'OCDE.*

Pour ces activités, ainsi que pour d'autres travaux connexes, l'AEN collabore étroitement avec l'Agence Internationale de l'Énergie Atomique de Vienne, avec laquelle elle a conclu un Accord de coopération, ainsi qu'avec d'autres organisations internationales opérant dans le domaine nucléaire.

FOREWORD

The process of uranium milling produces large volumes of waste, the radioactive content of which is small compared with that of wastes generated at other stages of the nuclear fuel cycle. It comprises uranium and its daughter products, of which thorium-230, radium-226 and radon-222 are of particular importance because their radiotoxicity.

Although the amounts of radionuclides contained in milling wastes are low, some of them, in the absence of proper management, may become sources of low level radiation exposure to local and more distant populations over very long periods of time. Techniques of stabilisation are applied to tailings to reduce their impact on the environment and public health, and monitoring programmes are in use so that tailings leakage via atmospheric and liquid pathways may be evaluated.

The meeting was intended to provide a forum for an exchange of information about current practices and technologies for stabilising mill tailings and to examine different aspects of their environmental impact. The discussions covered present and possible future R & D programmes and the contribution that international agencies could make in co-ordinating such programmes and developing guidelines and standards.

AVANT-PROPOS

Le traitement de l'uranium produit des quantités considérables de déchets dont la teneur radioactive est faible en comparaison de ceux produits à d'autres stades du cycle du combustible nucléaire. L'uranium et ses produits de filiation, dont le thorium-230, le radium-226 et le radon-222, présentent une importance particulière en raison de leur radiotoxicité.

Bien que les quantités de radioactivité contenues dans les résidus soient faibles, certains d'entre eux pourraient, faute d'une gestion convenable, constituer des sources d'exposition pendant des périodes très longues pour les populations locales et même les populations plus éloignées. Des techniques de stabilisation des résidus sont utilisées pour réduire leur incidence sur l'environnement et la santé publique, et des programmes de mesure sont mis en oeuvre pour évaluer les écoulements par voie gazeuse et liquide des terrils.

L'objet de la réunion était de constituer un forum d'échange d'informations sur les pratiques et les technologies actuelles utilisées pour la stabilisation des résidus et d'examiner des questions relatives à l'incidence sur l'environnement. Des programmes de recherche et de développement, en cours ou prévus pour l'avenir, ont été examinés, ainsi que le rôle que pourrait jouer des agences internationales afin de coordonner de tels programmes et de mettre au point des guides et des normes.

TABLE OF CONTENTS
TABLE DES MATIÈRES

SESSION 4 - POLICIES AND REGULATORY ASPECTS

SEANCE 4 - PRATIQUES REGLEMENTAIRES

Chairman - Président : Mr. J.B. MARTIN (United States)

SESSION 5 - CONCLUSIONS OF THE SEMINAR

SEANCE 5 - CONCLUSIONS DU SEMINAIRE

Chairman - Président : Mr. R.E. CUNNINGHAM (United States)

OPENING SESSION
SÉANCE D'OUVERTURE

Chairman — Président
Dr. E. WALLAUSCHEK

OPENING ADDRESSES - ALLOCUTIONS D'OUVERTURE :

Dr. E. WALLAUSCHEK

Head of Division of
the Radiological Protection and
the Radioactive Waste Management

on behalf of the Director General of the
OECD Nuclear Energy Agency

and

Mr. R.E. CUNNINGHAM

Deputy Director
Division of Fuel Cycle and Material Safety
US Nuclear Regulatory Commission

on behalf of
the United States Authorities

INTRODUCTORY PAPER

COMMUNICATION D'INTRODUCTION

ISSUES ON MANAGEMENT, STABILIZATION AND
ENVIRONMENTAL IMPACTS OF URANIUM MILL TAILINGS

R. E. Cunningham
U.S. Nuclear Regulatory Commission
Washington, D.C. U.S.A

ABSTRACT

Management and stabilization of uranium mill tailings has been controversial for over two decades. There are two basic issues: the nature of the risk to the public from tailings and what must be done to mitigate that risk. This paper provides an overview of the issues and sets some goals to be accomplished at the 1978 NEA Seminar on Management, Stabilization and Environmental Impacts of Uranium Mill Tailings that could be helpful in resolving the issues.

QUESTIONS LIEES A LA GESTION, A LA STABILISATION
ET AUX INCIDENCES SUR L'ENVIRONNEMENT DES
RESIDUS DE TRAITEMENT DE L'URANIUM

RESUME

La gestion et la stabilisation des résidus de traitement de l'uranium donnent matière à controverse depuis plus d'une vingtaine d'années. Deux questions fondamentales se posent en l'occurrence : la nature du risque que ces résidus comportent pour le public et les mesures à prendre pour atténuer la gravité de ce risque. Cette communication donne un aperçu des questions ainsi soulevées et détermine certains objectifs à atteindre lors du Séminaire AEN de 1978 sur la gestion, la stabilisation et les incidences sur l'environnement des résidus de traitement de l'uranium qui pourraient contribuer à la solution de ces questions.

Today, nuclear waste management is one of two major impediments to a more rapid growth in the nuclear power industry. The other impediment is the issue of nuclear weapons proliferation and safeguards. There are several subsets of problems within nuclear waste management. The one which has gained the greatest public attention is the disposal of high level waste. This might be due to a popular perception of hazard and because its disposal alternatives tend toward the exotic; disposal into deep geologic structures or the sea bed, or even into outer space. In the final analysis, however, disposal or long term stabilization of uranium mill tailings quite possibly will prove to be the most difficult technical challenge we face in solving our nuclear waste disposal problem.

Uranium mill tailings contain primordial radioisotopes which have always been part of man's environment. As found in nature, these radioisotopes contribute significantly to natural background radiation. When contained in mill tailings, they can pose some risk to man above that which would result from background radiation alone. Therefore, under radiation protection policies to which most nations subscribe, the risk from mill tailings must be controlled and minimized. Although most will agree with the objectives of these radiation protection policies, there is no clear agreement about what must be done to conform with the policies as applied to mill tailings for several reasons.

Because radioisotopes contained in tailings are also common in the natural environment, and since there is ordinarily only a small difference between levels of radiation in the environment and some slightly increased levels due to mill tailings, assessment of risk from tailings is difficult. The long half life of some of the radioisotopes in tailings, such as thorium 230 and radium 226, makes the control problem long term. One of the radioisotopes, radon, is a gas which increases mobility, thereby contributing to the dimension of the control problem. The volume of tailings is large. There are currently about 140 million tons of uranium mill tailings in the United States today with about 10-15 million tons being generated annually. This volume severely limits the options available to isolate the tailings from the environment over very long time spans. The concentration of radioisotopes in tailings is minute; about 500 micrograms of radium per ton of tailings for 0.2% ore. This extremely low concentration of radioisotopes contained in the tailings seems to make separation of the radioisotopes from the bulk of the tails impractical, at least with today's technology and costs.

Since the precise dimensions of the radiation risk from mill tailings has been elusive and control measures costly, management of mill tailings has been controversial for well over two decades and continues to be controversial today. The two basic issues are the nature of the risk to the public from tailings and what must be done to mitigate that risk. A brief review of efforts in the United States to come to grips with these issues, which have not been particularly successful until recently, might provide some insight about how to proceed both nationally and internationally.

Studies of airborne particulates and radon emission from mill tailings began in the 1950's. Although early studies showed elevated levels of radioactivity in the vicinity of tailings piles, the results of these studies were both inconsistent and inconclusive. However, it was generally believed that radiation exposure to the public resulting from these tailings were well within applicable radiation protection standards and that no immediate hazard existed.

Nineteen hundred and sixty-six was a vintage year for equivocal findings about mill tailings. An Atomic Energy Commission (AEC) evaluation of available data on mill tailings led to the conclusion early in 1966 that from a radiation safety standpoint, the use of the Commission's regulatory authority to effect long term control is not warranted but that the Commission should pursue its program for encouraging mill owners to stabilize tailings piles. About the same time, the United States Public Health Service (PHS) issued a report which recommended that measures to prevent erosion and spread of uranium mill tailings should be undertaken without undue delay; that distribution of mill tailings for use as construction materials should be halted until adequate precautionary review procedures could be instituted; and that binding agreements be reached as soon as possible regarding long term public and private responsibility for adequate maintenance of tailings piles. Toward mid-1966, the AEC told Congress that uranium mill tailings are not a radiation hazard at present and that it was difficult to conceive of mechanisms whereby they could become a hazard in the future. The debates continued and late in 1966 the AEC, as well as other federal agencies, agreed that tailings piles should be stabilized and contained to prevent water and wind erosion and that active tailings piles should be managed to minimize such erosion during use. Specific control measures, however, were not defined.

Further studies were undertaken. In 1969, the AEC and PHS issued a joint report which concluded that there was no significant exposure from radon and, therefore, no measures were needed to control radon emanation from tailings piles. By this time, however, the National Environmental Policy Act (NEPA) was passed by Congress. This was to become an important milestone for national direction in tailings management. It expanded the AEC's authority over mill tailings and added responsibility to assure that the environmental impacts of mill tailings were minimized. The NEPA requirement to analyze environmental impacts in connection with licensing mills revealed gaps in our knowledge about radiation dose from mill tailings and highlighted the need for stabilization performance objectives. Because of these deficiencies, the NRC began in 1976 to prepare a generic environmental impact statement (GEIS) on uranium milling which, when coupled with the supporting research and field studies, is intended to address these issues. A draft of this GEIS is scheduled for publication late this year and will be used to support standards governing tailings stabilization and management. The results of various research activities undertaken in support of the GEIS will be presented during this seminar.

As can be seen from this quick sketch, some effort was made during the 1950's and 1960's to come to grips with the problem but there was a great deal of vacillation on the issues. In my opinion, the problem remains today instead of being solved years ago because the data base on which to assess dose commitments and health effects was inadequate. Knowledge about source terms, critical pathways and appropriate dose models was insufficient to make informed decisions. We intuitively felt that tailings should be managed and stabilized for the protection of the present and future generations, but there was a division of opinion about how much effort this merited. Research to develop suitable options for management and stabilization was fragmented and inadequate. There was little incentive to spend significant resources on what was considered to be an inconsequential health and safety problem by most scientists. During the heydays of the 1950's and 1960's when tremendous programs were being undertaken to develop nuclear energy, diverting resources to mill tailings management was not particularly attractive.

There is another reason why we still have the mill tailings issues. We have been, and continue to be, faced with changing perceptions of what constitutes adequate protection of the public from radiation risk. A decade ago, prestigious organizations made statements before our Congress and peer review groups that uranium mill tailings would not cause persons to be exposed to radiation in excess of applicable standards for radiation protection. In so far as the criteria for radiation protection was based on dose limits in U.S. regulations, which in turn are based on ICRP recommendations, those statements were believed to be true then and seem to remain true today even though there is a downward trend in dose limits. However, few seriously believe today that permitting exposure up to the limits in present standards or even some fraction of those limits is acceptable without extensive justification. The application of the concept of "As Low As Reasonably Achievable" or ALARA is gaining national and international momentum. The January 1977 ICRP Publication 26 reflects this new added emphasis on the ALARA principle. Its application means that any exposure to radiation from mill tailings, however small, must be coupled with justification. It is in the context of the application of ALARA and changing perceptions of what constitutes adequate protection of our environment that a considerable effort has been expended in recent years both nationally and internationally to assess exposure from mill tailings and to determine how this exposure might be minimized both for the present generation and future generations.

During this seminar, we will evaluate progress in solving mill tailings issues and, hopefully, chart a course on how to proceed from here on. There are several observations which might be worth bearing in mind as we do so. The first is that the volume of mill tailings will grow exponentially regardless of how recycling of reactor fuel is resolved internationally. In the U.S. alone, we expect to have generated as much as 750 million tons of tailings by the year 2000. A typical mill being planned in the U.S. today will generate about 13 million tons of tailings over its lifetime. Because of these large volumes, decisions which are made at the outset of milling operations for management and stabilization of tailings are difficult to alter substantially except at enormous expense. The Department of Energy (DOE) is undertaking remedial actions to correct adverse conditions or improve stabilization at 22 abandoned mill sites. The cost estimates for these retroactive corrections range up to 130 million dollars of public money. The public will expect better performance in the future. For new milling operations, we must be in a position to reasonably demonstrate that the tailings will be adequately managed and stabilized.

A second observation is that mill tailings management and stabilization is an international problem. Even though relatively few countries are major uranium producers, tailings can reasonably be expected to contribute to the world population dose. Perhaps more important from an international standpoint is the prospect of mill tailings issues affecting the supply of uranium or substantially affecting the cost of uranium. It obviously could do either or both depending on our ability to provide solutions which are satisfactory to the public and which are provided at a reasonable cost to the industry.

The final observation is that the most difficult task before us is probably that of achieving a consensus on an acceptable level of risk for present and future generations which result from mill tailings. This is a social issue which cannot be settled by technology alone. As technologists, we can define the environmental impacts from mill tailings, establish what can be done to mitigate these impacts, and determine what it will cost. Our governmental institutions and international organizations must work toward achieving an agreement on what constitutes ALARA as applied to mill tailings. In addition to the ALARA issue, there are questions being raised in the United States today, and probably in other nations, about how we should perceive the matter of exposure from mill tailings. The exposure from radon production in tailings piles is extremely small as applied to any one generation but the sum of these exposures can be made large by counting far into the future, large enough in fact to be the dominant source of exposure from the nuclear fuel cycle. Whether it is meaningful to attach significance to radiation exposure thousands of years in the future, or conversely, whether it is justifiable to ignore them , are questions without easy answers. Science and technology alone will not provide the answers although they can go far in providing public institutions with the type of information needed to make informed judgments. This, however, will take time. The most satisfactory approach available to us today seems to be to make every reasonable effort to dispose of tailings in a way which minimizes radiation releases to the environment.

In the United States, the Nuclear Regulatory Commission, the Department of Energy and the Environmental Protection Agency, individual states such as New Mexico, and the industry have undertaken substantial coordinated programs to come to grips with uranium mill tailings problems. Our Congress has allocated resources to federal agencies to solve the problems and is considering legislation which will assure further protection of the public health and safety from mill tailings. Other countries, most notably Canada and Australia, have mounted similar efforts. Research and field studies have made good progress in identifying source terms and critical pathways, and in developing suitable dose models. In parallel with this, we believe considerable progress has been made in exploring alternatives for mill tailings management and stabilization. We have gone far in building a solid technical foundation for decision making, however, we must maintain our momentum until the task is completed.

The NEA has convened this seminar so that participants can bring their collective knowledge to bear on questions concerning environmental impacts of uranium mill tailings and what can be done to reduce these impacts. While technology can go only so far in resolving public issues concerning adequate protection and acceptable risk, this seminar can make a valuable contribution to the debates which will go on at least for the next several years. We can assess the technology which will be used to formulate national and international objectives for reduction of risk from mill tailings. I, therefore, view the goals of this seminar as follows:

1. To establish a record in these proceedings which accurately identifies the current status of our knowledge so that all nations can share this knowledge through the published proceedings.

2. To identify further areas where research, development, field investigations and standards might reasonably contribute to our knowledge and improve our ability to effectively manage, stabilize and reduce the environmental impacts from mill tailings.

3. To establish a basis for international cooperation in pursuing these goals.

In closing, I wish to commend the NEA for foresight in holding this seminar. I believe this meeting is timely and I look forward to many interesting papers and discussions over the next several days and trust that it will be beneficial to all.

Session 1

SOURCE TERMS

Chairman — Président
Dr. D.M. LEVINS
Australia

Séance 1

CARACTÉRISTIQUES DE LA SOURCE

SOURCE TERMS FOR AIRBORNE RADIOACTIVITY ARISING FROM

URANIUM MILL WASTES

M. C. O'Riordan   A. L. Downing,
National Radiological Protection Board   Binnie and Partners,
Harwell, Didcot, Oxfordshire,   Consulting Engineers,
England.   Artillery House, Artillery Row,
                                                   London, England.

ABSTRACT

One of the problems in assessing the radiological impact of uranium
milling is to determine the rates of release to the air of material from the
various sources of radioactivity. Such source terms are required for modelling
the transport of radioactive material in the atmosphere. Activity arises from
various point and area sources in the mill itself and from the mill tailings. The
state of the tailings changes in time from slurry to solid. A layer of water may
be maintained over the solids during the life of the mine, and the tailings may
be covered with inert material on abandonment. Releases may be both gaseous and
particulate. This paper indicates ways in which radon emanation and the
suspension of long-lived particulate activity might be quantified, and areas
requiring further exploration are identified.

TERMES-SOURCES DES SUBSTANCES RADIOACTIVES EN SUSPENSION
DANS L'AIR PROVENANT DES DECHETS DE TRAITEMENT DE L'URANIUM

RESUME

L'un des problèmes soulevés par l'évaluation des incidences
radiologiques du traitement de l'uranium consiste à déterminer les
vitesses de libération dans l'air de matières provenant des diverses
sources de radioactivité. Il faut connaître les termes-sources pour
modéliser le transport des substances radioactives dans l'atmosphère.
L'activité provient aussi bien de diverses sources ponctuelles et
diffuses existant dans l'installation de traitement elle-même que des
résidus de traitement. Avec le temps, les résidus passent de l'état
de boues à l'état solide. Une couche d'eau pourra être maintenue au-
dessus des solides pendant la durée d'exploitation de la mine et, lors
de la fermeture de cette dernière, les résidus pourront être recou-
verts de matières inertes. Les libérations peuvent se produire sous
forme de gaz ou de particules. Cette communication décrit les procé-
dés susceptibles d'être employés pour quantifier les émanations de
radon et l'activité des particules de longue période en suspension,
de même qu'elle définit les domaines méritant de faire l'objet d'un
complément d'étude.

1. INTRODUCTION

It may be wondered why there is a paper in this seminar from the United Kingdom, where we have no uranium mills and limited uranium prospects /1/. There are three reasons, two general and one particular. First, the UK has a substantial involvement in the later parts of the nuclear fuel cycle, so naturally,there is interest in the initial part. Then, some radiological problems are common to the various phases, and this exchange of information may help us elsewhere. The particular reason is the participation of my co-author and myself in the preparation of an environmental impact statement for an overseas mine and mill.

This last circumstance explains our approach to the problem of providing source terms for tailings. In such a situation, one wishes expeditiously to provide predictions that are as exact as possible. When gaps in essential data are discovered, it may not be practicable to mount experiments to fill them; rather, one must bridge them cautiously.

We shall suggest ways in which the emission of radon from tailings ponds might thus be estimated. Both wet and dry tailings will be considered. We shall also attempt to deal with the suspension of dry particulate activity and touch upon the effects of inert cover. The source terms are then linked to a dispersion analysis and some observations are offered.

2. MASS TRANSFER APPROACH

The kinetics of removal of radon from water may be described by the two-film theory of Lewis and Whitman /2/. This depicts gas exchange (Figure 1) as occurring by molecular diffusion through stagnant gas and liquid inter-facial films or layers /3/. The main body of each fluid is assumed to be well mixed.

Since the partial pressure of radon in the atmosphere is extremely low and its solubility slight, the transfer may be represented by the equation:

$$\frac{dM}{dt} = - \frac{K_L A M}{V_p} \qquad \ldots (1)$$

M = mass of radon at time t;

K_L = mass transfer coefficient;

A = area of air-water interface;

V_p = volume of water.

The parameter K_L is related to the diffusion of radon in the water medium and to the thickness of the hypothetical liquid boundary layer at the interface by:

$$K_L = \frac{D}{z} \qquad \ldots (2)$$

D = diffusion coefficient of radon in the water medium;

z = thickness of liquid boundary layer.

Consider a situation in which the pores of a tailings pile are kept saturated with water and inflow just balances evaporation and seepage losses. A differential equation may be set up to describe, in an approximate manner, the rate of change of the mass of radon in the pore water:

$$\frac{dM}{dt} = a\beta V - \lambda M - \frac{K_L A M}{V_p} - \frac{FM}{V_p} \qquad \ldots (3)$$

a = mass of radon per unit activity;

β = emanating power of tailings;

V = volume of tailings;

- 22 -

λ = decay constant of radon;

F = rate of inflow of water.

The quantity a is the inverse of the specific activity of radon. β is expressed in terms of radon activity emitted per unit volume of tailings in unit time. λ is the decay constant of radon.

Inflowing water is assumed to be free of radon. It is also assumed, as required by the boundary layer theory, that the water in the pores is uniformly mixed. This is unlikely to be fully achieved, although emission of radon into pore water will not vary greatly with depth, and a small horizontal motion of the water will destroy vertical gradients; nevertheless, the simplification is probably acceptable for predictive purposes.

At equilibrium, the rates of production and loss of radon are equal $\left(\dfrac{dM}{dt} = 0\right)$, and we may obtain the total mass M of radon in the tailings dam:

$$M = \frac{a\beta V}{\lambda + \dfrac{K_L A}{V_p} + \dfrac{F}{V_p}} \qquad \ldots \ldots (4)$$

Consider a hundred hectare dam with a 7m depth of tailings and a rate of water inflow of 10^4 m^3/d. Suppose the tailings had the properties shown in Table I. The value of a is 6.50×10^{-18} g/pCi and of λ 2.10×10^{-6}/s. An illustrative value of M could be computed if the value of K_L were known.

Emerson et. al. /4/ concluded that K_L for a small Canadian lake in summertime was in the range 0.6 to 2.5 cm/h with 1.6 cm/h as their best estimate. This value was obtained in light air (wind speed less than 1.5 m/s) with an assumed radon diffusion coefficient of 1.36×10^{-5} cm^2/s in water at 25°C. Since the boundary layer thickness is independent of gas species, and since the diffusion coefficient for oxygen at 25°C is about 2.4×10^{-5} cm^2/s /5/, the implied best estimate of K_L for oxygen is 2.8 cm/h. This compares reasonably well with a value of about 6 cm/h for oxygen in the Thames Estuary /6/ a body of water colder than 25°C and more exposed to the action of wind, which circumstances, on balance, would tend to increase K_L by reducing z.

Table I. Illustrative properties of tailings.

Specific activity of ^{226}Ra	500 pCi/g
Density	1.4 g/cm^3
Porosity	0.4
Emanation coefficient	0.2
Resultant emanating power*	3.0×10^{-4} pCi/cm^3.s

* refers to volume of tailings

The gas exchange rate is roughly proportional to the square of the wind speed /7/. One would therefore expect a much lower value of K_L for the still water near the surface of the tailings. The mass transfer coefficient for oxygen in quiescent water has been found to be 0.3 cm/h approximately /8/. In the experiments from which this result was obtained, small streaming currents due to slight thermal inhomogeneities appeared sufficient to eliminate gradients of dissolved oxygen even in columns of water several metres deep. On this basis, one might expect K_L for radon under comparable conditions to be about 0.17 cm/h.

In an extensive review of radon migration, Tanner /9/ cites a diffusion coefficient of 5.7×10^{-6} cm^2/s at 20°C for radon in a mud that roughly matches the tailings in physical properties. This value refers to the interstitial fluid rather than the bulk medium; it must therefore be reduced by the porosity to 2.3×10^{-6} cm^2/s. The ratio of this value to that for water suggests 0.03 cm/h as being an appropriate value of K_L for the saturated tailings.

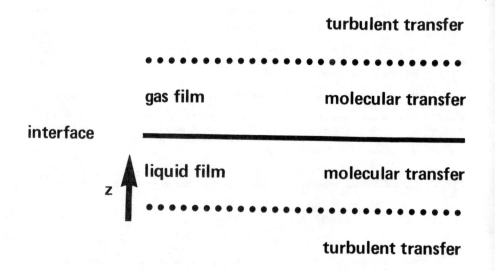

FIGURE 1   DEPICTION OF TWO-FILM THEORY

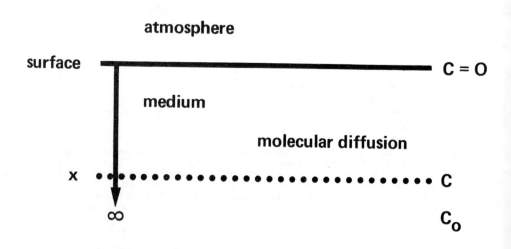

FIGURE 2   VISUALISATION OF DIFFUSION APPROACH

Using the foregoing values in equation 4, and assuming the area of the air-water interface to be the area of the dam multiplied by the porosity, we estimate the total mass of radon in the tailings under equilibrium conditions, M, to be 6.3×10^{-3}g. Furthermore, the activity concentration of radon in the pore water, C, (given by $\frac{M}{aV_p}$) is 3.5×10^{2} pCi/cm^3. It should be noted that the magnitude of C is quite insensitive to changes in F, the rate of inflow of water.

The implication of the simple model developed above is that the rate of emission of radon activity per unit surface area of wet tailings, E_W, is given by:

$$E_W = K_L C \qquad \ldots\ldots (5)$$

Substituting, we get $E_W = 8.33 \times 10^{-6} \times 3.5 \times 10^{2} = 2.9 \times 10^{-3}$ pCi/cm^2.s or 29 pCi/m^2.s. Normalised to unit specific activity of radium in the tailings, the emission rate becomes 5.8×10^{-2} pCi/m^2.s per pCi/g ^{226}Ra.

In round terms, therefore, we may say that the rate of emission of radon from deep saturated tailings is about 0.06 pCi/m^2.s per pCi/g ^{226}Ra for material with an emanation coefficient of 20%.

This model is obviously most appropriate for deriving emission rates of radon from fully liquid sources in mines and mills such as dewatering slots, evaporating ponds for mine water, and cooling ponds at the mill. Such sources exist during production only, a phase not of primary interest to this meeting. Nevertheless, the model yields a reasonable result for wet tailings and might also be considered for wet ore bodies. It would appear to merit further exploration.

3.   DIFFUSION THEORY APPROACH

Another, and more usual, approach to the calculation of emanation rates is to employ one-dimensional diffusion theory in the manner, for instance, of Kraner et al. /10/. This model envisions the radon moving solely by molecular diffusion through the pores in the tailings, the interstitial concentration of radon increasing from zero at the surface to a limiting value C_o at infinite depth (Figure 2). Porosity is deemed constant with depth.

The formula developed to describe the steady-state flux E as a function of depth x is:-

$$E = -DC_o \sqrt{\frac{\lambda}{D/S}} \; e^{-\sqrt{\frac{\lambda}{D/S}}\, x} \qquad \ldots\ldots (6)$$

The new symbol S is for the porosity.

This reduces at the surface, where x = 0, to:

$$E = -DC_o \sqrt{\frac{\lambda}{D/S}} \qquad \ldots\ldots (7)$$

The tailings pile considered above may be regarded as infinitely deep. Ignoring the effect of the water inflow (which is negligible in any case) we find that C_o is given by the product of the radium specific activity, the bulk density, and the emanation coefficient divided by the porosity; its value is 350 pCi/cm^3 also. The rate of emission of radon activity per unit area of wet tailings, E_W, is given by this method as:

$$E_W = 2.3 \times 10^{-6} \times 350 \sqrt{\frac{2.1 \times 10^{-6} \times 0.4}{2.3 \times 10^{-6}}}$$

$$= 4.86 \times 10^{-4} \text{ pCi/cm}^2.\text{s or } 4.86 \text{ pCi/m}^2 \text{ s}.$$

Normalised to unit specific activity of ^{226}Ra, the emission rate becomes 0.01 pCi/m^2.s per pCi/g ^{226}Ra. This is about a sixth of the value given by the uniform mixing model for the same saturated tailings.

We may also apply the diffusion model to dried out tailings, by which we mean air-dried of course. Assuming cautiously that the hygroscopic moisture content of tailings is 4%, we employ the diffusion coefficient for a building sand that also matches the dry tailings in terms of density and porosity /9/. This procedure completely ignores the fact that there will be a varying moisture profile in the tailings depending on climatic conditions in the area, on the actual physical properties of the tailings, and on the construction of the dam. The concentration of slimes at the surface due to their lower rate of sedimentation is also ignored.

Labelling the emission rate of radon per unit area of dry tailings as E_D, we obtain:

$$E_D = 2.2 \times 10^{-2} \times 350 \sqrt{\frac{2.1 \times 10^{-6} \times 0.4}{2.2 \times 10^{-2}}}$$

$$= 4.76 \times 10^{-2} \text{ pCi/cm}^2.\text{s or } 476 \text{ pCi/m}^2.\text{s}.$$

Normalised to unit specific activity of ^{226}Ra, this corresponds to 1 pCi/m^2.s per pCi/g ^{226}Ra, which is two orders of magnitude higher than the value for saturated tailings.

The foregoing estimates of surface emission rates per unit area and per unit specific activity of radium-226 are summarised in Table II.

Table II. Estimated rates of emission of radon from unit surface

area of deep tailings* per unit specific activity of incorporated ^{226}Ra

Model	Condition	pCi/m^2. s per pCi/g
Two-film	Saturated	0.06
Diffusion	Saturated	0.01
Diffusion	Air-dry	1

* Emanation coefficient 0.2

Although the discrepancy between the two estimates for the saturated tailings is discouraging, it is not altogether surprising that it should appear, when one considers the simplifications and the approximations that have had to be made.

There is some experimental support for the radon emission rate from air-dried tailings. Bernhardt et al. /11/ instance a value of about 0.6 pCi/m^2.s per pCi/g ^{226}Ra for bare tailings with characteristics similar to those considered above; these were situated in a semi-arid region. One might also cite an overall value of 0.6 for soils, which value can be deduced from an overall mean emanation rate of 42 aCi/cm^2.s and a specific activity for ^{226}Ra of 0.7 pCi/g /12/.

There appears to be no information on surface emission from saturated tailings, but it has been suggested /13/ that the disparity between saturated and dry tailings is an order of magnitude less than envisaged above. This suggestion was recently reinforced in a summary of the work of the same group of investigators /14/ where the surface rate of emission was calculated to be 0.3 pCi/m^2.s per pCi/g ^{226}Ra from deep wet tailings as opposed to a value of 1.2 for dry ones. The authors use a two-tier diffusion model to calculate the emission rate at the surface taking into account both the tailings and the soil below them. If one ignores the soil term and back-computes the bulk diffusion coefficient for tailings such as those described here, the result seems to point to the use of the coefficient for 17% moisture sand in Tanner /9/ whereas we use the value for 37% moisture mud. This divergence underlines the importance of the moisture content in this situation.

In view of the uncertainties surrounding the value for saturated tailings, we feel that is is prudent to adopt 0.1 pCi/m^2. s per pCi/g incorporated ^{226}Ra for

the surface emission rate. Perhaps the situation will be clarified at this
meeting.

The effect of inert cover on the emission rate from a tailings pile may be
estimated in a simple fashion by employing the half-value thickness of the material
in question. This is based on the exponential attenuation in concentration that
occurs with thickness of inert material over the tailings, the surface of which is
regarded as a plane source /9/. The attenuation coefficient, or diffusion length,
is $\sqrt{\frac{\lambda}{D/S}}$. The half-value thickness is thus $\ln 2 \div \sqrt{\frac{\lambda}{D/S}}$, which for the building sand
mentioned above, for instance, yields a value of 110 cm approximately.

It is not possible to quantify the effect of inert cover from the
experimental work of Bernhardt et al. /11/ but cover is seen to bring about a
sensible reduction. The chief merit of inert cover is that it prevents the
translocation of radioactive particulate material.

4. PARTICULATE SUSPENSION

It is difficult to imagine that anybody starting now would abandon an
uncovered tailings pile, so in the narrow sense of this conference, it is perhaps
unnecessary to analyse the situation. However, the surface may be allowed to dry
out during the filling of the dam thus creating a potential source. Moreover, in
assessing the radiological impact of a mine or mill, one encounters other
potential sources of airborne particulate material such as the ore body itself,
the ore stockpiles, and the waste rock. Since these situations throw up similar
problems, it is indeed worthwhile pursuing the topic. The radionuclides of
interest are the long-lived ones in the ^{238}U series.

The mechanics of wind transportation of dust particles is a complex
subject, and no generally acceptable model appears to be available in the
literature. Drastic simplifications and broad assumptions need therefore to be
made for predictive purposes. We shall ignore such properties of the tailings
as surface encrustation and moisture content and the leachability of the radioactive material, all of which tend to reduce the source term.

It is first assumed that particles other than respirable ones (moving
mainly by surface creep and saltation, but also by short-range transportation) are
mostly retained in the dam by the embankment or form a relatively narrow penumbra
of decreasing surface activity limited perhaps by a local topographical feature
such as a bund of drain. The respirable particles will however be selectively
sorted by the wind and be transported off-site; those in the top 1 mm of soil will
be deemed to be available for suspension.

The choice of 1 mm is traceable to Healy's suggestion for treating the
resuspension of inhalable particles (< 10 μm AMAD) from a low-vegetated semi-arid
area contaminated with plutonium /15/. This value, considered by him to represent
the "very surface of the ground" has been adopted also for evaluating the dust
likely to be blown off a waste ore dump in Australia /16/.

Although not in accord with strict physiological considerations /17/
particles having diameters less than 5 μm (less than 10 μm AMAD say) are regarded
here as respirable.

The next step is to ensure that data are available (or can be generated
from bench-scale or pilot-plant experiments) on the likely particle size
distribution of the tailings and the specific activity of the several radionuclides
in each size fraction. This will reveal the enhancement that occurs in the finer
fractions /18/. One can then compute the suspensible activity per unit surface
area, H, in, for instance, pCi/m^2.

Sehmel and Lloyd /19/ have provided suspension rates over a lightly
vegetated area on which inert submicrometer tracer was sprayed. These rates, R,
defined as the fraction of the total material on the ground suspended per unit
time, were measured for respirable particles, in this case particles with nominal
diameters for unit density spheres of less than 7 μm. The results for these

particles, obtained in three wind speed intervals, are given in Table III. The winds vary from light air to gale force. Suspension is seen to increase faster than linearly with wind speed.

Table III. Suspension rates of respirable particles

in various windspeed bands, fraction/s

Speed band, m/s	Rate, fraction/s
1.3 to 3.6	2×10^{-10}
3.6 to 5.8	9×10^{-10}
5.8 to 20.1	2×10^{-8}

The long-term fractional occurrence of winds in these velocity bands ($\emptyset$) is likely to be available from the local meteorological records; an annual basis is appropriate. The fractions will vary with wind direction, of course, but unless there is an unusually pronounced correlation between speed and direction, one might reasonably assume that they are independent. (Given the speculative nature of some of this assessment, we would have no hesitation in making such an assumption if the relative standard deviation of the occurrence in various sectors was around the 50% mark.) These fractional occurrences are used to weight the suspension rates, given for each speed band by the product HR, so that the long-term activity suspension rate is given by $\Sigma HR\emptyset$ pCi/m^2.s.

The suspension rate for respirable particulate material is analogous to the emission rate for radon gas discussed earlier. Both rates are used as input terms for atmospheric dispersion analysis.

A numerical example for tailings from an alkaline leach mill may clarify this approach to the generation of particulate source terms. Consider the bulk specific activity and density of the tailings described in Table I, that is 500 pCi/g ^{226}Ra and 1.4 g/cm^3. The respirable fraction might well constitute 10% by weight /20/ and its specific activity enhancement factor might well be 5 /18/. Therefore, the ^{226}Ra activity per unit area in the top mm of tailings, H, is given by;

$$H = 500 \times 1.4 \times 0.1 \times 5 \times 10,000 \times 0.1$$
$$= 3.5 \times 10^5 \text{ pCi/}m^2.$$

The remainder of the calculations is set out in Table IV, the values of $\emptyset$ being for a site for a mill /21/.

Table IV. Calculation of particulate source terms

Windspeed band, m/s	R, fraction/s	$\emptyset$, fraction	$HR\emptyset$, pCi/m^2.s*
1.3 to 3.6	2×10^{-10}	0.6	4×10^{-5}
3.6 to 5.8	9×10^{-10}	0.25	8×10^{-5}
5.8 to 20.1	2×10^{-8}	0.15	1×10^{-3}
			$\Sigma 1 \times 10^{-3}$

* H = 3.5×10^5 pCi/m^2

The dominating effect of strong to gale force winds is clearly evident in the table. That is why one might also wish to consider the probability of special meteorological phenomena such as cyclones and willy-willies, although it is not clear how one might quantify their effects.

It is difficult to evaluate how realistic the foregoing approach is because of the virtual impossibility of separating suspension from dispersion when examining field data. One is also hampered by the fact that data tend to reflect an interest in exposure estimates rather than source terms. There is, therefore, a degree of speculation in the method, but this topic is fraught with difficulty and information is scanty. Perhaps later papers in this seminar will help to measure the uncertainty. Because some phenomena tending to reduce suspension have been ignored, the estimate probably errs on the safe side.

5. DISPERSION ANALYSIS

It is not our brief to discuss dispersion modelling, but a few observations may be in order.

There are many computational models, but one that may be employed /21/ is that due to Start and Wendell /22/. This is based on the general gaussian model with an instantaneous point source, or puff release, equation for a ground level source. (The large tailings area may be reduced to a number of virtual point sources of equal strength.) The model disperses plume effluent through the advective transport of plume segment of puff centres and through the diffusion of effluent puffs about their individual centres. The transport of puffs is determined from a horizontal field of spatially and temporally varying winds. The diffusion of puffs is described by distance-dependent values of the crosswind and vertical diffusion coefficients.

The effect of a capping lid of stable air aloft, which limits vertical diffusion, is encompassed by the model, but removal mechanisms such as dry deposition are not. The contributions of the several sources to the activity concentrations at each receptor point in the model area are then combined, and isopleths of ground level concentration are constructed.

Of particular interest is the relative concentration of radon gas and radon daughters. Gas concentration at or near a source, as determined by the model, may be relatively high, but this is not too significant radiologically. More important is the radon daughter concentration measured in units of Working Level (WL). For young radon near sources, the ratio of daughter to gas concentrations is low, but as the radon ages during transport, the daughters grow towards equilibrium.

For times up to 90 min (but excluding the first several seconds) the increase in WL with time t in minutes is given, within 25%, by the following analytical expression /23/:

$$WL = 2.3 \times 10^{-7} \times \chi \ t^{0.85} \qquad \ldots\ldots (8)$$

Here, χ is the gas concentration in units of pCi/m^3. For times greater than 90 min, one may assume that $1 \ pCi/m^3 \ ^{222}Rn = 10^{-5}$ WL. This time-dependent expression may be incorporated in the model /21/ to yield WL isopleths which facilitate subsequent radiological assessment.

6. CONCLUSIONS

The conclusions we draw from our involvement in this field are that a simple yet cautious approach is adequate for predictive purposes, but that uncertainties do exist about source terms. These uncertainties have to be counteracted by a degree of overestimating, and it would be helpful if that were reduced so that radiological impact might be gauged more realistically.

The main difficulties appear to arise with waterlogged wastes. The utility of the mass transfer model for predicting radon emanation from aqueous sources cannot be disputed, but its application to saturated material involves a considerable degree of extrapolation. Faced also with the apparently contradictory indications from diffusion theory, the reviewer looks for field data. There appear to be none. He then has to take a line that may be unnecessarily cautious.

It might be argued that saturation is a transient phenomenon, lasting only for the life of the mill, and that one's concern should be concentrated on dry abandoned wastes. Nevertheless, radiological assessment of the operations phase is required, and field data linked to values of the principal properties affecting surface emission rates would be welcome.

There appears to be agreement on the rate at which radon is given off by so-called dry tailings, and there is supporting field evidence. Perhaps one should let well enough alone, but moisture content has such a singular effect that more complete information would be useful. Field studies involving the measurement of moisture at various depths as well as the other relevant properties would be of interest. For the person trying to predict the emission from a future dam, such information might be linked to a water balance model of the tailings.

The suspension of long-lived particulates from abandoned tailings is, hopefully, an historical problem, but there are similar sources in mining itself and there are analagous situations in other parts of the fuel cycle: we have in mind, for instance, contaminated seashore and estuarine sediments. Suspension data gathered from appropriate field experiments on tailings systems may therefore have wider applicability than one at first expects.

The papers that follow may however resolve all the difficulties mentioned here, and the Proceedings may well become the vade mecum of those engaged in tailings control.

7.   ACKNOWLEDGEMENTS

We wish to acknowledge all the advice and help given by G.N. Fernie and E. Englund of Maunsell and Partners, B.M.R. Green and J.A. Jones of the National Radiological Protection Board, and P. Dempsey of Binnie and Partners.

8.   REFERENCES

1.   Bowie, S.H.U.: "Global distribution of uranium ores and potential UK
     deposits", Geological aspects of uranium in the environment, pp. 12-20.
     London Geological Society, 1978.

2.   Lewis, W.K. and Whitman, W.G.: "Principles of gas absorption", Ind. Eng.
     Chem. Vol. 62, no. 12, p. 1215 (1924).

3.   Liss, P.S. and Slater, P.G.: "Flux of gases across the air-sea interface",
     Nature, Vol. 247, p. 181 (1974).

4.   Emerson, S., Broecker, W. and Schlindler, D.W.: "Gas exchange rates in a
     small lake as determined by the radon method", J. Fish. Res. Board Can.,
     Vol. 30, no. 10, p. 1475 (1973).

5.   Bruins, H.R.: "Coefficients of diffusion in liquids", International
     Critical Tables, Vol. 5., p. 63, McGraw-Hall, New York, 1929.

6.   Water Pollution Research Paper No. 11, "Effects of polluting discharges
     in the Thames Estuary", Chap. 13, HMSO London, 1964.

7.   Peng, T.H., Takahashi, T. and Broecker, W.S.: "Surface radon measurements
     in the North Pacific Ocean Station Papa", J. Geophys. Res., Vol. 79,
     no. 12, p. 1772 (1974).

8.   Downing, A.L., Melbourne, K.V., and Bruce, A.M.: "The effects of
     contaminants on the rate of aeration of water". J. App. Chem., Vol. 7
     p. 590 (1957).

9.   Tanner, A.B.: "Radon migration in the ground", The Natural Radiation
     Environment, pp. 161-190, University of Chicago Press, 1964.

10. Kraner, H.W., Schroeder, G.L. and Evans, R.D.: "Measurements of the effects
     of atmospheric variables on radon-222 flux and soil gas concentrations".
     ibid.

11. Bernhardt, D.E., Johns, F.B. and Kaufmann, R.F.: "Radon exhalation from
     uranium tailings piles", Technical Note ORP/LV-75-7 (A), USEPA, Las Vegas,
     1975.

12. UNSCEAR: "Sources and effects of ionizing radiations", United Nations, New
     York, 1977.

13. Kennedy, R.H., Deal, L.J., Haywood, F.F. and Goldsmith, W.A.: "Management
     and control of radioactive wastes from uranium milling operations",
     Nuclear power and its fuel cycle, Vol. 4, pp. 545-560, IAEA Vienna 1977.

14. Goldsmith, W.A., Haywood, F.F. and Leggett, R.W.: "Transport of radon which
     diffuses from uranium mill tailings", The natural radiation environment
     III, Book of Summaries, pp. 242-244, Houston, 1978.

15. Healy, J.W.: "A proposed interim standard for plutonium in soils", LA-5483-
     MS, Los Alamos Scientific Laboratory, 1974.

16. Clark, G.H: "An assessment of the meteorological data and atmospheric
     dispersion estimates in the Ranger I uranium mining environmental impact
     statement, AAEC, Lucas Heights, 1976.

17. ICRP Task Group on Lung Dynamics: "Deposition and retention models for
     internal dosimetry of the human respiratory tract", Hlth Phys, Vol. 12,
     p. 173 (1966).

18. Sears, M.B., Blanco, R.E., Dahlman, R.C., Hill, G.S., Ryon, A.D. and
     Witherspoon, J.P.: "Correlation of radioactive waste treatment costs and
     the environmental impact of waste effluents in the nuclear fuel cycle for

use in establishing as low as practicable guides - mining and milling of uranium ores", ORNL-TM-4903, Oak Ridge National Laboratory, 1975.

19.   Sehmel G.A. and Lloyd, F.D.: "Particle suspension rates", Atmosphere - surface exchange of particulate and gaseous pollutants (1974), pp. 847 - 858, USERDA Washington, D.C.

20.   Fernie, G.N.: Personal communication

21.   Steedman, R.K.: Personal communication

22.   Start, G.E. and Wendell, L.L.: "Regional effluent dispersion calculations considering spatial and temporal meteorological conditions" ERL ARL-44, p. 63, National Oceanic and Atmospheric Administration Idaho 1974.

23.   Evans, R.D.: "Engineers guide to the elementary behaviour of radon daughters", Hlth, Phys. Vol. 17, p. 229 (1969).

DISCUSSION

D.M. LEVINS, Australia

     Your estimate of two orders of magnitude higher for release into air as distinct from sub saturated tailings ; has this taken account of the variation in eminating power in water and air ?

M.C. O'RIORDAN, United Kingdom

     I think that we assume that the eminating coefficient by which we mean the percentage form that escapes is in fact independent of the physical state, so we use the value which is established in the laboratory for the dry material.

D.M. LEVINS, Australia

     What was that value ?

M.C. O'RIORDAN, United Kingdom

     We worked with a value of 20 %.

D.M. LEVINS, Australia

     We have made some measurements at the Atomic Energy Commission and our results seem to indicate that in water, the eminating power is highter than in air. I do not know if anyone else has got any comments on this ? Our experiments suggest a factor of two between the emination power in water compared to air.

EMANATING POWER AND DIFFUSION OF RADON
THROUGH URANIUM MILL TAILINGS

C.M. Jensen, R.F. Overmyer, P.J. Macbeth, V.C. Rogers,
T.D. Chatwin
Ford, Bacon & Davis Utah Inc.
United States

ABSTRACT

Radon emanation from uranium mill tailings particles,
radon exhalation from tailings, and methods that could be used to
reduce radon exhalation from tailings are discussed. Surface radon
flux and soil gas concentrations were measured in experimental
columns consisting of uranium tailings and various types and thicknesses of cover materials. The applicability of diffusion theory
was examined and effective diffusion coefficients were determined
for diffusion of radon gas through clay, soil, and sand covers over
tailings.

EMANATION ET DIFFUSION DU RADON PAR L'INTERMEDIAIRE
DES RESIDUS DE TRAITEMENT DE L'URANIUM

RESUME

Cette communication traite des émanations de radon en
provenance des particules de queue de traitement de l'uranium, de
la quantité de radon libérée par les résidus de traitement et des
méthodes susceptibles d'être utilisées pour la réduire. Les flux de
radon en surface et les concentrations de gaz dans le sol ont été
mesurés dans des colonnes expérimentales comportant des résidus de
traitement de l'uranium, ainsi que divers types et épaisseurs de
matériaux de recouvrement. On a examiné les possibilités d'appliquer
la théorie de la diffusion et déterminé les coefficients effectifs
de diffusion du radon gazeux dans de l'argile, de la terre et du
sable utilisés pour recouvrir les résidus de traitement.

1. INTRODUCTION

Radon gas, the first decay daughter of ^{226}Ra, is recognized as one of the significant potential health problems associated with uranium mill tailings. As a result of this potential health impact, there is interest in reducing radon exhalation from tailings piles. The process of production of free radon in the tailings and the transport of radon through the tailings are, therefore, important.

This paper contains a discussion of research being performed to characterize the radon source in the uranium mill tailings and the diffusion properties of radon in various cover materials. These investigations have included determination of the radium content and the emanating power of 16 different samples of uranium mill tailings and the diffusion coefficient of radon in 10 different soil samples.

The analytical and experimental procedures used to determine the radon source parameters and the diffusion coefficients are also given. Because these measured parameters are site specific, e.g., depend upon the ^{226}Ra content of the uranium ore and the milling process used to remove the uranium, only general equations and examples of the measurements performed will be presented.

2. EMANATING POWER OF URANIUM MILL TAILINGS

The emanating power of radon has been studied to parameterize models to predict radon exhalation from tailings. Emanating power is defined as that fraction of radon produced in some mineral matrix which escapes the matrix and is free to diffuse in the pore spaces. Values of the emanating power of uranium ores range from 1% to 91% and show that the emanating power is dependent on many parameters such as porosity, particle size, mineral species, radium mineralogy, etc.[1] Emanating powers of approximately 20% have been used to model radon sources from western sites. [2]

The principal method used to determine the emanating power is described by Scott et al. [3] Dry uranium mill tailings were deemanated to free the radon gas by evacuation in a bell jar. The evacuation produced no size separations and samples were otherwise untreated in any respect. The tailings were then sealed in cans to trap all radon that emanates from the material. After allowing equilibrium of the radon daughters to be established, the can of tailings was analyzed using a NaI detector and a pulse height analyzer to determine the initial activity of A_o of ^{214}Bi and hence the activity of the residual radon trapped in the crystalline structure of the minerals after deemanation. Waiting 30 days allows the radon to grow back into complete equilibrium with its radium progenitor. The additional amount of radon A_1 is equal to the amount that had been removed previously from the tailings by deemanation. The percent emanating power of the uranium tailings is then given by:

$$\% \text{ emanation} = 100 \times 1 - \frac{A_o}{A_\infty} \qquad (1)$$

where,

A_O = initial activity

A_1 = radon activity deemanated

A_∞ = $A_1 + A_O$ the activity after 30 days

A modification of this procedure was used. The activity was determined at several times after deemanation and the resulting data were fit by the method of least squares to the equation:

$$\text{Activity} = A_O + A_1 (1 - e^{\lambda t}) \tag{2}$$

to determine the parameters A_O and A_1. Using these best fit parameters gives:

$$A_\infty = A_O + A_1 \tag{3}$$

and the emanating powers can be determined.

Table I contains the emanating power and radium content of 22 mill tailings and soil samples collected in the Wyoming and New Mexico mining regions of the United States. The cover materials used were overburden which could conceivably be used to cover mill tailings. The radium content of the cover material was low and in the range of natural concentrations of ^{226}Ra in the earth's crust. Because of the small radium concentrations, emanating power determination varied greatly because of problems with counting statistics.

The radium content of the shine tailings samples were found to have a higher radium content than their corresponding sandy sample. The selection of sandy and shine tailings was accomplished by taking the sample either at the slurry discharge point from the mill to obtain the sandy tailings which settle faster or closer to the standing water in the pond and obtaining samples of shines which are carried farther into the pond upon discharge from the mill. No corresponding pattern was observed for the emanating power of the tailings. The deviations shown are from the variation between replicate samples. Those with no deviation shown either had no replicate determinations or no deviation observed.

3. RADON SOIL-GAS CONCENTRATION

The radon gas concentration in bare tailings and in a tailing-cover material combination with a finite source can be modeled using diffusion theory. The main limitation of the theory is a result of assuming that the flux is proportional to the concentration gradient as given by Fick's law,

$$J(x) = \frac{-D \; dC(x)}{dx} \tag{4}$$

where,

$J(x)$ = the radon flux (Ci/m^2-sec)

D = the effective diffusion coefficient (m^2/sec)

$\dfrac{dC(x)}{dx}$ = radon concentration gradient (Ci/m^4)

This limitation occurs because the flux is not necessarily proportional to the gradient of the concentration near or at inter-

TABLE I

PROPERTIES OF COVER MATERIAL AND MILL TAILINGS

Sample Identification	pCi Ra/g Tailings (dry)	Emanating Power %
Shirley Basin Clay Cover	$5.85 (\pm.25) \times 10^0$	16 ± 2
Shirley Basin Soil Cover	$5.99 (\pm.29) \times 10^0$	10 ± 5
Powder River Clay Cover	$2.69 (\pm.18) \times 10^0$	25 ± 6
Powder River Soil Cover	$1.18 (\pm.13) \times 10^0$	40 ± 15
Ambrosia Lake Soil Cover	$4.87 (\pm.85) \times 10^{-1}$	26
Ambrosia Lake Shale Cover	$1.17 (\pm.12) \times 10^0$	14 ± 3
Shirley Basin Sandy Tailings	$2.61 (\pm.08) \times 10^2$	12 ± 4
Shirley Basin Slime Tailings	$8.75 (\pm.15) \times 10^2$	8 ± 3
Powder River Sandy Tailings #1	$8.22 (\pm.46) \times 10^1$	19 ± 12
Powder River Slime Tailings #1	$1.29 (\pm.06) \times 10^2$	7 ± 1
Powder River Sandy Tailings #2	$1.45 (\pm.06) \times 10^2$	6
Powder River Slime Tailings #2	$1.63 (\pm.04) \times 10^2$	12 ± 4
Ambrosia Lake Sandy Tailings #1	$2.69 (\pm.07) \times 10^2$	19
Ambrosia Lake Slime Tailings #1	$8.50 (\pm.18) \times 10^2$	24 ± 10
Ambrosia Lake Sandy Tailings #2	$8.81 (\pm.35) \times 10^1$	10 ± 4
Ambrosia Lake Slime Tailings #2	$4.49 (\pm.10) \times 10^2$	--
Ambrosia Lake Sandy Tailings #3	$1.38 (\pm.04) \times 10^2$	20 ± 1
Ambrosia Lake Slime Tailings #3	$5.35 (\pm.12) \times 10^2$	18
Gas Hills Sandy Tailings #1	$6.32 (\pm.23) \times 10^1$	18 ± 1
Gas Hills Slime Tailings #1	$8.70 (\pm.44) \times 10^1$	8 ± 3
Gas Hills Sandy Tailings #2	$4.14 (\pm.69) \times 10^0$	11 ± 9
Gas Hills Slime Tailings #2	$4.11 (\pm.12) \times 10^2$	31 ± 4

faces and boundaries.   At the boundary of the system the atmosphere
acts as an infinite sink to radon gas.   This limitation becomes
apparent as the theoretical expressions for the flux are compared
to experiment.

The boundary conditions applicable to the experiments performed
for this study are summarized in Table II, and a schematic of the
system is given in Figure 1.   Only solutions in the x direction are
considered here.

Using these boundary conditions, the equations describing the
radon concentrations are:

$$C_t = C_o \frac{\cosh(\alpha x)}{\cosh(\alpha b)} + \frac{S}{\alpha^2} \left[1 - \cosh(\alpha(b-a)) \left(\frac{\cosh(\alpha x)}{\cosh(\alpha b)} \right) \right] \quad (5)$$

in tailings where C_o is the radon concentration at the top of the
cover material, and

$$C_c = C_o \frac{\cosh(\alpha x)}{\cosh(\alpha b)} + \frac{S}{\alpha^2} \left[\cosh(\alpha(a-x)) - \cosh(\alpha(b-a)) \left(\frac{\cosh(\alpha x)}{\cosh(\alpha b)} \right) \right] \quad (6)$$

in cover material.

The concentration C_o is a measured quantity while the source
S is determined by using the relationship

$$S = \frac{R \rho \lambda E}{D} \quad (7)$$

as given by Culot et al.[4]   where the constants are emanating power
E, density of tailings ρ, and ratio of grams or radium to grams of
tailings R.   The factor α is a characteristic length of radon in
the cover material given by $\alpha = x/D$.

The experimental configurations consisted of placing tailings
in two 55-gal test chambers to a depth of 2 ft.   Gas sampling was
accomplished with 25 mil ID stainless steel tubing which extended
through the chamber walls as shown in Figure 1.   A medium fritted
glass filter on the tubing prevented fine tailings particles from
entering the tubing.   For the radon concentration measurements, an
evacuated Lucas cell was attached to the sampling tube and soil gas
was withdrawn from the interior via the steel tubing and an additional filter to remove radon daughters.   The cell was counted as
described by Lucas[5]   to determine the radon concentration in the
extracted soil gas.   All penetrations into the drum were sealed
with epoxy and the tubing ends capped when samples were not being
taken to prevent radon loss through the tubing and holes in the
drum.

The test chamber containing sand as a cover material was 12-ft
high, accomodating 2 ft of tailings and up to 10 ft of sand.   The
test chamber containing the dry bentonite was 6-ft high with 2 ft
of tailings and up to 3.5 ft of bentonite.

Figures 2 and 3 show the experimentally and theoretically
determined radon gas concentrations in sand and bentonite cover
material.   Equations (5) and (6) were used to correlate the data.
The parameters for the source term are summarized in Table III.   The
porosities of sand and tailings were measured to be 0.30 and 0.28,
respectively, and were assumed to be equal in all calculations.   The
diffusion coefficient for uranium tailings has been measured by many
authors and ranges from 9×10^{-3} cm^2/sec to 5×10^{-2} cm^2/sec.
Matching the experimental data with diffusion theory predictions
gave an effective diffusion coefficient of 3×10^{-2} cm^2/sec for both

TABLE II

BOUNDARY CONDITIONS FOR DIFFUSION EQUATION

(1)	$J_t(o) = 0$	flux is zero at the bottom of the test chamber
(2)	$C_t(a) = C_c(a)$	concentration is continuous across inter-face of tailings and cover material
(3)	$J_t(a) = J_c(a)$	flux is continuous across the tailings-cover material interface
(4)	$C_c(b) = C_o$	the concentration must equal the experi-mentally determined concentration at the top of the cover material

TABLE III

SUMMARY OF CONSTANTS USED

p (porosity) 0.3

E (emanating power) 0.2

D (effective diffusion coefficient) . . . . . . 3.0×10^{-2} cm^2/sec

λ (radon decay constant) . . . . 2.097×10^{-6} sec^{-1}

R (ratio: radium to tailings) . . . . 0.78×10^{-9} gRa/g tailings

ρ (density of tailings) . . . . 1.6 g/cm^3

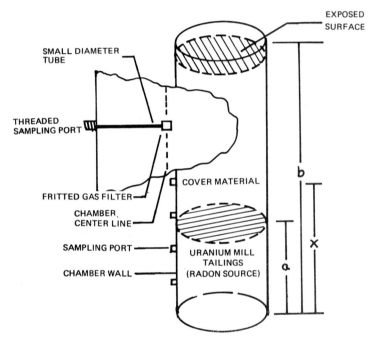

FIGURE 1 SCHEMATIC REPRESENTATION OF RADON SOURCE,
COVER MATERIAL AND TEST CHAMBER

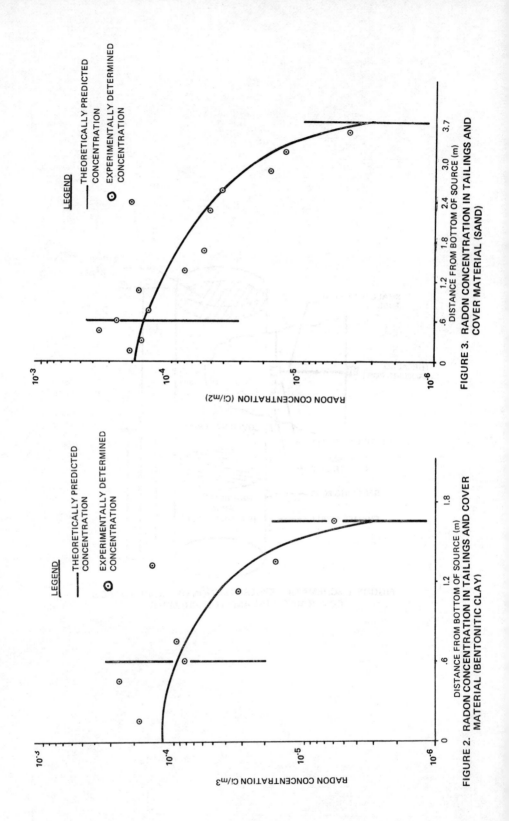

FIGURE 3. RADON CONCENTRATION IN TAILINGS AND COVER MATERIAL (SAND)

FIGURE 2. RADON CONCENTRATION IN TAILINGS AND COVER MATERIAL (BENTONITIC CLAY)

tailings and cover material. The radon gas concentration was measured at the interface of the cover material and air was used as a
boundary condition. Using these parameters, the radon gas concentration was very effectively modeled using diffusion theory.

The above experiments and analysis demonstrate the applicability
of diffusion theory in modeling radon gas concentrations. Its further utility can be seen by the relatively small amount of information needed to model the radon source.

4. SURFACE RADON FLUX AS A FUNCTION OF BARE URANIUM TAILINGS DEPTH

The surface radon flux as a function of uranium tailings depth
was also investigated and compared to the semi-empirical curve given
by Schiager. [6]

The following equations characterize the radon flux from bare
tailings:

$$J_t(a) = \frac{D_t}{\alpha} \tanh(\alpha a)\ (S - C_o\ \alpha^2). \tag{8}$$

The maximum flux, given in equation (8), is obtained when "a" increases to infinite.

$$J_t(\infty) = \frac{D_t}{\alpha}\ (S - C_o\ \alpha^2) \tag{9}$$

The ratio of the flux from tailings of a finite thickness to the
infinite thickness flux is:

$$\frac{J_t(a)}{J_t(\infty)} = \tahn\ (\alpha a) \tag{10}$$

The radon flux was measured using the charcoal canister technique for varying depths of coarse sandy and 200 mesh fine uranium
tailings and corrected for pressure variations. The results
plotted in Figure 4 were fit to Equation (8) using an iterative
least squares routine to determine the infinite flux and the effective diffusion coefficient. The flux values normalized to the infinite flux are plotted in Figure 5 and show that the flux for these
tailings reach saturation at a depth of 25 to 50% greater than predicted by Schiager. [6]

The curves are described by Equation (8) and the differences
are due to variations in the effective diffusion coefficient D.
Physically, these differences in flux can be attributed to variations in the porosity to which the effective diffusion coefficient
is related. The effective diffusion coefficient may, therefore, be
readily deduced from the flux attenuation as a function of depth of
tailings. [7]

5. THE EFFECTS OF COVER MATERIAL ON RADON EXHALATION

As can be seen from Figure 6, the theoretical prediction for
the surface flux:

$$J_c(b) = \frac{D_c\ S \sinh\ (\alpha a) - C_o\ \alpha^2 \sinh\ (\alpha b)}{\alpha \cosh\ (\alpha b)} \tag{11}$$

using experimentally determined modeling parameters is about a factor of 2 greater than the experimental measurements of the flux
through a sand cover. Determination of the source of the difference

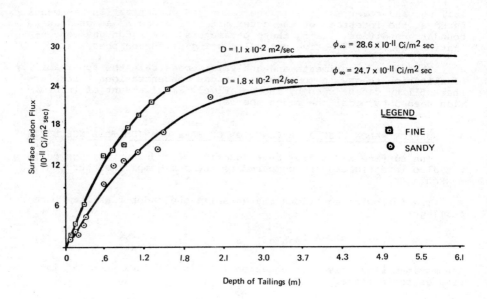

FIGURE 4. RADON FLUX AS A FUNCTION OF TAILINGS DEPTH

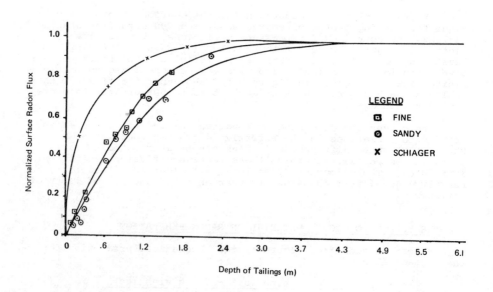

FIGURE 5. NORMALIZED RADON FLUX AS A FUNCTION OF TAILINGS DEPTH

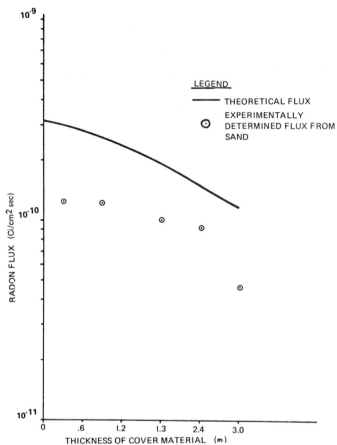

FIGURE 6. SURFACE RADON FLUX AS A FUNCTION OF COVER
MATERIAL THICKNESS

would require further examination. Also, the difficulty of making accurate measurements of C_o at the soil-air interface can introduce uncertainties in the theoretical flux curve.

The theoretical functional dependence is essentially exponential at a distance of 6 ft or more from the radon source and gives a relaxation length of 2.5 to 3.0 m. This value of relaxation length is in disagreement with the experimental data and is higher than the value reported by Schiager. [6] Assuming the flux is exponential beyond 6 ft and fitting the apparent slope of the data, a relaxation length of 1.4 m is observed, which is in good agreement with the value report by Schiager. [6] Because of the uncertainties, a range could be assigned to the relaxation length from 1.3 to 1.7 m. Except for the absolute magnitude of the flux, the model gives a fair approximation to flux behavior and some adjustment of the parameters in a phenomenological approach may provide a better fit of the data.

6. CONCLUSIONS

Empirical methods have been developed for correlating experimental measurements with parameters used in theoretical expressions for emanating power and diffusion coefficients for radon gas in cover materials. The methods incorporate well known procedures for measuring radon concentration, radon flux, and emanating power.

Emanating power is found to be dependent upon the mineralization and milling process of the uranium ore. Values are scattered and show little correlation with radium content, although high radium content can be correlated with the slimes.

Analytical models agree with the measured radon gas concentrations in the test chambers. They can be used with the data to determine the effective diffusion coefficient of the models cover material.

Radon flux from bare tailings can be described using analytical models if the effective diffusion coefficient can be determined. Radon fluxes through cover materials are not described well using diffusion theory, Fick's law, and diffusion coefficients determined from concentration experiments.

REFERENCES

1. Austin, S.R., "A Laboratory Study of Radon Emanation from Domestic Uranium Ores", Radon in Uranium Mining, IAEA-PL-565/8, 1975.

2. Schiager, K.J., "Analysis of Radiation Exposures on or Near Uranium Mill Tailings Piles", Radiation Data and Reports, EPA, Vol. 15, No. 7, July 1974.

3. Scott, J.H. and Dodd, P.H., "Gamma-Only Assaying for Disequilibrium Corrections", RME-135, Geology and Mineralogy, April 1960.

4. Culot, M.V.J., Schiager, K.J., and Olson, H.G., "Predication of Increased Gamma Fields After Application of a Radon Barrier on Concrete Surfaces", Health Physics, Vol. 30, pp. 471-478, June 1976.

5. Lucas, H.F.; "A Fast and Accurate Survey Technique for Both Radon-222 and Radium-226"; The Natural Radiation Environment; J.A.S Adams and W.M. Lowder, eds; University of Chicago Press; 1964.

6. Schiager, K.J., "Analysis of Radiation Exposures on or Near Uranium Mill Tailings Piles", Radiation Data and Reports, EPA, Vol. 15, No. 7, July 1974.

7. "Laboratory Research on Tailings Stabilization Methods and Their Effectiveness in Radiation Containment", Published by Ford, Bacon and Davis Utah for the Department of Energy, GJT-21, April 1978.

DISCUSSION

J.D. SHREVE Jr., United States

      My question relates to the experimental apparatus where you went in at different depths with tubing, you said. I would like to know the nature of the tubing, the material ?

R.F. OVERMEYER, United States

      It was a stainless steel tube with a 25 thousands ID diameter.

J.D. SHREVE Jr., United States

      It was metal then ?

R.F. OVERMEYER, United States

      It was metal, with a small fitted glass filter on the end so that it did not draw the fine tailings in when you put the evocuated lucas cell on the other end.

R.W. GILLHAM, Canada

      Your analysis considered the flexity of purely diffusive process, in fact it could be out of flexive components as a result of changes of barometric pressure or as a result of the gas phase beingdisplaced by infiltrating waste. I am wondering if these other components are being insignificant or if you are aware of any data in that regard ?

R.F. OVERMEYER, United States

      No, they were not at all considered insignificant, we did make some measurements, some corrections for pressure, and it was indicated in the paper, that was done, so we did consider barometric pressure differences. Secondly, water is very important, but we were trying to understand the diffusion process first. Most of the tailings that we have been working with are essentially dry in, the top meter or so. They are mostly in an arid region and they have been inactive for 10 to 15 years. Down below, we have found slimy tailings which have a wet clay appearance, but at least toward the top they are mostly dry and that is how we approach this particular problem because we are looking for cover material to reduce the radon, which in this case is a higher flux than one would normally see from saturated tailings from an operating mill.

A.B. GUREGHIAN, United States

      When you talk about Section 3, the last two sentences, you suggest limitations to fixed slope, implying that the continuity of flux is no longer valid. Ar you assuming that your system is non homogenous ?

R.F. OVERMEYER, United States

      It is not homogenous.

A.B. GUREGHIAN, United States

Why should not fluxes be continuous at any interface in the system ?

R.F. OVERMEYER, United States

I did not say that it was not continuous, in fact that was the boundary condition between the tailings and the soil, but the difficulty with the atmospheric interface is that it is almost like an infinite sink at that point and it becomes difficult to measure the gas concentration properly;if you can measure it a millimeter above that surface you may get some value different from a centimeter above and that is different from the value you measure a few centimeters.

A.B. GUREGHIAN, United States

I suppose that is the only difference in applying fixed slope or the continuity at interface between two different type of material, I appreciate the difference between the two, which would be to try to implement your equation by some reasonable values for "D" which probably would be remarkable in the case of the atmosphere.

R.F. OVERMEYER, United States

That is true. It is a complication at the interface, and while it is still as proportional to the gradient, the gradient at that point is very large, it is almost a step function, and so it is definitely a complication, but it could be handled I expect.

RADIUM BALANCE STUDIES AT THE BEAVERLODGE
MILL OF ELDORADO NUCLEAR LIMITED

V.I. Lakshmanan and A.W. Ashbrook
Eldorado Nuclear Limited
Ottawa, Canada

ABSTRACT

One major problem facing the uranium industry in Canada and elsewhere to-day is that of containment of radionuclides, specifically ^{226}Ra, in the extraction of uranium from its ores.

As part of a program within Eldorado Nuclear Limited to control, contain and isolate ^{226}Ra from the environment, a radium mass balance on its Beaverlodge uranium mill was undertaken early in 1978. The objective of this study was to determine the course taken by ^{226}Ra in the circuit, and provide basic data from which a rational approach will be taken in defining the best technology for eliminating the release of ^{226}Ra from this operation.

This paper first describes the mill circuit and the approach taken to sampling, sample preparation and analytical techniques for ^{226}Ra, involving some 27 sample stations. The results of the study are next presented, followed by an assessment of the data.

Finally, a comparison and assessment is made with data from other radium balance studies on both acidic and alkaline uranium mill circuits.

ETUDES DU BILAN MATIERE DU RADIUM EFFECTUEES A L'INSTALLATION
DE TRAITEMENT DE BEAVERLODGE DE LA SOCIETE
ELDORADO NUCLEAIRE LIMITEE

RESUME

L'un des principaux problèmes auxquels l'industrie de l'uranium est actuellement confrontée au Canada et ailleurs tient au confinement des radionucléides, et notamment du radium-226, pendant les opérations visant à dissocier l'uranium de ses minerais.

Dans le cadre d'un programme exécuté à la Société Eldorado Nucléaire Limitée en vue d'assurer le contrôle et le confinement du radium-226 et de l'isoler de l'environnement, on a entrepris au début de 1978 un bilan matière du radium portant sur l'installation de traitement de l'uranium de Beaverlodge. Cette étude visait à déterminer le cheminement du radium-226 dans le circuit et à établir des données fondamentales à partir desquelles on pourra suivre une démarche rationnelle pour définir la meilleure technologie permettant d'éliminer la libération de radium-226 au cours de ces opérations.

Cette communication décrit tout d'abord le circuit de traitement et la méthode adoptée pour procéder au prélèvement d'échantillons, à leur préparation et aux analyses du radium-226 qui fait intervenir quelque vingt-sept postes d'échantillonnage. On présente ensuite les résultats de cette étude suivis d'une éva-luation des données.

Enfin, on établit une comparaison et une évaluation à l'aide de données tirées d'autres études du bilan matière du radium sur des circuits de traitement de l'uranium par voie acide et alcaline.

INTRODUCTION

Due to the relatively high carbonate ($\approx$5%) and low sulphide ($\approx$1%) contents of the pitchblende-bearing ore, an alkaline leach process was chosen for uranium extraction at the Beaverlodge Mill[1]. The ore is leached in hot sodium carbonate/bicarbonate solution, and oxidised by the direct addition of oxygen to the leach pachucas. Uranium is precipitated from the leach solution as sodium diuranate. A schematic diagram of the milling process is shown in Figure 1.

One significant waste disposal problem for Eldorado, like other companies in the uranium mining industry, is the need for controls over the release of radioactive materials such as ^{226}Ra, either dissolved or in solid form, into the surrounding eco-system. It has been shown by various workers that most of the radium from the ore appears in the tailings[2,3]. This could mean either radium present in the ore is not leached, or is leached and precipitated at a certain stage in the mill circuit. Control of radium leaching from tailings will, however, depend on the nature of radium present in the tailings, precipitated or in the ore itself. The fact that most of the radium appears in the tailings can also probably mean that radium is bound within the crystal lattice of the mineral. Under such conditions, extensive grinding to liberate extra radium for fixation may not be successful. Any attempt made towards increased fixation of radium, under these conditions, can only be obtained from the mobile portion of radium present in the ore.

Radium content in the yellowcake obtained from Beaverlodge, before and after the lime circuit was introduced, indicates the possibility of radium leaching at a certain stage in the circuit. Thus radium concentrations before and after the introduction of a lime circuit were 1000 pCi/g and 50 to 80 pCi/g, respectively[4]. This can be due to the adsorption of radium by calcium salts. If radium is leached and precipitated, then testwork on topics such as leaching and precipitation kinetics, flocculation and settling rates should prove to be useful. However, if the portion of precipitated radium present in the tailings should prove to be only a small fraction of the whole, more attention is probably warranted in studying the containment of radium from tailings by pelletisation, selective leaching and precipitation techniques.

Hence, to study the fate of radium as it passes through the mill a systematic investigation was undertaken. The results obtained from the preliminary survey are presented here. Uranium, sodium and total alpha were also analysed on the samples taken during this study. In this report, the results obtained for radium and uranium only are included for discussion.

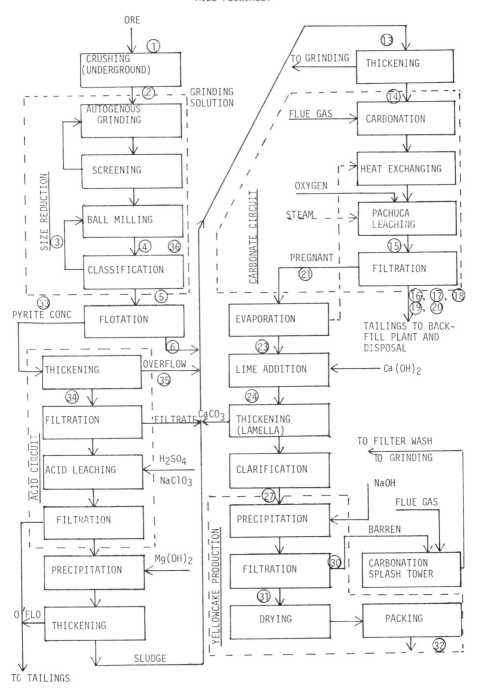

FIGURE 1

ELDORADO NUCLEAR LIMITED
BEAVERLODGE, SASKATCHEWAN
MILL FLOWSHEET

EXPERIMENTAL

Prior to the radium balance campaign, procedures to be followed in sampling, washing, etc. were established. Merits of grab and composite sampling procedures with respect to their representative nature, and the fate of radium while preparing composite samples, were considered. On the basis of preliminary results obtained for radium from two stations (pachuca leach feed and lime thickener underflow), with grab and eight hour composite sampling techniques, a grab sampling procedure was adopted for this campaign (Table I).

Methods such as alkaline fusion, *aqua regia* digestion and HF-HCl digestion to release ^{226}Ra from solids, together with washing procedures using water, and sodium carbonate/bicarbonate solutions, were tested (Table II). Based on the results, together with the fact that uranium and sodium analyses were to be included, water washes were accepted.

The campaign was carried out between 10 and 11am on January 11, 1978, and involved grab sampling from 27 pre-determined stations. Lime addition in the circuit during sampling was not operating due to low carbonate levels in the circuit. This condition had existed for about two days prior to sampling. The autogenous mill and acid leach circuits were not functioning during this sampling period.

During sampling, five litre slurry and one kilogram solid cuts were taken. The specific gravity of the slurrys was determined using pulp density meters. Slurrys were filtered using 3 x 12.5 cm Buchner funnels with Whatman No.3 filter paper, and filtrates from the three flasks were combined and mixed. Three clean filtrates of 0.5 L each were retained from each sample station for analysis. Filtrates, except the ones for immediate analysis, were stabilised by the addition of HNO_3 (10 ml L^{-1}) and stored in polyethylene containers.

Filter cakes from low percent solids slurrys were washed with 4 x 300 mL of hot distilled water. Filter cakes from high percent solids slurrys were treated to secure representative (≈500 g) samples. The samples were subsequently repulped with 4 x 300 mL of hot water for ten to twenty minutes, filtered and oven dried. Plus and minus 200 mesh fractions from these solids were then obtained by wet screening. The plus 200 mesh fractions obtained from stations 1, 3, 4, 5, and 36 were reduced in size (≈100% -150M) to make them suitable for elemental analysis. Plus 200 mesh fractions obtained from other stations were considered acceptable with respect to particle size for analytical treatment. The samples were thoroughly mixed after drying by tabling, and four equivalent samples were secured for analysis.

Radium in solids was determined after HF-HCl dissolution and water washing by the de-emanation technique. Solution samples were analysed for radium by the de-emanation procedure. Uranium was analysed fluorimetrically.

The tonnage of ore processed on January 11, 1978 was 748.8 tons. In the absence of automatic flowrate measurements in the mill circuit, certain assumptions were made in calculating the flows. Because of the selectivity of the leaching process for uranium compounds and the low concentration of uranium in the ore, it was assumed that the tonnage of undissolved ore solids did not vary significantly. In percent solid calculations a specific gravity of 2.66 for the ore, 5.90 for the yellowcake, and measured solution specific gravities were used. All solid tonnages were based on the mill flowsheet, ML59 MISC 10. All solution flows, except those for stations 21, 22, 23, 29, and 30, were calculated using specific gravities. Solution flows for stations 21, 22, 23, 29, and 30 were obtained from the flowsheet.

TABLE I

COMPARISON OF 8 h COMPOSITE AND GRAB SAMPLING

Sample	Solids ^{226}Ra pCi/g	Filtrate ^{226}Ra pCi/L
Leach Feed - Grab Nov. 7	710	1760
Leach Feed - 8 h comp. Nov. 7	650	1740
Lime Thickener U/F Nov. 7 - Grab	620	-
Lime Thickener U/F Nov. 7 - 8 h comp.	570	-

TABLE II

EFFECT OF ALTERNATIVE SOLID DISSOLUTION AND WASHING METHODS

Sample Treatment	pCi/g ^{226}Radium - Sample dissolution		
	Alkaline Fusion	Aqua Regia digestion	HF-HCl digestion
#1 Filtered #42. As is - no wash wet wt = 480 g dry wt = 410 g filtrate - 300 mL	630	570	590
#2 Filtered #42. Washed with 4 x 300 ml hot water	615	580	580
#3 Filtered #42. (1) Washed with 2 x 300 ml of hot carb solution (Na2CO3, NaHCO3). (2) Washed with 3 x 300 ml hot water	620	550	530

RESULTS AND DISCUSSION

Table III presents certain process stream characteristics for the stations sampled. Station identification, specific gravity of the slurry, percent solids and specific gravity of the solution are shown for each sampling station.

Analytically determined concentrations of ^{226}Ra in dissolved and undissolved form at the various stations are given in Table IV. Undissolved radium, as quoted here, is that portion retained on a Whatman No.3 filter, while dissolved radium represents that contained in the filtrate.

One of the primary purposes of this survey was to determine the amount of ^{226}Ra in the process at various locations and the amount in the dissolved and undissolved form at various stations. It was also part of the scope of the study to determine the distribution of radium with +200M and -200M portions of the solids. To this end, a radium balance or distribution is shown in Tables V, VI, VII.

An initial test of the results was made on the basis of the uranium analyses (Table VIII) and the assumption of secular equilibrium between uranium and radium in the raw ore. This situation need not be true for all ore samples. The equilibrium could have been disturbed due to selective migration or weathering. However, if the assumption of secular equilibrium is valid there would be 0.758 curies of ^{226}Ra in the process, which compares favourably with the experimentally determined value of 0.756 curies. If an extension of this hypothesis to include all solids prior to pachuca leaching is made, the ^{226}Ra content drops significantly and remains at 0.525 to 0.568 curies at stations 5, 6, 13, 14, and 36. This compares with the experimentally determined value of 0.499 to 0.629 curies in solids. Such a drop in radium (and uranium) values in solids at stations 5, 6, 13, 14, and 36 compared to the primary ball mill feed at station 1 suggests either the grab sample obtained at station 1 is different from samples obtained at stations 5, 6, 13, 14, and 36 (where equilibrium values can be compared) or substantial leaching of uranium and radium has taken place. It is known that grinding usually leaches up to three percent of the contained uranium and, hence, the latter hypothesis may not be correct. It should also be noted that uranium analyses carried out at the mill during the month of January on flotation feed varied from 0.152% to 0.344% U_3O_8, and on January 10 and 11 were 0.293% and 0.232%, respectively. The results also show, on the basis of rougher tails entering the alkaline leach circuit, that the ^{226}Ra associated with flotation concentrates and entering the acid leach circuit as ATO (acid thickener overflow) thickener underflow is only 1.3%.

Total radium at stations 6, 13, 14, 15, and 16 does not change significantly. However, a significant drop in radium content from 0.57 curies to 0.39 curies at the tailings tank is observed. The radium content in solution did not change between stations 16, 17, and 18. It is not known whether samples obtained at stations 17 and 18 were different to that at station 16. The reason for the drop will be studied during the next campaign.

Distribution of radium between the +200M and -200M fractions in the leach circuit shows a higher percentage of radium associated with the -200M fraction. Tests carried out by Khoja[3] with Beaverlodge tailings showed that the bulk of the radium in tailings is concentrated in the fines, and the -200M tailings fraction contained 86.2% of the total radium. The results obtained from the present campaign show that 83.6% of the radium in solids is associated with the -200M fractions of the tailings tank discharge.

The maximum dissolved radium in solution at any station in the Beaverlodge alkaline leach circuit, as shown by this preliminary survey, is 1.2% which compares with the results obtained from American mills using alkaline circuits. The dissolved portion of the radium decreases to <0.1% when discharged through the tailings tank. This could mean that radium precipitates as it passes through the circuit.

It should be noted, however, that the feed to the ATO thickener ($\approx$2% total feed) after flotation showed 5.2% of the radium in solution, while the underflow after thickening showed only 0.1% in solution. Activities of radium at station 33 (feed to ATO thickener), 34 (ATO underflow) and 35 (ATO overflow) were 2870, 2150, and 1680 pCi/L, respectively.

A significant drop in the radium values in solution between station 23 (2850 pCi/L) and station 27 (50 pCi/L) was observed. These values were obtained when liming in the circuit was not in operation. Such a drop may mean either that sufficient calcium salts were present in the lime reaction tank to adsorb radium or, that caustic addition caused radium to precipitate. Yellowcake obtained during this campaign also contained only 48 pCi/g ^{226}Ra. As mentioned earlier, radium content in the yellowcake before and after the introduction of a lime circuit was 1000 pCi/g and 50 to 80 pCi/g, respectively. If the yellowcake sample taken is representative, and corresponds to the process conditions as they existed during the campaign, it supports the determined loss of radium from solution in the precipitation circuit.

Results obtained from similar campaigns in acid leach processes with resin-in-pulp, CCD ion exchange and CCD solvent extraction circuits have indicated that from 0.2% to 6.7% of the radium from the feed is dissolved[2,5]. In the acid leach mills most of the soluble radium does not follow the soluble uranium values which are selectively recovered in ion exchange and solvent extraction circuits, but rather it reports in the effluent or raffinate tailings. The amount of radium contained in the yellowcake produced from the acid plants varies from 0.06% to 0.26% of the radium in the ore feed[5]. In alkaline leach circuits the available literature indicates the dissolution of radium at 1.5% to 3.0%[6]. In one mill 2.3% of the total radium from the feed dissolved and nearly all the dissolved portion (2.2%) reported with the yellowcake[5].

SUMMARY

The results obtained from this preliminary campaign show that for the process described, the maximum soluble portion of radium observed in the alkaline leach circuit is 1.2%, and falls to <0.1% at the tailings tank discharge. The results also show that the radium content entering the acid circuit is not significant. In the tailings discharge, most of the radium is associated with the -200M fraction, and is 83.6% of the total radium at that station. Even in the absence of lime addition in the precipitation circuit, a substantial drop in the radium level in solution, from 2580 pCi/L to 50 pCi/L, was observed. Radium balance on the basis of the feed to the primary ball mill is poor.

ACKNOWLEDGEMENTS

The support, assistance and contributions of Mr. D.H. Bell and Mr. A.B. Cron (Beaverlodge) to this campaign are gratefully acknowledged, as is Mr. D.Y. Shimano for doing much of the experimental work. Our thanks are extended to Mr. H.H. Wirch (Beaverlodge), Mr. T. Szaplonczay (Ottawa) and Mr. R.D. John (Port Hope) for the analytical work involved.

REFERENCES

[1] "Brief to Cluff Lake Inquiry", Eldorado Nuclear Limited, August, 1977.

[2] "Process and Waste Characteristics at Selected Uranium Mills", U.S. Dept. of Health, Education and Welfare, Public Health Service. Technical Report W62-17.

[3] Khoja, Z.M. M.Sc. Thesis, Carleton University, Ottawa, 1977.

[4] Lakshmanan, V.I., Macdonald, D.E., and Ashbrook, A.W. Workshop on Radium Control, Canadian Uranium Producers' Metallurgical Committee, Ottawa, 1977.

[5] "Summary Report on: I The Control of Radium and Thorium in Uranium Milling Industry. II Radium-226 Analysis Principles, Interference and Practice. III Current Winchester Laboratory Projects". National Lead Co., Inc., WIN-112. 1960. 97 pages.

[6] Merritt, R.C., "The Extractive Metallurgy of Uranium", Boulder, Colorado, Johnson Publishing Co., 1971.

TABLE III

SAMPLE STATIONS AND PROCESS STREAM CHARACTERISTICS

Sample Station		Slurry		Solution
Number	Location	sg	solids, %	sg
1	Primary Ball Mill Feed	2.622	97.5	n/a
2	Grinding Solution	1.085	0.0	1.091
3	Secondary Ball Mill Feed	1.960	72.1	1.086
4	Classifier Feed, Secondary Ball Mill	1.730	59.6	1.092
5	Surge Tank	1.300	26.2	1.089
6	Rougher Tails	1.185	12.7	1.092
13	Carbonate Thickener Feed	1.210	16.3	1.088
14	Carbonate Thickener U/F	1.635	53.8	1.091
15	Pachuca Discharge	1.600	50.8	1.097
16	Filtration Repulp Slurry	1.830	62.8	1.119
17	S. Filtration Repulp Slurry	1.280	34.9	1.004
18	Tailings Tank	1.155	21.1	1.004
19	Dorrclone O/F	1.050	7.2	1.003
20	Dorrclone U/F	1.700	66.4	1.002
33	A.T.O. Thickener Feed	1.140	7.2	1.089
34	A.T.O. Thickener U/F	1.750	65.7	1.091
35	A.T.O. Thickener O/F	1.105	0.0	1.084
36	Classifier Feed, P.B.M.	1.580	50.4	1.087
21	P. Filtrate after Carbonation	1.100	<0.1	1.102
22	S. Filtrate	1.100	2.1	1.086
23	Leach Slurry, Steam Stripped	1.100	0.0	1.115
27	Precipitation Feed	1.180	5.9	1.137
28	Reseed Cone U/F	1.200	7.7	1.123
29	Reseed Cone O/F	1.100	<0.1	1.121
30	Filtrate, Barren	1.090	0.0	1.117
31	Yellowcake Slurry	2.360	65.7	1.072
32	Yellowcake, Dry	5.9	100.0	n/a

n/a: not available

TABLE IV

CONCENTRATIONS OF RADIUM IN SOLID AND SOLUTION SAMPLES

Station Number	Solid +200M, pCi/g	Solid -200M, pCi/g	Solution pCi L^{-1}
1	1040	2300	n/a
2	n/a	n/a	1490
3	950	2180	2870
4	1060	1275	2810
5	690	1080	2180
6	450	950	1380
13	425	1060	1940
14	460	1090	2130
15	207	1285	1020
16	305	1300	1120
17	260	820	70
18	190	910	70
33	375	755	2870
34	625	1140	2150
35	n/a	n/a	1680
36	560	1260	1970
21	n/a	1350	2230
22	n/a	1030	1320
23	n/a	n/a	2580
27	n/a	205	50
28	n/a	155	40
29	n/a	310	31
30	n/a	n/a	28
31	n/a	71	50
32	n/a	48	n/a

n/a - not available

TABLE V

RADIUM DISTRIBUTION IN THE BEAVERLODGE MILL

Station Number	^{226}Ra distribution[a] % +200M	% -200M	Solution, %	Total Curies in Solids & Solution D-1*	Theoretical ^{226}Ra distribution Total CiD-1
1	88.2	11.8	–	7.56×10^{-1}	7.58×10^{-1}
5	29.5	69.8	0.7	6.33×10^{-1}	5.61×10^{-1}
6	13.6	85.3	1.1	5.60×10^{-1}	5.32×10^{-1}
13	18.5	80.3	1.2	5.63×10^{-1}	5.25×10^{-1}
14	21.5	78.3	0.2	5.67×10^{-1}	5.33×10^{-1}
15	11.1	88.8	0.1	5.47×10^{-1}	–
16	16.2	83.7	<0.1	5.72×10^{-1}	–
17	18.4	81.6	<0.1	3.95×10^{-1}	–
18	16.4	83.6	<0.1	3.91×10^{-1}	–
33	10.4	84.5	5.2	5.84×10^{-3}	8.06×10^{-3}
34	12.8	87.1	0.1	8.42×10^{-3}	8.46×10^{-3}
35	–	–	100	3.4×10^{-4}	–
36	57.2	42.6	0.3	5.00×10^{-1}	5.68×10^{-1}
21	–	–	100	1.1×10^{-3}	–
22	–	–	100	7.4×10^{-4}	–
23	–	–	100	8.8×10^{-4}	–
27	–	99.6	0.4	2.7×10^{-4}	–
28	–	99.3	0.7	1.2×10^{-3}	–
29	–	99.8	0.2	6.7×10^{-4}	–
30	–	–	100	1×10^{-5}	–
31	–	99.9	<0.1	1.5×10^{-4}	–
32	–	100	–	1.2×10^{-4}	–

*D-1 - per day

a: from Analytical determinations

TABLE VI

RADIUM DISTRIBUTION IN THE LEACH CIRCUIT

Station Number	226Ra in +200M solids (Ci D-1)*10	226Ra in -200M solids (Ci D-1) 10	Total 226Ra in solids (Ci D-1) 10	226Ra in Sol'n. (Ci D-1) 10^4	Total 226Ra in solids & solution (Ci D-1) 10
36	2.86	2.13	4.99	13.2	5.00
5	1.87	4.42	6.29	41.9	6.33
6	0.76	4.78	5.54	63.8	5.60
13	1.04	4.53	5.57	66.9	5.63
14	1.22	4.44	5.66	12.3	5.67
15	0.61	4.86	5.47	6.6	5.47
16	0.93	4.78	5.71	4.5	5.72
17	0.73	3.22	3.95	0.9	3.95
18	0.64	3.27	3.91	1.8	3.91

*D-1 - per day

TABLE VII

RADIUM DISTRIBUTION IN THE PRECIPITATION CIRCUIT OF THE BEAVERLODGE MILL

Station Number	226Ra in -200M solids (Ci D-1) 10^4	Total 226Ra in solids (Ci D-1) 10^4	226Ra in solution (Ci D-1) 10^4	Total 226Ra in solids & solution (Ci D-1) 10^4
21			11.0	11.0
27	2.7	2.7	<0.1	2.7
29	6.6	6.6	0.1	6.7
30	-	-	0.1	0.1
31	1.5	1.5	<0.1	1.5
32	1.0	1.0	-	1.0

*D-1 - per day

TABLE VIII

CONCENTRATION OF URANIUM IN VARIOUS SOLIDS AND LIQUID SAMPLES

Station Number	Solid % U, +200M	Solid %U, -200M	Liquid U, g/l
1	0.315	0.53	-
2	-	-	1.28
3	0.30	0.51	1.28
4	0.30	0.36	1.10
5	0.19	0.28	1.41
6	0.15	0.26	0.70
13	0.15	0.28	1.28
14	0.16	0.28	1.51
15	0.015	0.01	3.00
16	0.015	0.01	0.39
17	0.015	0.01	<0.01
18	0.015	0.01	<0.01
33	0.16	0.32	1.425
34	0.215	0.33	1.16
35	-	-	0.98
36	0.24	0.26	1.52
21	-	-	2.81
22	-	0.02	0.67
23	-	-	3.05
27	-	66.3	0.17
28	-	66.5	0.065
29	-	67.1	0.065
30	-	-	0.05
31	-	66.9	0.05
32	-	68.7	-

DISCUSSION

F.G.H. JANTZON, Canada

To what extent is this knowledge you so describe useful in an active metallurgic plant ?

V.I. LAKSHMANAN, Canada

Well, this helps us in a few ways. This will help us if there is any place in the mill cycle where we could contain the radium before it reaches the tailings treatment system. We wanted to find where there is any selective leaching any place in the mill cycle where there could be a lot of radium going into the solution and we could isolate that portion of radium so that the amount of radium getting into the tailings system could be reduced. But we found through this whole exercize, that the maximum amount of radium getting into the solution is only 1.2 %. This means that we have to concentrate our efforts down the stream and see how we could stabilize and reduce the radium content in the tailings.

AIRBORNE PARTICULATE CONCENTRATIONS AND FLUXES AT AN
ACTIVE URANIUM MILL TAILINGS SITE

G. A. Sehmel
Pacific Northwest Laboratory
Richland, Washington, U.S.A. 99352

ABSTRACT

Direct measurements of airborne particulate concentrations and fluxes of transported mill tailing materials were measured at an active mill tailings site. Experimental measurement equipment consisted of meteorological instrumentation to automatically activate total particulate air samplers as a function of wind speed increments and direction, as well as particle cascade impactors to measure airborne respirable concentrations as a function of particle size. In addition, an inertial impaction device measured nonrespirable fluxes of airborne particles. Calculated results are presented in terms of the airborne solid concentration in g/m^3, the horizontal airborne mass flux in $g/(m^2 \text{ day})$ for total collected nonrespirable particles and the radionuclide concentrations in dpm/g as a function of particle diameter for respirable and nonrespirable particles.

CONCENTRATIONS ET FLUX DE PARTICULES EN SUSPENSION DANS L'AIR

AU-DESSUS D'UN DEPOT DE RESIDUS RADIOACTIFS DE TRAITEMENT DE L'URANIUM

RESUME

On a procédé, au-dessus d'un dépôt de résidus radioactifs de traitement de l'uranium, à des mesures directes des concentrations et des flux de particules en suspension dans l'air résultant du déplacement de ces résidus. Le matériel de mesure expérimental était composé d'une instrumentation météorologique permettant de déclencher automatiquement des dispositifs d'échantillonnage de la teneur globale en particules de l'air en fonction des accélérations et de la direction des vents, ainsi que des impacteurs en cascade destinés à mesurer les concentrations respirables en suspension dans l'air en fonction de la granulométrie. En outre, on a employé un système à impacts par inertie pour mesurer les flux non respirables de particules en suspension dans l'air. Les résultats calculés sont présentés en fonction de la concentration des solides en suspension dans l'air exprimée en g/m^3, du flux massique horizontal en suspension dans l'air exprimé en $g/(m^2\text{-jour})$ correspondant à la totalité des particules non respirables recueillies et des concentrations de radionucléides exprimées en dpm/g d'après le diamètre des particules respirables et non respirables.

This work was prepared for the Nuclear Regulatory Commission under Contract EY-76-C-06-1830 with Pacific Northwest Laboratory.

INTRODUCTION

The general topic of this afternoon's discussion is "Source Terms for Uranium Mill Tailings Areas." The work reported is directed towards determining the source term for airborne particulates. For these areas, there are at least four source-term parameters that may influence wind erosion of the uranium mill tailings piles. These parameters are: the "flat" surface of a pile, which can be partially covered with water, the sloping sides of a pile, the coarse sands in a pile and the fines in the pile. The coarse sands are preferentially deposited to form the pile sides and retaining dike, while slimes consisting of fine particulates are deposited on the flat surface. Although all piles do not have identical characteristics, studies of these four parameters can lead to significant insight into the wind erosion processes at any mill tailings site.

The site selected for the initial wind erosion research was at the United Nuclear-Homestake Partners mill tailings pile near Grants, New Mexico. The mill tailings pile is approximately 700 by 1300 m in area and is about 35 m in height above the surrounding flat terrain. At this pile, all four parameters are studied for a carbonate leach mill tailings separation process.

The purpose of this paper is to present an overview of current research results from airborne particulate measurements begun in August 1977 at this site. Selected overview results will be discussed within the framework of the allotted time. To be discussed are equipment used in collecting airborne particulates, the experimental sampling array for measuring airborne particulates, results in terms of the dpm/g of airborne radionuclide, airborne solids concentrations in g/m^3, and the average airborne fluxes of nonrespirable particles in $g/(m^2 \, day)$. In all cases, results are reported as average values for a selected wind direction and wind speed increments.

AIRBORNE PARTICLE SAMPLING EQUIPMENT

Airborne particle sampling equipment was either electrically powered or operated by wind forces. Electrically powered samplers were activated automatically as a function of wind direction and wind speed increments. These wind parameters were automatically determined with a specially designed meteorological sensing instrument by which both the wind direction and wind speed increments for particle sampling could be selected. Thus, when the wind direction and wind speeds were within selected ranges, the electrical control signals from the cup and vane anemometer were used to activate selected air sampling equipment. Selected wind speed increments for sampling airborne particulates were 3 to 5 m/s, 5 to 7 m/s, and 7 to 11 m/s for the selected wind direction or for all wind speeds for a selected wind direction.

The control signal from the meteorological sensing instrument was transmitted to activate either isokinetic air samplers for measuring total particulate loading or to particle cascade impactors to measure airborne solid concentrations as a function of respirable particle diameter. For the isokinetic airborne particle sampler, airborne concentrations of both "respirable" and nonrespirable particles were measured as a function of wind speed increments.

The isokinetic air sampler inlet is shown in Figure 1. A cutaway view of the sampler inlet is shown with the inlet attached to a filter holder. The inlet cover is shown opened in this cut-away view of the sampler. A solenoid operated closure uncovers the inlet when the air sampler is activated by the meteorological sensing instrumentation. The inlet is 25 cm wide, and its height is adjusted by matching isokinetic flow with the selected wind speed increment to be sampled. The inlet cross-sectional area is held constant for the first 7.5 cm towards the filter, after which the top of the 16 ga aluminum inlet is angled upward to fit over the filter holder. The total distance between the inlet and the filter is 25 cm. The inlet is attached to a standard 20 x 25 cm (8 x 10 in.), high-volume air sampler operating at 1.13 m^3/min (40 cfm).

Also shown schematically are the paths of motion for "large" and "small" particles. In this case, large refers to those particles that settle on the inlet bottom, and small refers to those particles collected on the filter. Airborne concentrations in g/m^3 are calculated from the solid collection on the filter. The airborne flux of nonrespirable particles in g/(m^2 day) is calculated from the solid collection on the inlet bottom, the inlet cross-sectional area, and sampling time.

Airborne fluxes of nonrespirable particles were also collected with air impact flow particle collectors. This air impact flow collector is schematically shown in a cutaway view in Figure 2. Particles enter the 7.6-cm-diameter inlet due to the combined effects of particle inertia driving large particles into the inlet as well as the airflow rate permitted by the collector air flow resistance. The internal resistance is decreased by expanding the outlet diameter to 25 cm. The outlet diameter is sealed with a 25 μm nylon screen, which retains all large particles and tends to collect some smaller particles. That is, this collector collects airborne nonrespirable particles greater than 25 μm in diameter. In interpreting data from this collector, nonisokinetic sampling errors are assumed minimal. This is a reasonable approximation for particles greater than 25 μm in diameter and for a field instrument operated by wind impact pressure.

The third sampler for collecting airborne particulates was a particle cascade impactor (Sierra Instrument Company) sampling at 1.13 m^3/min (40 cfm). A cyclone preseparator on the particle cascade impactor tended to prevent larger than respirable particles from entering the particle cascade impactor. Although large particle collection was minimized, some large particles of an unknown amount did pass through the cyclone preseparator and were collected within the cascade impactor. Thus, these data are qualified in that some large particles may have been collected within the particle cascade impactor and were subsequently reentrained through the particle cascade impactor to the backup filter. Nevertheless, the data to be presented are the only data at present that have been analyzed radiochemically to show the radionuclide concentrations as a function of respirable particle size. Respirable particle sizes are those particles less than about 3.5 μm in diameter.

Only limited radionuclide results have been determined to date. For the radionuclide data to be shown, the dpm/g of airborne solid are given as a function of particle diameter for three different types of airborne particulate samples. The first samples were collected on stages of cascade particle impactors. Data will be shown as a function of the cascade particle impactor stage 50% cutoff diameter. The second samples were collected on the cascade particle impactor backup filter. Particles collected on the backup filter are nominally less than the cutoff diameter of the last stage. The third samples are for large particles collected by inertial collection. These large particles were subsequently sieved into eight size increments between 37 and 420 μm.

ONSITE AIR SAMPLING ARRAYS

The relative locations of airborne solid sampler equipment on and around the active mill tailings site are shown in Figure 3. As shown, north is towards the top of the figure. In meteorological terminology, north is termed 0° while winds from the southwest are termed 225° winds. At this location, prevailing winds are from the southwest, while the secondary principal wind direction is from the northwest. The sampling array was set up to determine the mill tailings pile movement.

In this figure, three samplers types are shown in the symbol key. The isokinetic samplers are shown by diamonds, particle cascade impactors are shown by squares, and air impact flow particle collectors are shown by circles. These air samplers were located at sampling tower heights from 0.3 to 15 m above the tower base.

There are six principal onsite sampling areas shown in this figure. These are site A, site B1, site B2, site C, the air impact flow collectors along the northern retaining dike, and the samplers inside the north fence. The purpose of these sites was to determine the increased airborne particle and radionuclide concentration as the air blew from the background site and across the pile. Between site A and sites B1-B2, these B sites were to determine increased airborne concentrations caused by wind erosion from the sloping sides of the pile. Between sites B and sites C, the effect of a flat mill tailings surface was investigated. Between site C and the north fence, the decreased airborne concentration with distance was studied. Thus, between sites A and B, the wind erosion of a coarse source particle material constituting the banks of a sloping mill tailings pile was studied. Between sites B and C, the wind erosion of fines on a flat surface of a uranium mill tailings pile was determined.

Site A was a background site at which wind speed and direction instrumentation was located. When wind was blowing from the southwest, selected air samplers were activated automatically to determine airborne solid concentrations in background air approaching the pile before air was contaminated by wind-eroded material from the pile.

At site A, there were three sets of isokinetic air samplers, shown by the three orientations of the triangle symbol. One set of airborne solid collectors (triangles) was activated for wind speeds between 3 and 5 m/s, the second between 5 and 7 m/sec, and the third between 7 and 11 m/sec for the selected wind direction, which was centered around $211°$. Three cascade impactors are shown by the three box symbols. One cascade impactor was turned on for each of the three wind speed increments.

Control signals for activating samplers at site A were transmitted to sites B1, B2, and C for sampling at those sites for the three wind speed conditions at site A. However, after control signals were received at site C, control signals were combined before sending a single control signal to samplers inside the north fence. The single signal was combined so that the north fence air samplers sampled the same wind direction increment, but sampled all winds between 3 and 11 m/sec.

These four sites, A, B1, B2, and C, determine airborne concentrations below 15 m and give little information about the crosswind airborne concentrations variations due to wind erosion. Consequently, there are nine sites at 91 m spacing along the north fence to show crosswind variations in airborne concentrations. Similarly, between site C and the north fence, there are impact flow particle collectors located (52 m spacing) along the retaining dike ridge to determine crosswind variations caused by both local terrain (pile) variations as well as meteorological variations across the pile.

Changes in nonrespirable particle airborne fluxes were determined by collection on both isokinetic air sampler inlets as well as within air impact flow particle collectors across the pile.

OFFSITE AIR SAMPLERS

Since there is a road immediately north of the north fence, samplers north of the road are termed offsite sampler locations in this report. There were four principal offsite air sampling stations. At these sites, air samplers were activated as a function of wind direction while sampling all wind speeds from that direction. At these four sampling sites, wind directions increments of $90°$, $60°$, and two at $15°$ were sampled. The wind direction increments were selected to sample most winds blowing across the pile. Thus, at the closest sampling station, a $90°$ wind direction increment centered on the pile was sampled, while at the two outermost sampling locations only $15°$ wind direction increments were sampled. Air impact particle collectors were also located at each of these four offsite locations.

RESULTS

To estimate an airborne source term for this uranium mill tailings pile, airborne radionuclide concentrations in dpm/g must be known, as well as airborne solids concentrations in g/m^3 and nonrespirable fluxes in $g/(m^2 \text{ day})$. Some of these pertinent values will be discussed.

Airborne radionuclides concentrations in dpm per gram

Radionuclide concentrations in dpm/g were determined for ^{238}U, ^{230}Th, ^{226}Ra, and ^{210}Pb as a function of particle diameter. These results are shown in Figures 4-7 for two different time periods. Concentrations on airborne respirable particles were determined between August 10 and September 12, 1977, while concentrations on airborne nonrespirable particles were determined from samples collected between November 19 and December 8, 1977. Data in dpm/g are shown with one sigma radiochemical counting limits as vertical bars. Otherwise, the data symbol encompasses the one sigma range. For nonrespirable particles above 37 μm diameter, horizontal limits around each data point show sieve size limits for each data point.

The dpm/g of airborne solids is shown for ^{238}U in Figure 4. For all particle diameters collected within the particle cascade impactor, the ^{238}U concentration was nearly constant at about 140 dpm/g of airborne solid. For larger nonrespirable particles, the dpm/g decreased. The decrease was inversely proportional with particle diameter for particle diameters between 90 and 320 μm. This inverse relationship, with a minus-one slope shown in the Figure, suggests that ^{238}U was attached as small particles to the surface of these larger particles. For the data collected to date, the ^{238}U ranged between 140 and to 32 dpm/g.

Similar data are shown in Figure 5 for ^{230}Th, which is a subsequent daughter product of ^{238}U. In this case, the ^{230}Th dpm/g in the respirable range shows some nonuniformity with particle diameter. For nonrespirable particles, the dpm/g decreases. Above 90 μm, the decrease with diameter again shows a minus-one slope. This decrease suggests that ^{238}Th is also attached as small particles to larger host solid particles in this size range. The ^{230}Th ranges from 2300 dpm/g in the respirable range down to 230 dpm/g in the nonrespirable range.

The dpm/g for ^{226}Ra is shown in Figure 6. In this case, the dpm/g shows a nonuniform distribution as a function of particle diameter in both the respirable and nonrespirable size ranges. By comparison, with the earlier members of the radioactive decay chain, the succeeding members of the decay chain show a more pronounced variation in dpm/g for all particle diameter ranges. The maximum is approximately 3300 dpm/g for a 0.5 μm particle diameter. Again, for particle diameters above 90 μm, the minus-one slope suggests small ^{226}Ra particles are attached to larger host solid particles.

Pb-210 was the last decay chain member for which radiochemical data were obtained. These data, shown in Figure 7, indicate a maximum concentration of 2500 dpm/g for a 0.5-μm-particle diameter . The maximum dpm/g for both Figures 6 and 7 is for 0.5-μm-diameter particles.

One might question why the backup filter dpm/g is less than for the 0.5-μm-diameter impactor stage. Possibly, larger particles have reentrained through the cascade impactor to be retained on the backup filter. However, it is unknown how much reentrainment occurred within the impactor. Another possibility might be related to particle diameter. Possibly, it is the 0.5-μm-diameter particle that becomes attached most readily to all other particle sizes. For smaller particles sizes represented by the backup filter collection, these 0.5-μm-particles cannot become attached to these smaller particles without changing the agglomerate size. Whatever the attachment mechanism is, it is presently unknown and results are still being obtained to determine the radionuclide distribution on airborne solids. Nevertheless, the dpm/g data are consistent and show that radionuclides are transported on all particle diameter ranges. The maximum radionuclide concentrations are on the respirable particle size ranges. For particle diameters above 90 μm, the radionuclide concentrations appear to indicate attachment of small radionuclide particles to the surface of larger inert particles.

The observation that small particles are attached to larger host particles is important in modeling and in proper characterization of tailings materials for wind erosion. Mechanisms that dislodge these fine particles from larger host solid particles will potentially be a source of many fine particles for airborne transport.

Airborne solid concentrations in g/m^3

Airborne solid concentrations reported here were for four different time periods beginning August 1977 and ending in April 1978. Airborne concentrations for the first sampling period are shown in the upper portion of Figure 8. For this initial sampling period, sampling was for all wind speeds for a wind direction of $211° \pm 30$ at site A. Average airborne particulate concentrations increased one to two orders of magnitude.

In the next time period from November 19 to December 8, sampling was a function of wind speed increments. Airborne particulate concentrations in g/m^3 are shown in Figure 9 at sampling sites A1, B1, B2 and along the north fence. As expected, airborne solid concentrations were least at the background site A and increased across the sloping sides of the pile towards site B. Airborne concentrations along the north fence were one to two orders of magnitude higher than at the background site A. Background airborne concentrations at site A ranged from 2×10^{-5} to 3×10^{-4} g/m^3. At site B, airborne concentrations were up to four orders of magnitude higher than background. These increases in airborne mass concentrations at site B and along the north fence show wind erosion was occurring from the mill tailings area. It is surprising, however, when one looks at airborne concentrations as a function of wind speed. Airborne concentrations did not always increase with increasing wind speed as was expected. Wind speed effects are still being studied.

Similar results are shown in Figure 10 for the succeeding time period from February 21 to March 27, 1978. In this case, data are also shown for site C as well as offsite locations. At site B, maximum airborne concentrations in g/m^3 increased up to three orders of magnitude above background. This increase indicated material was rapidly transported from the bluff sides of the mill tailings pile. In contrast, at site C, airborne concentrations did not show a rapid increase across the flat pile surface of the slimes area. Although airborne concentrations increased towards site C, concentrations at breathing height along the north fence were only up to one order of magnitude greater than background. As wind blew eroded material offsite to distances of 4,000 m from the central north fence sampling site, airborne solid concentrations were within the same order of magnitude as background. Similar results are shown in Figures 11 and 12 for succeeding sampling time periods. Airborne solid concentrations increased across the pile and tended to decrease toward background at 4,000 m from the mill tailings site.

Airborne nonrespirable mass flux

The airborne mass flux is defined in terms of a unit square meter. The unit square can be visualized in terms of a one-square meter flat board. The board is oriented with the plane of the board perpendicular to the earth's surface. In this case, a horizontal wind blows through the unit area when the horizontal wind is perpendicular to the unit area.

Airborne mass fluxes were calculated for the air impact flow particle collectors and the total time collectors were in the field. Mass fluxes in $g/(m^2 \text{ day})$ are shown in Figure 13 for the time period from November 19 to December 8, 1977. Mass fluxes for the background site A, sites B1 and B2, and site C are shown separately for collection in the impact flow particle collectors: either on the 25-μm screen filter or this filter plus inlet collection. Also shown are mass fluxes for the particle cascade impactor cyclone inlet. These samplers were all facing $211°$. At the background sampling site A, the airborne mass flux of nonrespirable particles averages about 0.4 $g/(m^2 \text{ day})$ for sampling heights from 0.3 to 3 m. Across the sloping bluff sides of the bluff uranium mill tailings pile, airborne fluxes increased between sampling site A towards sites B1 and B2. The increase at site B1 was up to four orders of magnitude above background, while at site B2 the increase was up to five orders of magnitude larger than background. At site C, the maximum airborne flux was up to four orders of magnitude greater than background. At site C, the flux at 15 m was up to two orders of magnitude larger than background.

Along the north fence, the airborne flux leaving the site was about two orders of magnitude larger than background. However, unexpectedly, the airborne flux transported from the north across the road toward the pile was also up to two orders of magnitude greater than background. This unexpected increase from the north might be attributed to road material suspension caused by vehicular traffic. Offsite, at distances of 4,000 m, the mass flux from the south tended towards the background flux at site A.

Similar airborne mass flux data are also shown in the lower portion of Figure 8. However, in this case, sampling was for all wind speeds. In Figure 14, airborne mass fluxes are shown which were determined with air impact flow particle collectors. In all cases, an increase in mass flux is shown.

Nonrespirable mass fluxes were determined with the isokinetic samplers for time periods during which the inlet closures were open. Mass fluxes were determined from the material collected at the bottom of the isokinetic sampler inlet. Inlet airborne mass fluxes in $g/(m^2 \ day)$ are shown in the bottom portion of Figure 12. Between site A and site B, across the bluff sloping surface, the mass flux increased one to two orders of magnitude. Across the flat fines area between sites B and C, airborne mass fluxes remained within the same order of magnitude. In contrast, along the north dike, the mass flux determined with the air impact flow particle collectors was up to twice the mass fluxes determined at site C from the isokinetic air sampler inlets. However, along the north fence, mass fluxes decreased to within the same order of magnitude as background.

Offsite north of the road, the mass flux at the first sampling site was one order of magnitude greater than along the north fence. This increase may have been caused by either soil movement caused by vehicular traffic suspension from the road or possibly by the maximum airborne plume concentration of nonrespirable particles passing above the north fence sampling sites and finally touching ground near the first offsite sampling location. It is unknown if either road suspension or plume height are more valid in explaining these airborne mass flux data. Nevertheless, at distances 4,000 m offsite, airborne mass fluxes are comparable to those measured at the background site A.

For comparison purposes, airborne mass loadings in g/m^3 are also shown in the upper portions of Figure 12. Airborne concentrations also increased across the pile edge from site A to site B and further increased up to site C. Along the north fence, airborne concentrations were still within one order of magnitude greater than background concentrations measured at site A. Concentrations 4,000 m offsite approached background.

CONCLUSIONS

These data on airborne solids concentrations in g/m^3 and airborne mass fluxes in $g/(m^2 \ day)$ show that mill tailings material is eroded by wind stresses and is removed from the mill tailings pile. Erosion occurs both from the mill tailings pile sides, as well as the flat pile surface. Concentrations and fluxes decrease with distance north of the pile and tend to be within the same order of magnitude as background at 4,000 m from the pile. Although airborne mass fluxes in g/m^3 and airborne mass fluxes in $g/(m^2 \ day)$ tend toward background at the 4,000 m distance, it is unknown for respirable particle size ranges how airborne radionuclide concentrations decrease with distance. The direct evaluations of airborne radionuclides concentrations in dpm/g have been only for sampling locations on the pile. It is unknown how far particles in each particle size range are transported downwind. Thus, one must proceed with caution in applying either a single or a range of radionuclides concentrations in dpm/g for all reported airborne mass loading and mass flux data.

The data presented at this time are in terms of airborne concentrations in g/m^3, airborne mass fluxes in $g/(m^2 day)$ and radionuclide concentrations in dpm/g) of airborne solids. These terms are needed for estimating particulate source release terms. In a subsequent paper, the airborne solid source term will be estimated by integrating the vertical and crosswind data. After the radionuclide dpm/g data are more adequately defined, the total airborne particulate radionuclide source term will be calculated from the integrated airborne solid source term and the selected dpm/g.

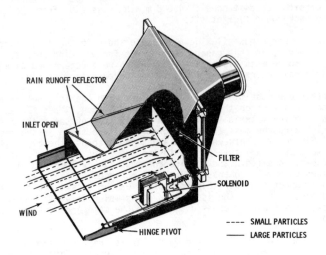

FIGURE 1. Isokinetic Sampler - Inlet Open

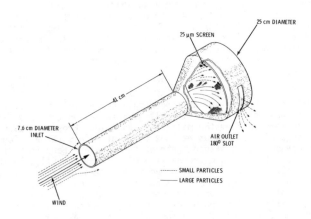

FIGURE 2. Inertial and Impact Pressure Sampler

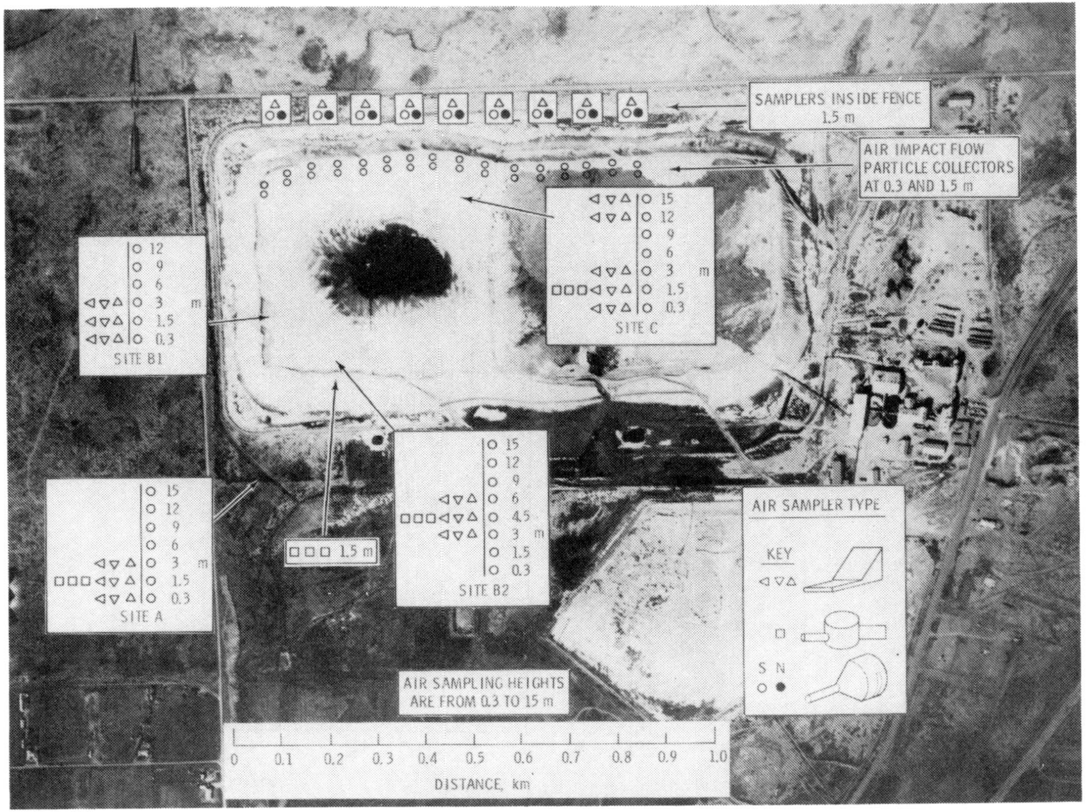

FIGURE 3. Air Sampling Arrays

Labels within figure:

SAMPLERS INSIDE FENCE 1.5 m

AIR IMPACT FLOW PARTICLE COLLECTORS AT 0.3 AND 1.5 m

KEY
AIR SAMPLER TYPE
S   N

SITE C
m
15
12
9
6
3
1.5
0.3

SITE B2
m
15
12
9
6
4.5
3
1.5
0.3

SITE B1
m
12
9
6
3
1.5
0.3

SITE A
m
15
12
9
6
3
1.5
0.3

AIR SAMPLING HEIGHTS ARE FROM 0.3 TO 15 m

1.5 m

DISTANCE, km
0   0.1   0.2   0.3   0.4   0.5   0.6   0.7   0.8   0.9   1.0

- 73 -

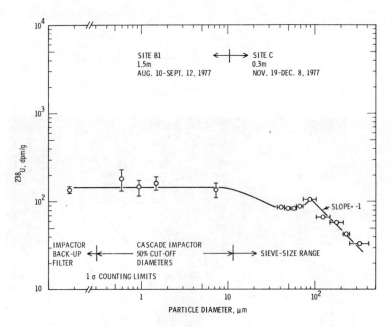

FIGURE 4. ^{238}U Concentration on Airborne Solids as a Function of Particle Diameter

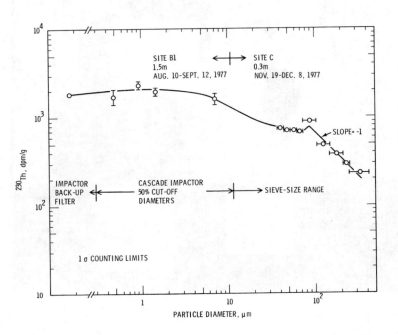

FIGURE 5. ^{230}Th Concentration on Airborne Solids as a Function of Particle Diameter

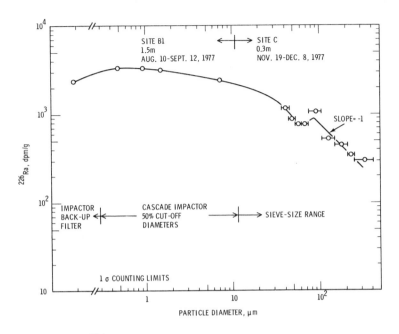

FIGURE 6. ^{226}Ra Concentration on Airborne Solids as a Function of Particle Diameter

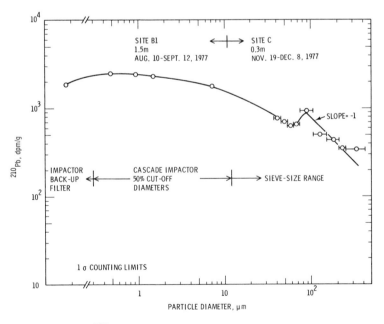

FIGURE 7. ^{210}Pb Concentration on Airborne Solids as a Function of Particle Diameter

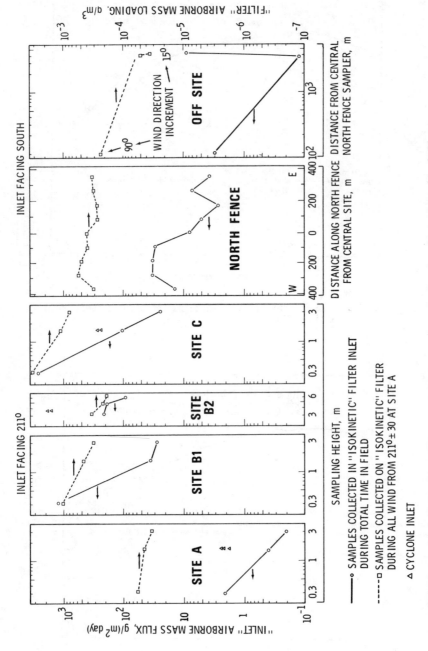

FIGURE 8. Average Airborne Fluxes and Concentrations During Aug 10 to Sept 12, 1977

- 76 -

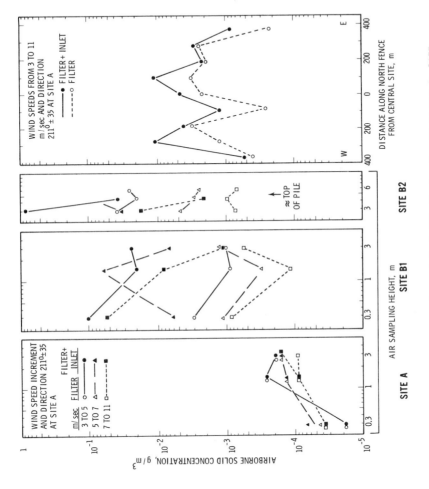

FIGURE 9. Airborne Solid Concentrations at Each Site During Nov 19 to Dec 8, 1977

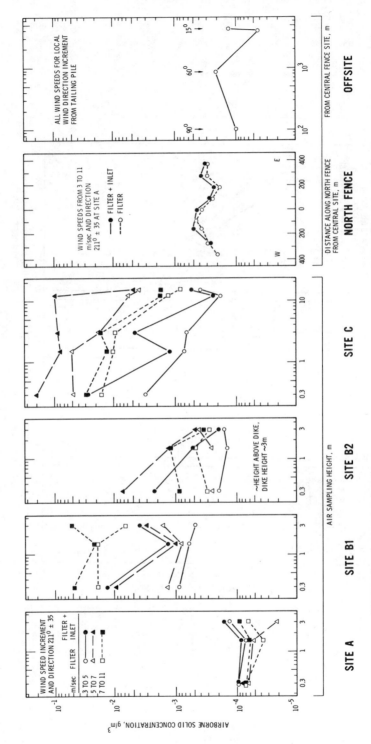

FIGURE 10. Airborne Solid Concentrations at Each Site During Feb. 21 to March 27, 1978

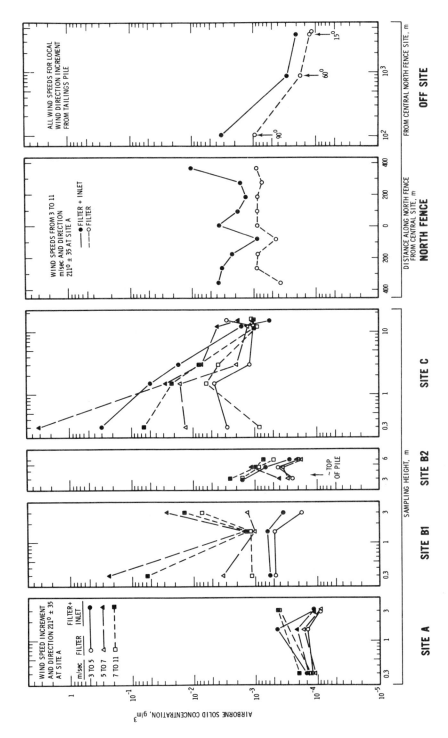

FIGURE 11. Airborne Solid Concentrations at Each Site During March 30 to April 17, 1978

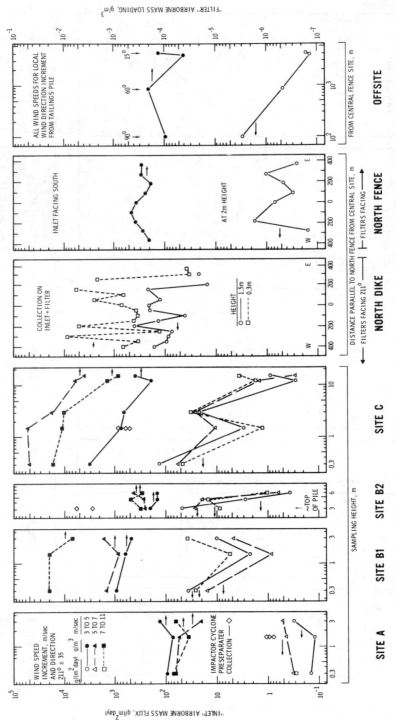

FIGURE 12. Average Airborne Fluxes and Concentrations During Feb. 21 to March 27, 1978

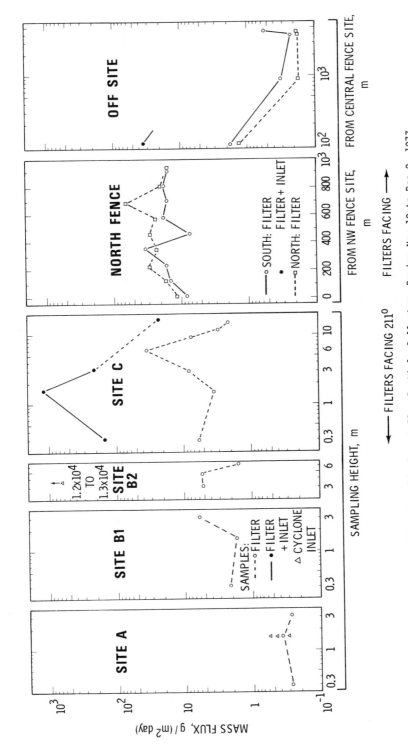

FIGURE 13. Airborne Mass Flux From Air Impact Flow Particle Collectors During Nov 19 to Dec 8, 1977

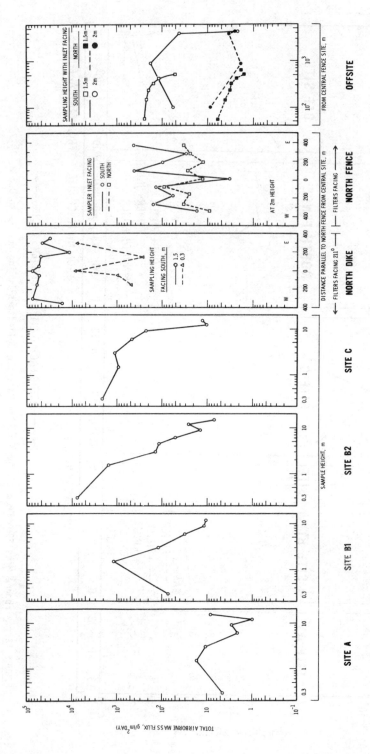

FIGURE 14. Total Airborne Mass Flux from Air Impact Flow Particle Collectors During March 30 to April 17, 1978

DISCUSSION

A.B. GUREGHIAN, United States

Are you contemplating to define a sort of diffusion co-efficient function of the velocity of the wind ?

G.A. SEHMEL, United States

I think part of the question is : are we going to look at a diffusion coefficient - square centimeters per second or whatever. I think that would be a very ambitious project. I guess personally I would be more interested in determining a release rate. I think that would be more applicable to many people.

A.B. GUREGHIAN, United States

A term like diffusion coefficient which probably includes what you are thinking of - is it possible to do that ?

G.A. SEHMEL, United States

It has not been done yet. At the present time this is where we are right now and the succeeding papers based on what is presented right now and other data that we have.

J.D. SHREVE Jr., United States

I am interested in comparing the size spectre you see in collected samples with surface material on the pile, and I would like to hear a comment or two on the degree to which such measurements are, a little bit, about middle or highly site specific ?

G.A. SEHMEL, United States

Any overall model we finally come up with, there should be relation between what is air born and what is on the ground and meterology, and the terrain. As I have the literature, this particular pile is a carbonate leach and I think it is about a 400 or 500 micron upper cutoff, and the largest particle we have seen - we have seen a few particles about 420 microns, so from that standpoint, our upper limit corresponds with something like the upper limit on the ground. There is other work being done at P&L by Jack Thomas and Dick Perkins, who have been characterizing the mill tailings area and the surface characteristics in the cross comparison between the air born material and the ground data has not been done at the present time, but will be done in the future.

D.C. McLEAN, United States

To what extent would you expect these curves to be modified by various degrees of coverage ? Would a cover make a significant decrease or something out of proportion to what you expect just by a fractional change ?

G.A. SEHMEL, United States

In another program of this investigation is the effects of surface coverage, and as the first speaker quoted, the resuspension rate data that I have got published so far. But we are looking at the effect of coverages and we should have in the future new resuspension rates, which I hope will be less as the vegetation has grown. In essence I agree that as the coverage of a pile increase, vegetation or whatever you want to put on the pile, that should decrease the release of material from the pile and I think it is more than a feeling I mean, it is sort of obvious. I think when you look back in the agricultural literature in Chapell, and so on, they have got predictions of wind erosion equation and they have numbers and curves for, say there is so much vegetation per square meter or whatever on the ground, tons per acre per year of erosive decreases. So on that, yes indeed. So the coverage will decrease erosion but as far as a mill tailings pile, I do not think there is data that I personally know of, that predicts how much that decrease will be, and it also should be a function of the water content. But where we are sampling and they are pumping more water on that side of the pile, maybe we can determine the effect of the increase of water content on the decreased airbourne concentrations that ought to come out of it - I hope it does come out of it.

J.D. SHREVE Jr., United States

Sorry to preempt the second question, but I was wondering why you were hesitant to say that if you covered the whole pile there would be zero airbourne radionucleides if indeed the cover had none in it ?

G.A. SEHMEL, United States

I am sorry, I did not mean to imply that at all. I thought the question was based more on if you grew trees or grass - I thought that was the question. If you cover it completely, as I understand at many mills that is going to be done in the future, at the present time I really cannot see how anything is going to come up from a covered area. But, after it has geen covered, I do not know what will happen. Something is going to come up. I do not know how much will.

"RADIOLOGICAL AND ENVIRONMENTAL STUDIES AT URANIUM MILLS:
A Comparison of Theoretical and Experimental Data"

M. H. Momeni, W. E. Kisieleski, Y. Yuan, C. J. Roberts

Division of Environmental Impact Studies
Argonne National Laboratory
Argonne, Illinois   60439   USA

ABSTRACT

Evaluation of radiological risk of uranium milling is based on identification and quantification of sources of release and consideration of dynamic
coupling among the meteorological, physiographical, hydrological environments
and the affected individuals. Dispersion pathways of radionuclides are through
air, soil, and water, each demanding locally tailored procedures for estimation
of the rate of release of radioactivity and the pattern of biological uptake and
exposure. The Uranium Dispersion and Dosimetry Code (UDAD), a comprehensive
method for estimating the concentrations of the released radionuclides, dose
rates, doses, and radiological health effects, is described. Predicted concentrations are compared with experimental data obtained from field research at
active mills and abandoned tailings.

ETUDES DE RADIOLOGIE ET D'ENVIRONNEMENT CONCERNANT

LES INSTALLATIONS DE TRAITEMENT DE L'URANIUM :

Comparaison des données théoriques et expérimentales

RESUME

L'évaluation du risque radiologique découlant du traitement de
l'uranium est fondée sur l'identification et la quantification des
sources de rejet et sur l'examen du couplage dynamique entre les
facteurs météorologiques, physiographiques et hydrologiques, d'une
part, et les individus affectés, de l'autre. Les voies de dispersion
des radionucléides sont l'air, le sol et l'eau ; or, il faut prévoir,
pour chacune de ces voies, des procédures adaptées aux conditions locales afin d'évaluer la vitesse de libération de la radioactivité et
le mode d'absorption et d'exposition biologique. On décrit le programme de calcul relatif à la dispersion et à la dosimétrie de l'uranium (UDAD), qui permet d'évaluer de façon globale les concentrations
de radionucléides libérés, les débits de dose, les doses et les effets
du rayonnement sur la santé. Les valeurs prévues des concentrations
sont comparées aux données expérimentales obtenues au cours de recherches effectuées sur le terrain dans des installations de traitement
en service et sur des résidus abandonnés.

Introduction

Evaluation of radiological impacts of uranium mining, milling, and mill tailings is based on identification and quantification of sources of release and consideration of the dynamic coupling among the meteorological, physiographical, hydrological environments and the affected individuals. Among the sources of release are mining, transportation, ore storage, ore crushing and grinding, leaching and extraction, product drying and packaging, and tailings. Dispersion of effluents from these sources is dependent on local parameters such as the wind field and vertical wind profile, and source characteristics such as elevation and area, and, for stack emissions, ejection velocity and temperature of the effluent. Particle deposition rate is dependent on their size, shape, and density. Interaction of airborne particles with the earth surface may result in deposition, or partial or total reflection, depending on surface structure and roughness. Thus the concentration of airborne and ground-deposited activity depends on characteristics of the specific mill site and region. Theoretical estimation of these concentrations is dependent on the choice of parameters and the validity of the assumptions utilized. Any field study designed to validate or calibrate a transport model should include real-time field measurements of all the input variables to the model as well as the ambient air concentrations. Only by measuring all of the input variables, and estimating measurement errors can a discrepancy between a prediction and actual measurement be resolved. Because of diurnal and other cyclic variations, field measurements should be conducted for rather long duration and under a wide range of conditions. One of the objectives of our on-going radiological and environmental field studies has been the provision of a data base for the evaluation of models for predicting airborne concentrations of radioactivity resulting from mining and milling operations. The principal parameters being measured are (a) source terms, (b) effluent size distribution and size-activity relation, (c) meteorological variables, i.e., windspeed, direction, and diffusion coefficients, (d) airborne concentrations of radioactivity as a function of time and distance relative to source locations.

Theoretical Model

The Uranium Dispersion and Dosimetry (UDAD) Computer Code was developed* to provide estimates of potential radiological impacts on individuals and on the general population within an 80-km radius of a mine-mill complex. In the UDAD code, only atmospheric transport is treated. Exposure resulting from inhalation, external irradiation from airborne and ground-deposited activity, and ingestion of food raised within the region is calculated. Detailed theoretical description of the logics utilized in UDAD, default values, user's guide, a test example and the code statements are reported elsewhere [1]. In the following section only a summary of the code methodology is described.

Continuous dispersion models describing spatial and temporal variations in the distribution of radioactive concentrations in plumes have been based on the Fickian classical differential equation of diffusion. Air dispersion is calcu-lated in the UDAD code using a sector-averaged Gaussian diffusion equation corrected for dry deposition, rainwash, and radioactive decay. Wind speeds in each of 16 directions or sectors are grouped in six categories (0-3, 4-6, 7-10, 11-16, 17-21, and > 21 knots). The atmospheric stabilities are classified into six categories (S = 1, 2, 3, 4, 5, and 6) based on criteria by Pasquill [2]. The values of $\sigma_z(x,S)$ used in the program are those of references 3 and 4.

Concentration of the radionuclide $\chi(x,D,W,S,Q,)$ at a distance x from a source Q and direction D, wind speed W, and stability S is computed from the appropriate equations. The average annual concentration χ from the source Q is calculated by summing each concentration $\chi(x,D,W,S,Q,)$ weighted by the frequency f (D,W,S) for the particular wind speed and stability class:

$$\chi(x,Q) = \Sigma_D \ \Sigma_W \ \Sigma_S \ f(D,W,S) \ \chi(X,D,W,S,Q) \ . \tag{1}$$

---

*This development was partially supported by the U.S. Nuclear Regulatory Commission.

The total annual average concentration from all sources is calculated from Equation (1) by summing the contributions from each source.

For area sources, the model converts the area into equivalent squares of width "d." It is assumed that a "virtual point source" is located at a distance of (d/2) cot (Δθ/2) upwind from the center of the source area, where Δθ is the 22.5-degree sector used to subtend the area width. For near receptors which cannot "see" the whole source area within a 22.5-degree sector, the source emission rate is multiplied by a correction factor. The correction factor is the ratio of that portion of the source area lying within a 22.5-degree sector located upwind from the receptor to the total source area.

The release of tailings particles by wind action is estimated in UDAD from theoretical and empirical wind erosion equations, according to the wind speed, particle size distribution, and surface roughness. Atmospheric concentrations of radioactivity from all sources are calculated using a dispersion-deposition-resuspension model. Source depletion as a result of deposition, plume tilting for the heavier particulates, and radioactive decay and ingrowth of radon daughters are included in a standard, sector-averaged, Gaussian plume model. The average air concentration at any given receptor location is assumed to be constant during each annual release period, but to increase year to year due to resuspension. Surface contamination is estimated by including buildup from deposition, ingrowth of radioactive daughters, and removal by radioactive decay and weathering processes. Deposition velocity is estimated based on particle size, density, and certain physical and chemical processes. The calculation of resuspension of previously deposited particles assumes a resuspension factor which decreases as a function of time to account for reduced availability as a result of natural processes.

Calculation of the inhalation dose to an individual is based on the ICRP Task Group Lung Model [5]. Following this model, the fraction of inhaled activity deposited in the lung compartments is determined by the aerodynamic properties of the particles. The rates of clearance from the lung are dependent on the solubility of the deposited materials. Estimates of inhalation dose due to radon and its short-lived daughters are calculated separately to distinguish from those due to the longer-lived daughters Pb-210 and Po-210.

External radiation doses from air immersion and exposure to contaminated surfaces are assumed to be proportional to the radionuclide concentration at the point of exposure. Doses are calculated using conversion factors that are provided as explicit input to the code [6]. Terrestrial food pathways included in the model are vegetation, meat, milk, poultry, and eggs. Calculation of internal organ or tissue doses is based on ICRP recommendations [7] with the option of using either a single or multiple exponential retention model.

Individual doses, dose rates, dose commitments, and population dose commitments may be computed at the end of an arbitrary number of years of mill operation. Also environmental dose commitments [8] are estimated as a measure of long-range potential radiological impact on the local population.

Field Study

A major part of the field study has been conducted since June 1977 at Anaconda Uranium Mill, Bluewater, New Mexico, in cooperation with William E. Gray, Director, Environmental Affairs, Uranium Mining and Refining. Since 1955 this mill has extracted uranium using an acid leaching, resin-in-pulp process. At present the ore is exclusively supplied from the Anaconda openpit Paguate Mines some 80 km east of the mill on the Laguna Indian Reservation. Previously the average ore quality was 0.34% (this is the % by weight of U_3O_8 in the ore) but in recent years this has decreased to about 0.25%. Before January 1978 the throughput of this mill was about 3500 tonnes per day; it was increased to about 5400 tonnes per day after alteration of the mill circuit and extensive increase in the tailings evaporation area. The tailings from the mill operation are pumped to a retention area of about 8×10^5 m^2 located 3 km from the plant. Figure 1 shows the Anaconda mill property and the field monitoring stations. Some phases of the study at this mill have been conducted jointly in cooperation with the Environmental Protection Agency (Radiation Programs, Las Vegas, Nevada).

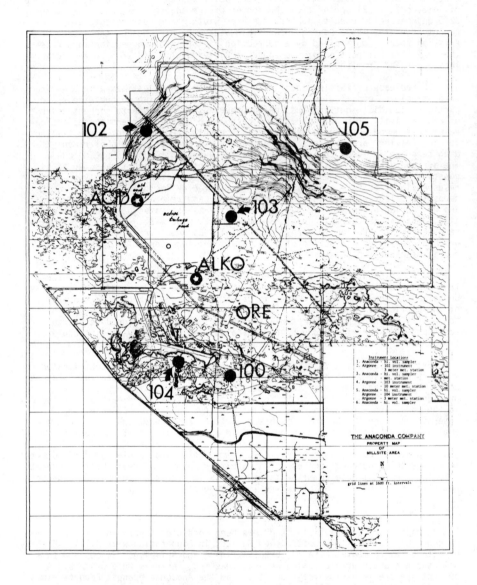

Figure 1. Anaconda Uranium Mill property depicting location of monitoring
stations. ALKO and ACID are abandoned tailings. ORE and T are
ore pad and Argonne National Laboratory field laboratory.

Present efforts at this mill have been to measure the following important parameters:

1. Source Terms
   a. ore pad and ore feed
   b. ore crushing and grinding
   c. product drying
   d. tailings: radon exhalation and fugitive dust movement
   e. radon diffusion through soil cover

2. Environmental Concentrations
   a. airborne particulates
   b. radon-222 and working level
   c. surface and subsoil contamination
   d. food-chain contamination

3. External Gamma Exposure Rates

The emphasis in these studies has been to provide a data base covering a long period of time, utilizing more than one technique of measurement, and if possible using continuous sampling procedures. Preliminary site selection for each station was based on a UDAD prediction, and practical considerations such as (a) cost of each station, (b) necessity of obtaining representative data, (c) accessibility, (d) availability of electrical power, and (e) security. In this paper selected data on radon concentration and working level using continuous monitors are presented.

Radon and Radon Daughter Monitors

Radon and radon daughter concentrations were measured using continuous monitors.** A block diagram of the monitor is shown in Figure 2. Air at a constant flow rate of about 30 LPM is passed through a 47-mm membrane filter where the airborne particulates and ions are removed. A fraction (1 LPM) of the radon-free air is passed through a 1.4-liter scintillation cell. The total system is enclosed in a 92 cm × 76 cm × 33 cm weatherproof container. The air inlet is 76 cm from the ground. The temperature inside the system is regulated by recirculation of the warm air generated by the pump and electronics during the winter period, and cooled in summer with filtered outside air.

A silicon diffused junction detector (490 mm^2) is used for measurement of alpha particles from Po-218 and Po-214 on the membrane filters. The scintillation cell, zinc sulfide coated lucite, is used for measurement of radon by detecting the alpha particles of Rn-222, and its daughters Po-218 and Po-214 which are formed within the cell. The detection limit of the instrument is about 1×10^{-4} working level and 0.1 pCi/L for Rn-222. The silicon diffused junction detector was calibrated with a Pu-238 electroplated source. Efficiency of this detector is 33%. The scintillation cell was calibrated with a Rn-222 source generated from a Ra-226 standard, and cross calibrated against a National Bureau Standard radium source. The system is periodically flushed with nitrogen gas and background is determined. The background for the silicon detector is about 5 counts/hour. During the past year of operation this detector has not shown any change in efficiency or background. The background in the scintillation cell is about 120 counts/hour. The increase in background has been less than 10% during the past year. Weathering of the zinc sulfide coat results in degradation of efficiency. This decrease in efficiency has been about 15% during the last year. The background buildup in the scintillation cell is only due to Po-210 from decay of Pb-210 (22-year half-life). A recommended procedure for continuous operation is to replace the scintillation cell after six months of normal operation.

**This system was developed in cooperation with M. R. Marley, H. Beene, and E. L. Geiger and was constructed by Eberline Instrument Corporation.

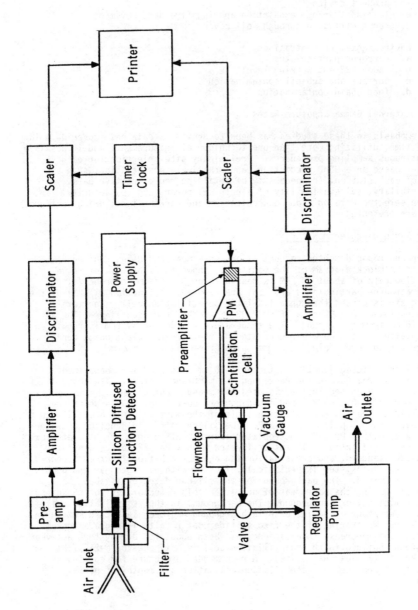

Figure 2. Continuous radon and radon daughter monitor.

Working level (WL) in air was estimated from the total alpha counting rate from Po-218 and Po-214 based on the relationship: WL = FC/LE where F is the calculated conversion factor (in WL-liter/alpha disintegration), C is the hourly count rate, L is the volume of the air sampled during one hour, and E is the efficiency of the detector. The conversion factor for air with only Po-218 or only Po-214 is respectively, $F = 5.9 \times 10^{-5}$ and $F = 5.3 \times 10^{-5}$ based on the working-level definition (1.3×10^5 MeV/WL-L). Since air contains both Po-214 and Po-218 depending on the equilibrium conditions, an average $<F> = 5.6 \times 10^{-5}$ WL-L/alpha disintegration has been used. The error associated with this choice of conversion factor is at most 5.4%.

Meteorology

Dispersion of pollutants is dependent on meteorological parameters--wind speed, direction and atmospheric stability. Atmospheric stability was estimated from the standard deviation of the wind direction. Figure 1 shows the location of the three meteorological stations. Wind speeds in each of the 16 directions Θ (north corresponding to $\Theta = 0$) and atmospheric stabilities were grouped in six categories each as noted above.

Results: Theoretical Estimations of Radon and Working Level

Figure 1 shows the location of the tailing retention area, ore pad, and monitoring stations. The coordinate system center ($x = 0$, $y = 0$) chosen for these calculations is located at the product dryer stack. The rate of release of radon from the ore pad, ORE in Figure 1, and the ore crushing and grinding operation is estimated to be 100 Ci/year. The estimated area of the tailings beach used in these calculations is 0.64 km^2, exhaling 1.4×10^4 Ci/year. For the analysis this area was divided into 16 equal sections (0.04 km^2), each releasing 8.75×10^2 Ci radon/year.

Figures 3 and 4 show the annual average radon concentration and working level as a function of distance for five angles ($\Theta = 0$ north, 90, 180, 270 west, and Θ_{max}) as predicted by the UDAD code [1]. The direction of maximum radon dispersion corresponding to the most prevalent wind direction is shown as Θ_{max}. The maximum predicted annual average radon concentration at 5 km from the mill is 0.26 pCi/L and occurs in the north-east direction ($\Theta_{max} = 45$). Table I gives total annual average atmospheric radon concentration for boundary locations, a nearby village and at the monitoring locations.

Table I. Total Predicted (UDAD) Annual Average Radon Concentration and Working Level at Selected Anaconda Locations

Identification	Radon Concentration (pCi/L)	Working Level
North Boundary	0.42	2.3×10^{-3}
East Boundary	0.36	2.1×10^{-3}
South Boundary	0.23	1.3×10^{-3}
West Boundary	0.05	4.6×10^{-4}
Bluewater Village	0.10	5.0×10^{-4}
Station #102	0.40	2.3×10^{-3}
Station #103	4.10	4.8×10^{-3}
Station #104	0.50	1.8×10^{-3}

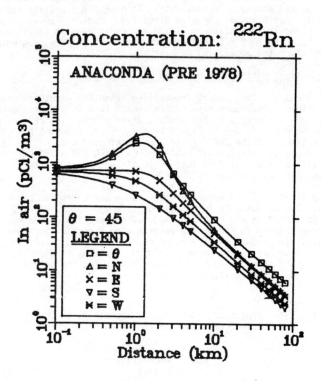

Figure 3. Annual average radon concentration predicted by the UDAD Code as a function of direction and distance from the mill.

Working Level

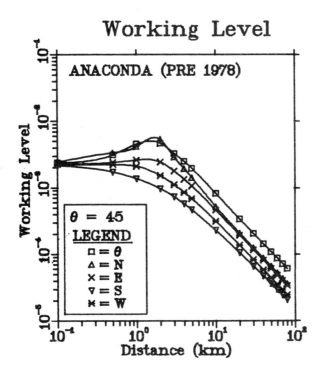

Figure 4. Annual average working level predicted by the UDAD Code as a function of direction and distance from the mill.

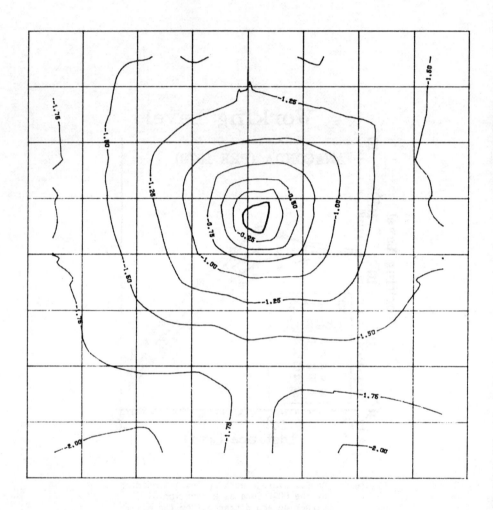

Figure 5.   Predicted radon concentration isopleths around
the Anaconda Mill. Grid spacing is 2 km.
Numbers are log (pCi radon per liter/suggested
concentration limit of 3 pCi per liter).

The predicted radon concentration is 0.42 pCi/L at the northern boundary, higher than any other boundary values, and is only 14% of 3 pCi/L, a suggested limiting radon concentration at a mill boundary within a low-density population region. At the south boundary, adjacent to Anaconda village, the radon concentration is only 0.23 pCi/L. An isopleth of radon concentration normalized to the suggested 3 pCi/L radon concentration standard is shown in Figure 5. Working levels at the boundaries of the Anaconda mill (Table I) are less than or equal to 2.3×10^{-3}. At monitoring station #103 the predicted working level is 4.8×10^{-3}.

Results: Experimental Estimation of Radon and Working Level

Figure 6 shows working level continuously measured at station #102 during September of 1977. The peak in working level is observed between 4 and 5 o'clock in the morning and is followed by a shallow minimum between noon and 4 o'clock in the afternoon. The radon concentration shows a similar pattern of diurnal variation.

Daily average radon concentrations for stations #102 and #103 are shown in Figures 7 and 8 for the month of September, 1977. Corresponding working levels are shown in Figures 9 and 10. Daily averages show patterns of periodicity of 3 to 4 days, presumably corresponding to regional frontal air mass movement.

Comparison between Predicted and Measured Values

Monthly averaged radon concentrations and working levels measured during the months of July, August, and September, 1977, are given in Table II. Each individual average is from about 2160 measured concentrations. The average radon concentrations from the mill operation for the three-month period are 0.4 pCi/L, 1.2 pCi/L and about 0.4 pCi/L for stations #102, #103, and #104, respectively. The predicted value for these three stations are 0.4 pCi/L, 4.1 pCi/L and 0.5 pCi/L, respectively. At stations #102 and #104 the predicted and the measured values are the same or nearly equal, but the predicted value for station #103 is larger than the measured value by a factor of about three.

Table II. Monthly Average Radon Concentrations and Working Levels
Derived from Hourly Data

Period	Radon Concentration (pCi/L)		Working Level $\times\ 10^3$		Working Level/ Radon Concentration
	Gross	Net	Gross	Net	
July, 1977					
Station #102	0.6	0.3	2.9	1.4	4.7×10^{-3}
Station #103	1.4	1.1	5.0	3.5	3.2×10^{-3}
Station #104	0.7	0.4	2.9	1.4	3.5×10^{-3}
August, 1977					
Station #102	0.8	0.5	3.6	2.1	4.2×10^{-3}
Station #103	1.6	1.3	5.8	4.3	3.3×10^{-3}
Station #104	0.6	0.4	2.8	1.3	3.3×10^{-3}
September, 1977					
Station #102	0.7	0.4	3.3	1.8	4.5×10^{-3}
Station #103	1.5	1.2	6.7	5.2	4.3×10^{-3}
Station #104	0.7	0.4	3.0	1.5	3.8×10^{-3}

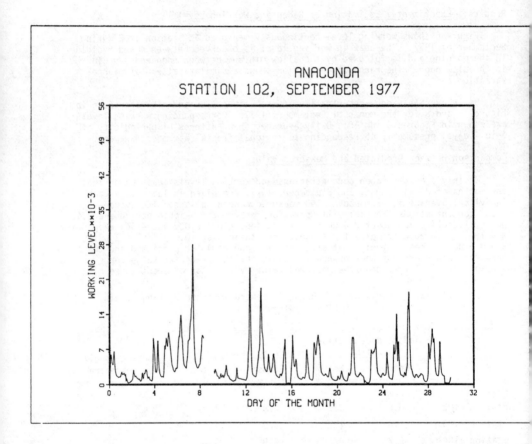

ANACONDA
STATION 102, SEPTEMBER 1977

Figure 6.   Average hourly working level as a function of
time measured at station #102.

Average radon concentration measured in air grab samples (10 samples--August 10 to September 19, 1977) at locations marked "Acid" and "Alko" (Fig. 1) are 5.2 pCi/L and 7.8 pCi/L, respectively. These sampled areas are located on tailings abandoned about a decade ago. Subsequently these tailings were covered with about 60 cm of silt-clay soil cover. Average radon concentration thereafter was decreased to 2.2 pCi/L (Acid) and 3.5 pCi/L (Alko) (12 measurements--October 1977 to January 1978). The values predicted by UDAD are (Acid) 2.0 pCi/L and (Alko) 2.5 pCi/L.

Working level was also measured using the American National Standard ANSI Technique [9]. The average of about 30 measurements (background plus increment contributed from the mill) at each station, all collected between 8 a.m. to 3 p.m. during July to October, 1977, are respectively (Table III) 0.2×10^{-3} (#100), 1.1×10^{-3} (#101), 0.8×10^{-3} (#102), 1.8×10^{-3} (#103), 0.3×10^{-3} (#104) and 1.8×10^{-3} at station #105. Station #101 is located about 100 m to the northwest of station #102 (Fig. 1). Individual measurements for each station vary about five times above and below the average values. The predicted average working levels are 1.7×10^{-3}, 2.3×10^{-3}, 2.3×10^{-3}, 4.8×10^{-3}, 1.8×10^{-3} and 3.3×10^{-3} working level, respectively for station #100 to #105. The working levels were measured when changes in radon concentration with time were moderate and lower than daily average. Thus, the working levels measured by this technique represent the low end of the range of daily average working levels.

Comparisons between the measured and predicted values of radon and working level concentrations (Table III) are encouraging. Improvement in the predicted values of source terms and adjustment for micrometeorological characteristics of each station will further improve the estimates. Work in this direction is in progress.

Table III. Comparison between Predicted and Measured
Radon Concentration and Working Level

| Station # | Radon (pCi/L) | | Working Level $\times 10^3$ | | |
	Predicted	Continuously Measured[a]	Predicted	Continuously Measured[a]	Grab Sampled
100	---	---	1.7	---	0.2
101	---	---	2.3	---	1.1
102	0.4	0.4	2.3	1.8	0.8
103	4.1	1.2	4.8	4.3	1.8
104	0.5	0.4	1.8	1.4	0.3
105			3.3		1.8

[a]Values listed are net concentrations after subtraction of estimated average background levels of 0.3 pCi radon per liter and 1.5×10^{-3} WL.

Aknowledgment

The authors acknowledge the assistance of Dr. A. Zielen in preparation of the reported computations. Also, we appreciate the encouragement and support of these studies by Paul Magno (EPA) and Harry Landon (NRC).

References

1. Momeni, Michael H., Yuan, Y., Zielen, A., with assistance of T. Beissel, "Argonne National Laboratory Uranium Dispersion and Dosimetry (UDAD), Volume 1 Dosimetry," in preparation, Argonne National Laboratory, Argonne, Illinois, 1978.

2. Pasquill, F., "Atmospheric Diffusion," 2nd edition, Halsted Press, New York, 1974.

3. Briggs, G. A., "Diffusion Estimation for Small Emissions, in Environmental Research Laboratory," Annual Report, USAEC Report ATDL-106, National Oceanic and Atmospheric Administration, 1974.

4. Gifford, F. A., "Turbulent Diffusion Typing Schemes: A Review," Nuclear Safety, V. 17-1, 1976.

5. ICRP Task Group on Lung Dynamics: Deposition and Retention Models for Internal Dosimetry of the Human Respiratory Tract; Health Physics 12, 173-207, 1966 and subsequent revision at ICRP meeting in Oxford, 1969.

6. Killough, G. G., and McKay, L. R., "A Methodology for Calculating Radiation Dose from Radioactivity Released to the Environment, Oak Ridge National Laboratory, ORNL-4992, 1976.

7. "Recommendations of the International Commission on Radiological Protection, Report of Committee II on Permissible Dose for Internal Radiation," Health Physics, Volume 3, pp. 1-380, 1960.

8. "Environmental Radiation Dose Commitment: An Application to the Nuclear Power Industry," U.S. Environmental Protection Agency, EPA-52014-73-002, 1974.

9. American National Standard, Supplement to Radiation Protection in Uranium Mines and Mills (Concentrators) ANSI N7-1a-1969, Supplement to N7.1-1960, American National Standards Institute, New York, N.Y.

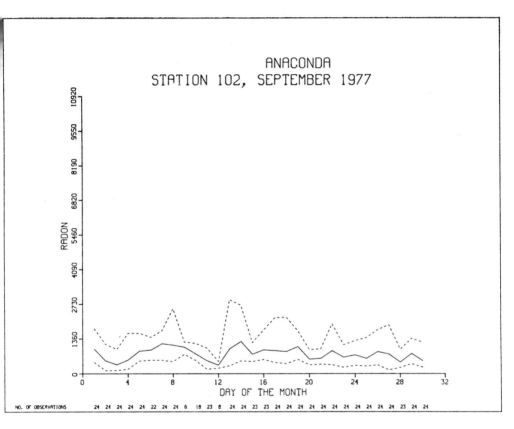

Figure 7.   Average daily radon concentration in pCi/m^3
(solid line) as a function of time.   The
upper and lower curves are the observed
daily maxima and minima.

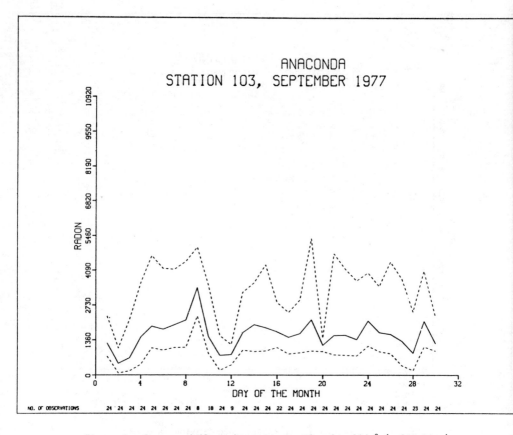

Figure 8.   Average daily radon concentration in pCi/m^3 (solid line)
as a function of time.   The upper and lower curves are
the observed daily maxima and minima.

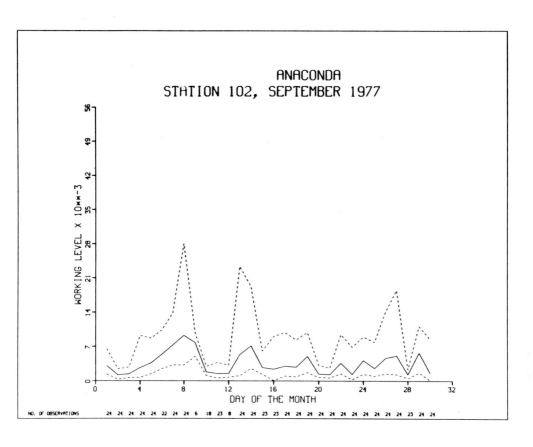

Figure 9. Average daily working level as a function of time. The upper and lower curves are the observed daily maxima and minima.

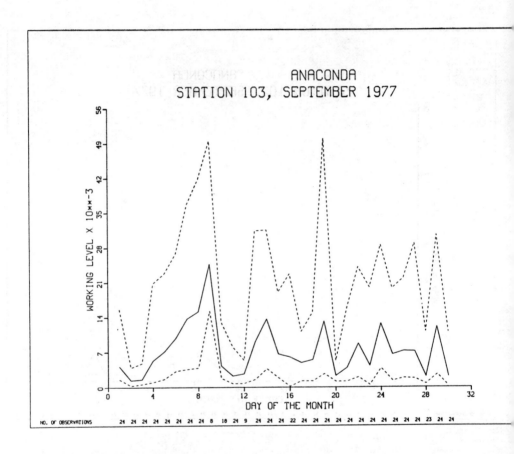

Figure 10. Average daily working level as a function of time. The upper and
lower curves are the observed daily maxima and minima.

DISCUSSION

J. HOWIESON, Canada

      I am wondering if the UDAD code is available or will be available to other laboratories ?

M.H. MOMENI, United States

      Yes, the UDAD code was reviewed not only by NRC but also by our colleague at EPA, and we have received permission to release after it has gone through editorial and clearing. It is my best estimate that it will be off the press by the end of August. And it will be available through the Argonne National Laboratory or the NRC. It is public property and it is available.

J. HOWIESON, Canada

      Could I ask just a little question about the code. Does it predict the cumulative long term effect as well ?

M.H. MOMENI, United States

      UDAD code is in two packages - package n° 1 is dosimetry and package n° 2 is risk estimate, which you are asking, the proposed risk estimate is on the basis of a competative risk analysis. In a sense the code tries to take into account probability of natural causes of death - such as falling off a bridge, dying of childbirth, dying from all other causes. Reduced effective population to this population which is a risk at every time. Also population vanishes by just natural mortality, by age. So, in that in a sense, the UDAD code, the risk estimate is closer to the reality of the situation, that part of the code, unfortunately, has not been brought into the main UDAD code, so that in the August release, that will not be incorporated.

G.L. MONTET, United States

      Looking at your Table 3, a comparison of the predicted and observed results, I noticed that those sample that were monitored continuously, agreed better with the prediction, which is I guess to be expected from your remark. But I noticed that the grab samples, rather than sort of randomly being higher or lower as you might impart from the variations, are consistently lower. Is that a feature of the grab sample or did you just happen to take them all during the minima ?

M.H. MOMENI, United States

      The grab samples we obtained - our attempt was to obtain it at that time of the day where variation in the radon environment was showing the least amount of variation, so we obtained between 12 o'clock and 5 o'clock. However, by choice those were the times when radon concentrations and working level are the lowest because of the property of a maximum dispersion at that time, so they show the lower end of the measurement, and this again points to one observation, those who obtain grab samples, one 4 o'clock in the morning, and one 4 o'clock in the afternoon, and then in between, they should not expect to find any correlation between their measurement and the reality of the situations.

THE RADIOLOGICAL IMPACTS OF URANIUM MILL TAILINGS - A REVIEW
WITH SPECIAL EMPHASIS ON THE TAILINGS AT RANSTAD IN SWEDEN

By Jan Olof Snihs   and   Per-Olof Agnedal
   National Institute of   Studsvik Energiteknik AB,
   Radiation Protection,   Nyköping
   Stockholm

ABSTRACT

        The environmental impact of uranium mill tailings can be
expressed in collective dose commitment and corresponding detriment
per $MW_e \cdot y$ energy produced by the uranium which corresponds to the
amount of waste of interest. The methods of dose commitment calculations are discussed and it is suggested for the purpose of estimation
of the detriment to limit the commitment to 10,000 years.

        The external radiation from the tailings is easily reduced
by covering but in case of a future settlement on the tailings the
collective dose commitment will be some hundreds to thousands of
$manrad/MW_e \cdot y$ depending on the quality of the uranium ore.

        The dispersion of dust from uncovered tailings is mainly a
local problem and the collective dose commitment for critical tissues
in the lung will be less than a $manrad/MW_e \cdot y$.

        In the long run the cover may be lost and the resultant
collective dose commitment for the critical tissues in the lung will
be a few tens of $manrad/MW_e \cdot y$ depending on the quality of the tailings. The effects of global dispersion of the material in the tailings are also discussed.

INCIDENCES RADIOLOGIQUES DES RESIDUS DE TRAITEMENT DE L'URANIUM - EXAMEN AXE NOTAMMENT SUR LES RESIDUS SE TROUVANT A RANSTAD, SUEDE

RESUME

Les incidences, sur l'environnement, des résidus de traitement de l'uranium peuvent être exprimées en fonction des doses collectives engagées et du détriment correspondant par MWe-an d'énergie produite à partir de l'uranium auquel équivaut la quantité de déchets en cause. On examine les méthodes de calcul des doses engagées et il est suggéré, afin d'évaluer le détriment, de limiter la dose engagée à 10.000 ans.

Il est facile de réduire l'irradiation externe en provenance des résidus de traitement en recouvrant ces derniers mais, en cas d'implantation à l'avenir d'un établissement sur le dépôt de résidus, la dose collective engagée sera comprise entre quelques centaines et quelques milliers d'homme-rads/MWe-an suivant la qualité du minerai d'uranium.

La dispersion des poussières provenant des résidus non couverts est un problème essentiellement local et la dose collective engagée au niveau des tissus critiques du poumon sera inférieure à un homme-rad/MWe-an.

Il se peut qu'à long terme la couverture des dépôts de résidus disparaisse et la dose collective engagée qui en résultera pour les tissus critiques du poumon atteindra quelques dizaines d'homme-rads/MWe-an suivant la qualité des résidus de traitement. On examine également les effets de la dispersion globale des matières contenues dans les résidus.

1.   INTRODUCTION

Mining and milling of uranium result in the formation of waste products formed in the various stages of the process. These affect the environment to varying degrees and they must be suitably disposed of. They consist of solid wastes, mainly from the mining process, liquid wastes which are residues from the milling processes or leakage during the milling, and airborne waste in the form of dust, chemical contaminants and radioactive substances, principally radon. The final stage in the milling process involves storage of the mill tailings and it is this which represents one of the main waste disposal problems.

The consequences for the environment depend on the working and milling techniques, on the protective measures applied and on the appearance and nature of the environment with regard to the vegetation, buildings and general land usage. During the decades during which uranium mining has been in progress, increased insight has been obtained on the composition and environmental impact of the wastes and stricter requirements have been imposed for purification plant, retaining systems and supervision.

The releases and leakage which can adversely affect the environment contain radioactive substances, metals and different chemical compounds. The radioactive substances, above all uranium, radium and radon, are dispersed in the environment via water or air and they can irradiate humans either internally after uptake in the body via water, foodstuffs or air or externally from deposits on the ground. However, due to radioactive decay and dispersal in the environment the environmental impact of the radioactive substances will be limited in time.

The radioactive substances may become enriched in plants and animals and the equilibrium levels reached after many years' release may constitute one reason for instituting release limitation and supervision. A great deal is now known concerning radiological uptake and enrichment processes and - by using pessimistic values in the assessment - it is possible to predict the consequences for the environment, including man, of radioactive releases over a long period of years. When estimating the resultant radiation doses and their consequences, the dispersion and uptake must be taken into account even at points far remote from the point of release. In principle it is the consequences of global dispersion which are calculated if such dispersion can be anticipated. The resultant collective dose, i.e. the sum of all the individual doses, expressed in man-rads is calculated in relation to the energy produced expressed in $MW_e \cdot year$ (MW_e means megawatts of electric power) which the practice has given rise to.

Since a release of radioactive substances with long half-lives will remain in the environment for many years, the collective dose is calculated taking into account this residence effect. It is then called the collective dose commitment. In principle, the collective dose commitment can be calculated for the whole period for which the released pollutant remains in the environment and can affect humans. For radioactive substances this means in practice for a finite time of the same order of magnitude as the half-life and for stable substances it means an infinite time. However, the longer the time perspective considered, the greater the uncertainty when predicting the behaviour of a pollutant in the environment. Periods of time with an order of magnitude exceeding 10,000 years may include one or more ice ages and are therefore extremely speculative and uncertain. In such cases calculations are therefore probably meaningless.

To estimate the collective dose commitment and the consequences in terms of the probable number of cases of cancer resulting from a particular practice, for instance uranium mining, the effects for this purpose have been calculated for a maximum period of 10,000 years.

Strictly speaking, the consequences of the releases of non-radioactive substances should be calculated in the same way. The environmental impact of these substances, however, is usually reported in the form of concentrations in air, water, plants and animals where these concentrations are related to the toxic limits. The results are usually also handled in a different way with regard to the effects on man. For radioactive substances and ionizing radiation the possible effect on man is calculated even for very low radiation doses. In the case of other substances, however, it is common to refer to threshold values below which the risk of injury can be neglected. A comparison between the risks on this basis is therefore unsatisfactory. In spite of this fact it has been found that it is often the effects of the non-radioactive substances which determine the limits which must be set. When the releases and leakage of these substances is limited to comply with these limits the conditions for the radioactive substances may be fulfilled automatically, possibly with the exception of the leakage of the gas radon and the external radiation from the piles of mill tailings both of which necessitate special remedial measures.

2.   THE MILL TAILINGS PILES

2.1 General

The wastes from uranium production which have attracted most attention are those known as mill tailings. They consist of the uranium-depleted ores, sometimes with additives such as limestone to achieve chemical neutralization. Initially, the consistency of mill tailings is semi-liquid but after evaporation and drainage they gradually become solid. The quantities are of the same order of magnitude as those of the ore taken from the mine and are therefore greater when the uranium content of the ore is smaller. In the USA the quantities are now of the order of 100 million tons. If 1 300 tons of uranium were mined annually in Ranstad, 6 million tons of mill tailings would be obtained each year.

Mill tailings are laid out in piles, the height of these piles varies from country to country, from roughly 5 to 30 metres. An annual production has an area of 0.1 - 1 km^2. The area can also be expressed in terms of the energy produced from the extracted uranium. For the USA this value has been given as 8 $m^2/MW_e \cdot year$ and for Sweden 25 $m^2/MW_e \cdot year$, due to different uranium contents and techniques for making the piles. The leaching process gives a uranium yield of 70 - 95 per cent. Thus 5 - 30 per cent of the initial content of uranium is present in the tailings. Furthermore, the tailings contain almost 100 per cent of the other radioactive substances, including thorium-230 (half-life 83,000 years) and radium-226 (half-life 1,600 years).

The mill tailings piles can affect the environment due to leakage from the piles of heavy metals, chemical compounds and radioactive substances via water, via airborne dispersal of radon and solid particles. In addition, in the immediate vicinity of the piles there is a weak gamma radiation field.

The environment can be protected by isolating and stabiliz-
ing the mill tailings. The piles are impounded in earth walls, stabilized chemically with neutralizing substances and mechanically by
the admixture of non-radioactive rock waste and covered with moraine
and topsoil. This prevents more than a very limited seepage of precipitation through the tailings with concomitant leaching out of substances deleterious to the environment.

The experience gained from modern installations shows that
the leach-out of radioactive substances is normally very slight. However, the chemical properties of the mill tailings and the filler
material affect the leach-out process. For example, the presence of
Cl-ions increases the leach-out of radium (1). The pyrities content
of certain ores may lead to the formation of sulphuric acid in the
tailings due to chemical and bacteriological oxidation and this may
lead to crumbling of the tailings and increased leach-out of deleterious substances. This effect can be reduced by the addition of limestone.

2.2 External radiation

The gamma radiation from mill tailings originates mainly
from the progeny of radon. The precursor of radon is radium-226 which
- as in the case of the precursor of radium, thorium-230 - is present
in the same quantities in the tailings as in the original uranium ore.
This gamma radiation will be practically unchanged for the whole foreseeable future. It will be 80,000 years before the level has decreased to half.

However, as an environmental hazard from a short term point
of view (hundreds of years) the gamma radiation from the tailings is
a fairly limited problem. It affects mainly the actual area occupied
by the tailings and results in an unacceptable exposure level in the
case of long-term occupation of the area. The tailings from the milling of uranium ores with contents of 0.03 - 0.3 per cent give about
0.2 - 2 mrad/hour on the surface, i.e. some 40 - 400 times the level
of the normal natural background radiation from the ground. By covering the tailings with layers of moraine and soil several metres thick
the gamma radiation can be effectively shielded off. Decisions concerning the future use of the area covered by the mill tailings must
therefore take into account the risk that the covering layer of moraine may be removed with the result that the gamma radiation again
becomes unacceptably high.

If the covering layer of moraine should be removed, either
by wind and erosion or by some human action, and persons unaware of
the radiation risks should take up residence in the area, they would
be exposed to gamma radiation levels of the order of 0.2 - 2 mrad/
/hour. However, one can assume some degree of shielding from the
layer of topsoil added to make the building site attractive. In addition, the building materials in a house provide some shielding.
Therefore, a more reasonable value of the gamma radiation level would
be about a tenth of the values above.

The annual individual risk of injury (cancer) from this
cause would be 0.003 - 0.03 per cent. If the presence of a community
of 10 000 persons is assumed, the collective dose commitment over a
period of 10,000 years would be 400 - 1 200 manrads/MW_e·year, i.e.
a mathematical expectation of 0.08 - 0.2 cases of cancer over a
10,000 year period per MW_e·year. These two different values represent
Swedish and American conditions respectively. However, a future community living directly on mill tailings piles is the worst conceivable case and the risk of this occurring cannot be other than speculative.

On an even longer time scale it is possible that the tailings will be spread by winds and erosion far beyond the original area. In the extreme case, global distribution must be assumed. However, the extra irradiation to which man would be exposed would be very marginal and would also be relatively limited in time due to covering by other dust which is also spread by the winds. It can be shown, with reasonable assumptions, that in this case the natural background radiation from the ground would be increased by less than 0.0002 per cent and that the resultant collective dose commitment would be some tens of manrads/MW_e·year. It can therefore be ignored in comparison with the previous, equally speculative, example.

2.3 Dust hazards

Airborne dispersion of dust particles from the mill tailings involves a risk that airborne radioactivity will be inhaled by humans. Long-term measurements in the immediate vicinity (about 100 metres) of non-stabilized piles of mill tailings in the USA (1) showed an average concentration of about 3 per cent of the maximum permitted concentration for members of the public. If the airborne radioactivity had been in the form of insoluble dust particles, this 3 per cent corresponds to a lung dose of not more than 45 mrem/year. Due to deposition of the airborne particles, the concentrations in the air, and thus the radiation doses, will decrease rapidly with distance. The radiation doses to persons in the locality will therefore be much less than 45 mrem/year. The collective dose commitment in the locality can be estimated to be less than one manrad per year, thus giving a mathematical expectation of lung cancer of less than 10^{-4} per year for 10 000 persons. It cannot be estimated how long this would continue but on a purely hypothetical basis the mathematical expectation would be less than 10 cases of lung cancer during a period of 10,000 years. The mill tailings in the example above correspond to about 20.000 MW_e·y produced energy. The collective dose commitment for critical tissues in the lung will therefore be less than 0.5 manrad/MW_e·y.

2.4 Radon

The problem which has attracted most attention and which is most realistic in connection with the mill tailings piles is that of radon leakage. Radon is formed on the radioactive decay of radium-226. Both radium and its precursor thorium-230 are present in approximately the same quantities in the mill tailings as in the original uranium ore and their half-lives are very long (83,000 years for thorium-230). Although it is technically feasible to remove the radium and thorium, up to now this has not been regarded as a practical possibility due to the concomitant costs and the new problems which such a measure would involve.

The radon leakage from a pile of mill tailings is the result of a number of factors: the radium content of the tailings and their radon emitting capacity, the radon diffusion properties of the radon in the tailings, the area and thickness of the pile and its covering etc. The radon leakage is usually given in picocuries per m^2·sec (pCi/m^2·s) and it can vary from 1 - 4 pCi/m^2·s (Ranstad) to 500 pCi/m^2·s (USA) (2). These values apply to dry uncovered piles of tailings. If the tailings have a high water content the radon leakage is considerably reduced, to 1/10 or less. If the piles are covered with moraine and topsoil the radon leakage is also reduced. 0.5 - 1 metre can give a reduction to half, five metres to a tenth or less. At the experimental installation at Ranstad, a reduction to one thousandth has been obtained by covering with bentonite, with 1 metre of moraine and then with 30 cm of topsoil. Covering methods can therefore give extremely low radon leakage - sometimes even lower than that from the surrounding ground.

If it is assumed that the tailings piles loose their cover-
ings after a very long period of time due to the effects of wind and
erosion, the radon leakage reverts to the value without covering.
The future consequences of this can be estimated from values for the
radon leakage per $m^2 \cdot sec$ and how many m^2 of tailings are formed in
order to produce uranium for 1 $MW_e \cdot year$. This latter value depends on
the uranium content of the uranium ore and on the thickness of the
tailings pile. If a typical value for American tailings piles is used,
8 $m^2/MW_e \cdot year$ (2), the result obtained is 44 manrads/$MW_e \cdot year$ calculated over 10,000 years. For Ranstad it is estimated that for one
of the planning alternatives the tailings piles will have an area of
about 25 $m^2/MW_e \cdot year$. If the radon leakage is assumed to be 10 pCi/
/$m^2 \cdot s$ the resultant value is 3 manrads/$MW_e \cdot year$ calculated over
10,000 years. The mathematical expectation of the number of lung
cancer cases per $MW_e \cdot year$ is then about 0.01 for the American conditions and about 0.001 for Ranstad over a period of 10,000 years.

In the hypothetical case in which the tailings are also
spread outside the original tailings area, the effective area of the
pile is increased, as is therefore the radon leakage. However, there
are a number of limiting factors. Often only a fraction of the radon
formed has any physical possibility of escaping from the particle in
which it is formed. The value of this emanation factor varies from
1 - 2 per cent (Ranstad) to 20 - 25 per cent (USA). This means that
even if all the tailings were laid out in a thin layer, not more than
1 - 25 per cent of the radon formed would be able to leak out into
the environment. Another limiting factor is the containment of the
dispersed tailings in other dust which is dispersed and spread on the
ground. This results in the radon leakage being less over a long time
scale after dispersion than when the tailings lay collected in piles
without covering. The resultant radon leakage and radiation doses
after dispersion of the tailings is therefore not considered in this
connection.

2.5 Water

The environmental impact of releases in water may be of
significance only for the local area and population. Leakage of uranium, thorium and radium may increase the concentration of these nuclides in fish and the consumption of fish may be the most important
pathway to the population.

As uranium, thorium and radium are natural radioactive nuclides normally found in ground water and in lake water, the increase
of the concentration in fish is not expected to be higher than the
increase of the concentration in the water. By large dilution of the
leakage from the tailings in ground water and in the lake the increase of the concentration in the water is expected to be moderate.
However, the chemical composition of the leakage may be a factor of
significance as regards the uptake in fish.

There are not many data on environmental consequences of
leakage of tailings and they are by the reasons given above of specific, local nature. The experiences from Ranstad are very limited,
but there is no indication that the leakage will be a serious radiological problem. However, and this is a general remark, because of
the very long-time perspective of the environmental problems of tailings, it is seriously recommended that the environmental consequences
of leakage to water will be carefully studied and analyzed.

3.   CONCLUSION

     The environmental impact of tailings from mining and milling of uranium has been discussed and estimated. The results are summarized in the table below. The radiological consequences have been calculated for a period of 10,000 years per an energy production of $1.000 \text{ MW}_e \cdot y$. Even if these environmental impacts can be considered to be low they are found to be much higher than corresponding environmental impacts of the mining and milling operations.

Table I

Sources	Mathematical expectation of number of cancers per $1.000 \text{ MW}_e \cdot y$ over 10,000 years
External radiation in case of settlement on uncovered tailings	80 - 200
Dust spread from the tailings: external radiation	< 10
inhalation	< 0.1
Radon from the tailings: covered	0
uncovered	1 - 10

References

1. Goldsmith, W.A. : Radiological Aspects of Inactive Uranium - Milling Sites: An Overview, Nuclear Safety, Vol. 17, No 6, Nov-Dec 1976.

2. United States Environmental Protection Agency. Environmental Analysis of the Uranium Fuel Cycle. Part II. Fuel Supply. Report EPA--520/9 - 73 - 003 - B, 1973.

DISCUSSION

W.E. KISIELESKI, United States

You have made a point of giving equal and comparable
attention to non radioactive substances, I wonder if you have
factored in considerations about the synergistic potential like
in the case of radon about the adsorbtion on other inhaled dust
particular or other aerosols ?

J-O. SNIHS, Sweden

The risk factors are considering these. They are taken
from the epidemiological studies in mines.

Session 2

ENVIRONMENTAL ASPECTS

Chairman — Président
Mr. M.C. O'RIORDAN
United Kingdom

Séance 2

INCIDENCES SUR L'ENVIRONNEMENT

OVERVIEW OF THE ENVIRONMENTAL IMPACT
ASSESSMENT FOR THE PROPOSED EXPANSION OF
THE ELLIOT LAKE ONTARIO URANIUM MINES

D.M. Gorber, R.G. Graham and B.G. Ibbotson
JAMES F. MacLAREN LIMITED
Toronto, Canada

ABSTRACT

As a result of the resurgence in the demand for uranium, Denison
Mines Limited and Rio Algom Limited began preparations, in 1973
and 1974 respectively, for expansion of their facilities at
Elliot Lake, Ontario. These programmes involved not only the
expansion of facilities currently in operation in the area, but
also the rehabilitation of non-operating properties that were
previously used during the 1960's.

This paper reviews the methodology employed during the environmental assessment study of the proposed expansion and highlights the
long and short-term strategies recommended.

RÉSUMÉ

Suite à l'accroissement des besoins en uranium, Denison Mines
Ltée et Rio Algom Ltée ont commencé à préparer en 1973 et 1974
respectivement, les plans d'expansion de leurs installations à Elliot
Lake, Ontario. Ces programmes impliquaient non seulement l'expansion des installations présentement en usage, mais également la
réhabilitation des installations utilisées au cours des années
60 et abandonnées depuis.

Ce mémoire explique la méthodologie utilisée au cours de l'étude
d'impact sur l'environnement de l'expansion proposée et les
points saillants des strategies recommandées à court et long
terme.

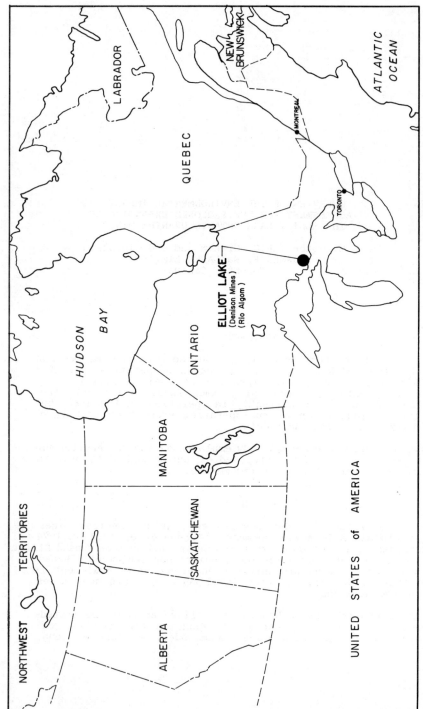

FIG. 1 — KEY PLAN

- 118 -

1.   INTRODUCTION

        The discovery of uranium deposits in the Elliot Lake
area of northern Ontario in 1953 initiated a crash programme
whereby twelve mines were brought into production over a span of
four years (1955 to 1958). By 1959, a town in excess of 25 000
people had been located at the eastern shore of Elliot Lake.
Figure 1 shows the location of Elliot Lake in relation to the
surrounding area of northern Ontario.

        In 1959, the principal customer for the production, the
United States Atomic Energy Commission, announced that it would
not extend the contracts beyond 1962. As a result, 10 of the
original 12 mines and mills closed down, and by the early 1960's
the population of the Town of Elliot Lake dropped to about 7000.
Denison Mines Limited and Rio Algom Limited were the only two
companies to continue operations from the early 1960's to the
present time.

        In response to the resurgence in the demand for uranium,
Denison Mines Limited and Rio Algom Limited began preparations in
1973 and 1974 respectively, for expansion of production at their
Elliot Lake properties. In view of this expansion, the Ontario
Minister of the Environment recommended that an environmental
impact assessment study be prepared and that the Environmental
Assessment Board be authorized and directed to conduct public
hearings in respect to the expansion.

        James F. MacLaren Limited was subsequently retained by
both companies in December 1976, to prepare a detailed environmental assessment of the proposed expansions.

        The findings of these studies were outlined in four
separate volumes submitted between March, 1977 and April, 1978,
and entitled "Background Information"; Background Information
Update"; "Community Assessment"; and "Environmental Assessment".

2.   VOLUME 1 - BACKGROUND INFORMATION

        Volume 1 summarized the need for expansion and intro-
duced the programmes proposed by the mining companies. Generally,
Denison Mines Limited proposed to expand their existing milling
capacity of 6350 metric tons per day (MTPD) to 13610 MTPD by
1982, and Rio Algom Limited proposed to expand their existing
milling capacity of 4085 MTPD to 9350 MTPD by 1980. This would
be accomplished by expanding existing operations as well as by
reactivating several of the mines and mills previously in operation
during the 1950's. Figure 2 outlines the location of the mines,
mills and tailings impoundments in the Elliot Lake area.

        The report reviewed the existing mining, milling and
waste management operations, and some insight was given to the
structure of the Elliot Lake community. The main body of the
report inventoried the existing terrestrial, aquatic and air
environments, outlined potential sources of radioactive emissions,
and noted where deficiencies in data were apparent.

        Process options were presented, and general comments
were included relative to the state of the art of these processes,
their environmental sensitivities and their applicability to the
Elliot Lake situation.

        Preliminary public hearings were held by the Environmental
Assessment Board in April, 1977 to review the findings of Volume
1 and determine the scope and content of additional reports to be
submitted for the final hearings. The Assessment Board ruled

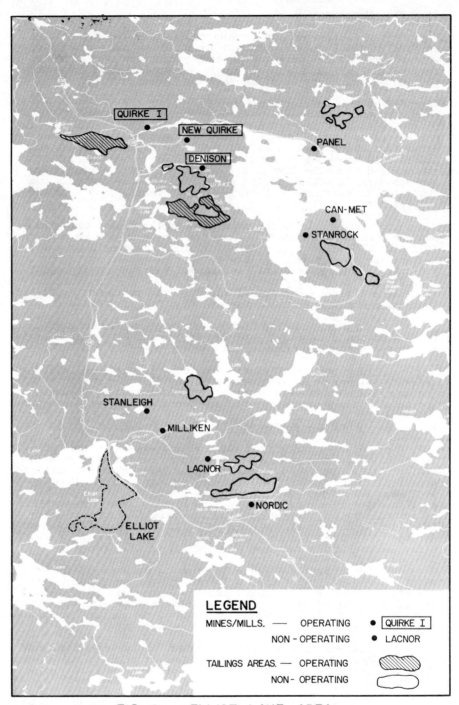

FIG. 2 — ELLIOT LAKE AREA

LEGEND

MINES/MILLS. — OPERATING   ● QUIRKE I

NON - OPERATING   ● LACNOR

TAILINGS AREAS. — OPERATING

NON- OPERATING

on the type of studies to be carried out and identified which
studies were to be undertaken by the mining companies, and which
were the responsibility of specific municipal, provincial and
federal agencies.

3.   VOLUME 2 - BACKGROUND INFORMATION UPDATE

    Volume 2, published in February, 1978, outlined addi-
tional background data gathered since Volume 1 had been submitted,
and addressed specific requests of the Environmental Assessment
Board. Results of field monitoring programmes dealing with surface
water hydrology and quality, meteorology and air quality, and
aquatic and terrestrial biology were presented.

    A major part of field programme was devoted to measuring
suspended particulate, dustfall, gamma radiation, radon exhalation
and radionuclide uptake of representative plant, animal and fish
species. Measurements were taken upon and adjacent to operating
and non-operating tailings areas and in the vicinity of the mines,
mills and townsites in the study area. In addition, a survey was
carried out by the Ontario Ministry of Natural Resources to determine the population of fish in the Serpent River watershed and to
assess the distribution, stock, species, age and growth rates in
the vicinity of the mining industry.

4.   VOLUME 3 - COMMUNITY ASSESSMENT

    In March, 1978, Volume 3 was released. This document
identified the social and economic effects of the proposed expansion on the Town of Elliot Lake, and examined such issues as employment, housing demands and supply, hard services (water supply,
sewage treatment, etc.), soft services (safety, education, etc.)
and health and social services. Data were collected through meetings and interviews with government agencies, municipal officials
and social service agencies, and through a social attitude survey
on a sample of over four hundred residents of the Town of Elliot
Lake.

    An assessment was then made of the existing quality of
life within the Town, and the effects that the proposed expansion
would have on it, the adjacent communities, and the Region as a
whole.

5.   VOLUME 4 - ENVIRONMENTAL ASSESSMENT

    Volume 4, the results of the environmental impact assess-
ment, was published in April, 1978. As well as assessing the
effects of the proposed expansion plans of both mining companies, a
number of additional operational alternatives were reviewed in
detail and the environmental consequences of incorporating these
process options into the expansion programmes were presented.

    Because the scope of the study encompassed such a broad
range of considerations, it was necessary to categorize these
according to the three principal operations of the industry, namely:
mining, milling and waste management. Typical subjects dealt with
under the mining phase of the operation included the production and
disposal of waste rock, the use of hydraulic backfill of tailings
in underground stopes, the feasibility of water reduction and
recycling underground, and the potential for reduction of ammonia
and nitrate levels in the mine water.

    The review of the milling operation concentrated upon
processes which would reduce potentially harmful discharges to the
environment. Specifically, these included methods to reduce ammonia

and nitrate, radium and other radionuclides in discharges to tailings, and to reduce fresh water usage and promote water recycling within the system.

With regard to waste management, three major components of the operation were reviewed, namely: tailings containment, effluent treatment and stabilization. Methods of tailings deposition and containment were investigated in terms of storage capacity, pyrite oxidation and acid production, seepage potential, radon release and the adaptability to various stabilization techniques. Methods of increasing the effectiveness of radium removal and pH control, and methods of achieving adequate stabilization of tailings were discussed.

The assessment of the foregoing alternatives, as well as the proposed expansion programmes, was presented in terms of the potential effects upon the aquatic, air and terrestrial environments. A computerized model was developed for evaluating the effects of the mine/mill processes and waste management alternatives upon the water quality and hydrology of the Serpent River watershed. Similarly, a model for dispersion of dust from tailings areas was developed and used to assess the air quality implications of various alternatives. In both cases, the models were calibrated using results of the field monitoring programmes. The radiological consequences of these actions were established by estimating radiation exposures to individuals in selected population groups.

In predicting environmental consequences it was important to look not only at the magnitude of the changes anticipated, but also at the rates of change and the degree of irreversibility of the proposed actions. Decisions on the suitability of these actions were therefore made in light of their short-term and long-term effects.

Short term in this study referred to the period when facilities were still in operation, or when some form of control or maintenance of the facilities could be undertaken. Long term referred to the period during which all operations had ceased and periodic maintenance would not be possible. While improvements and modifications to operations can be implemented and monitored during the operating life of a facility, it is desirable in the long term, that facilities can be safely shut down without further control or maintenance.

With this in mind, actions selected to provide benefits during the operating life of a facility must be chosen with care to ensure that they are compatible with long-term goals. The report suggested that sound management of operations in the short term should take the following into account:

i) Containment of tailings and control of seepage - minimizing uncontrolled seepage from tailings areas to reduce discharges of untreated effluent to receiving water bodies;

ii) Maximization of the use and capacity of tailings areas - maximizing the capacity of existing tailings areas to limit the areal extent of surface tailings during the operating period of a site and limit the potential for pollution in other watercourses or watersheds;

iii) Interim surface stabilization of tailings wherever necessary - several methods can be utilized to reduce dusting and radon emission from exposed tailings. The selection of a given remedial measure is site specific, and depends on such factors as the degree of reduction necessary, the stability of containment structures, and the surface characteristics of the tailings;

  iv) <u>Minimization of freshwater usage</u> - encouragement of internal water recycling within the bounds of health, process or economic restrictions;

   v) <u>Control of radium in tailings effluents</u> - Although the use of barium chloride does present a viable method of separating radium from the tailings effluent as a co-precipitate, improvements in retention time, the use of flocculants, removal of the radium sludge from the settling ponds, would provide a more complete removal system. The use of these measures is again site specific and is primarily governed by the physical constraints of the area.

  vi) <u>Control of pH in tailings effluents</u> - control of pH of the effluent leaving the tailings area by over-neutralizing in the mill, or where this is not successful, by neutralizing the final effluent.

The selection of a course of action to be taken to address long-term concerns must be made on the basis of current research data and practical technology. The two concerns associated with the long term are the potential oxidation of pyrite in tailings and the possible generation, release and migration of radionuclides to the natural environment.

Based on present technology, vegetative stabilization appears to offer the most practical long-term approach for the reduction of pyrite oxidation and radionuclide movement. Although actual operating experience is limited and more answers will be provided through continued research, this form of stabilization has fewer foreseeable problems than other methods evaluated. Another option which may warrant further research is deep lake disposal. While the scheme appears to present considerable long-term advantages, additional field studies and research are necessary to verify and refine details of the receiving water body characteristics, geology of the basin, and the long-term water quality effects anticipated.

An alternate approach to long-term control could involve the systematic removal of the problem constituents contained in the tailings, namely: pyrite, radium and thorium. While the technology for pyrite removal exists, concurrent with its removal are problems of separate containment and/or disposal. At present, there is no practical radium removal process within the milling circuit. Radium removal alone presents only a partial solution to the long-term radionuclide problem without the simultaneous removal and isolation of thorium. The removal and isolation of thorium is not practical at this time.

6.   SUMMARY

In August of this year, Volume 5 will be published, approximately two years after this study began. This report will present results of all radiological analyses of samples taken during the 1977 monitoring programme. It is anticipated that the final Environmental Assessment Board hearings will commence in September of this year.

The undertaking of a major environmental impact assessment of this nature requires the assembly and cooperation of a multidisciplinary team of engineers, geologists, biologists, sociologists and planners. It also requires the assistance and contributions of staff of the proponents and numerous government agencies as well as the involvement of experts from a wide range of disciplines.

To be effective in the development of sound management strategies and the decision-making process, environmental factors must be integrated with all engineering, social and economic aspects of the proposed programme. Although the initial role of this environmental assessment was to bring the environmental issues into focus for approval of the proposed programmes, it was evident that the real challenge was to fit the assessment findings into the design process early enough to achieve a real and meaningful contribution, while at the same time respecting the governmental review and decision process.

DISCUSSION

<u>R.S. DANIELS</u>, United States

Could you comment on methods that you were using for measuring exhalation of radon ?

<u>R.G. GRAHAM</u>, Canada

I am not too involved in the details of how that was done. I know it is outlined in the report in Volume II.

<u>D.C. McLEAN</u>, United States

What about the availability of those volumes ? Can they be purchased or obtained in some way ?

<u>R.G. GRAHAM</u>, Canada

All the reports that we submitted today are in the hands of the Assessment Board. On what the availability is right now of these reports, I am not sure. If you could leave your name with us, we can direct those inquiries to the Assessment Board. It is in their hands as to how the reports are distributed and whether they do have enough copies. I am not really sure what the status is right now. I know we published quite a few copies. In fact, I have gone back with republishing some of the volumes.

<u>J.D. SHREVE Jr.</u>, United States

I was curious about how you got all that pretty vegetation to grow. Are not the tailings very acid ? The only thing I could think is that because you have so much more rainfall than we have in New Mexico that maybe you dropped that pH and, of course, feed the plants at the same time.

<u>R.G. GRAHAM</u>, Canada

Well, I think that is true. The rainfall situation is one major factor. First of all, it is not us that does that, it is the mines that do that, just to make that distinction. As I understand it right now, it is seeded and it is fertilized, and there is a heavy addition of lime.

SEEPAGE FROM URANIUM TAILING PONDS
AND ITS IMPACT ON GROUND WATER

Perry H. Rahn
Department of Geology and Geological Engineering
South Dakota School of Mines and Technology
Rapid City, South Dakota   57701   USA

Deborah L. Mabes
Division of Environmental Impact Studies
Argonne National Laboratory
Argonne, Illinois   60439   USA

ABSTRACT

A typical uranium mill produces about 1800 metric tons of tailing per day.
An assessment of the seepage from an unlined tailing impoundment of a hypothetical
mill in northwestern New Mexico indicates that about 2×10^5 m^3/yr of water will
seep over a period of 23 years. The seepage water will move vertically to the
water table, and then spread out radially and ultimately downgradient with ground
water.

The principal dissolved contaminants in the tailing pond liquid are radium,
thorium, sulfate, iron, manganese, and selenium; in addition, the liquid is
acidic (pH=2). Many contaminants precipitate out as neutralization of seepage
water occurs. At the termination of mill operation, radium will have advanced
about 0.4 m and thorium no more than 0.1 m below the bottom of the tailing pond.

LES PHENOMENES DE SUINTEMENT DUS AUX BASSINS D'EVACUATION

DES RESIDUS DE TRAITEMENT DE L'URANIUM ET LEURS

INCIDENCES SUR LA NAPPE PHREATIQUE

RESUME

Une installation type de traitement de l'uranium produit
journellement de l'ordre de 1.800 tonnes métriques de résidus. Il
ressort d'une évaluation des suintements en provenance d'un bassin
sans revêtement renfermant les résidus d'une installation hypothé-
tique située dans la région nord-ouest du Nouveau-Mexique que le
débit des eaux de suintement sera d'environ 2×10^5 m^3/an sur une
période de vingt-trois ans. Ces eaux se déplaceront verticalement en
direction de la nappe phréatique, puis se répandront transversalement
pour atteindre finalement en bas de pente ladite nappe.

Les principaux produits de contamination dissous dans le
liquide de ce bassin sont le radium, le thorium, le sulfate, le fer,
le manganèse et le sélénium ; de plus, le liquide est acide (pH = 2).
De nombreux produits de contamination se séparent par précipitation
parallèlement à la neutralisation des eaux de suintement. A l'issue
de la période d'exploitation de l'installation, le radium aura avancé
de 0,4 mètre environ et le thorium, de 0,1 mètre au maximum en-
dessous du fond du bassin contenant les résidus.

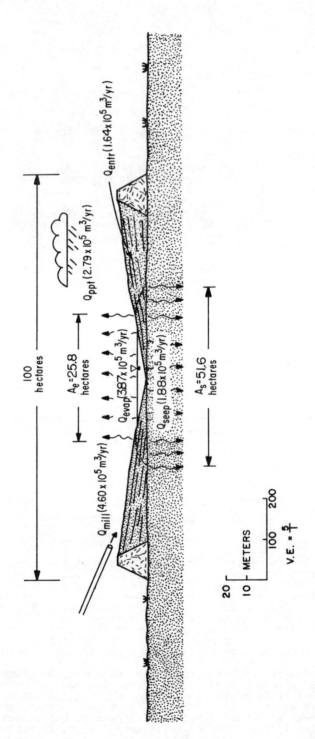

Figure 1. Water budget for tailing pond area.

INTRODUCTION

This paper describes a quantitative analytical model developed to predict the impacts to ground water caused by a "typical" uranium tailing pond. The model was developed for the generic environmental impact statement for uranium milling.[1]

The typical uranium mill in the United States processes 1800 metric tons of ore per day and produces 1800 MT of tailing. The ore averages 0.15% U_3O_8 and an acid leach process is used. The tailing are crushed to a gradation of 70% sands (> 70 microns) and 30% slimes (< 70 microns). The specificed gravity of the tailing is 2.65. The uranium, thorium, and radium remaining in the tailing are about 7%, 95%, and 99.8% of the original concentrations in the ore. Tailing are slurried in water at about 50% solids by weight and discharged into an impoundment. The tailing are deposited within a 100 hectare area surrounded by a very low permeability embankment. The bottom of the tailing disposal area is not lined. For this paper it is assumed that the subsoil consists of 125 m thick deposit of alluvium on top of bedrock of very low permeability; the water table is at 25 m depth. It is assumed that the alluvium has average hydraulic conductivities of 10^{-2} cm/sec horizontal and 10^{-5} cm/sec vertical. The average annual precipitation and evaporation rates for the area are 31 cm/yr and 1.5 m/yr respectively.

CALCULATION OF SEEPAGE RATES AND GROUND WATER CONTAMINATION

The following model calculates (1) the rate of seepage from the tailing impoundment, (2) the seepage water velocities, and (3) the dispersion and concentration of toxic contaminants at selected locations downgradient.

1. Seepage Discharge from an Unlined Tailing Pond

Assuming no surface overflow, seepage from an operating tailing pond can be calculated as the difference between the liquid inflow and liquid evaporated from the pond or entrained in the tailing. In general, the area through which seepage will occur is not known. Seepage can be directly estimated, however, by the method explained below wherein a hydrologic budget for the tailing area is established, and the seepage and resulting pond size are dependent upon subsoil permeability.

The water budget for the tailing pond can be expressed as:

$$Q_{in} = Q_{out} + Q_{entr} \tag{1}$$

where Q_{in} is rate of water discharge into the pond, Q_{out} is the rate of water discharge out of the pond and Q_{entr} is the rate of consumption of water that is entrained in the tailing.

Q_{in} is the sum of precipitation (Q_{ppt}) and the net inflow from the mill (assume Q_{mill} = 4.60 x 10^5 m^3/yr). It is assumed that all of the precipitation that falls on the surface area within the centerline of the embankment (roughly 90 ha) contributes to the water in the pond. Thus Q_{ppt} = 2.79 x 10^5 m^3/yr.

Q_{entr} can be expressed as a discharge, and is determined from the tailing porosity and rate of discharge. Assuming that the tailing are completely saturated at a porosity of 40%, the ratio of water entrained to the tailing production is 1:4. Thus Q_{entr} is 1.64 x 10^5 MT/yr. Shortly after cessation of mill operation, gravitational water in the tailing can be expected to drain out, leaving only hygroscopic and capillary water retained as thin films on the soil particles. On the basis of published data[2] for the "specific retention" of water in unconsolidated deposits of the grain size typically found in tailing (fine sand), 60% (0.98 x 10^5 MT/yr) of the water in the pores can be expected to drain out, leaving 40% (Q_{entr} = 0.66 x 10^5 m^3/yr entrained in the tailing.

Q_{out} is the sum of the seepage discharge (Q_{seep}) and evaporation rate (Q_{evap}). Both are a function of the area of the water surface and the wetted sands (see Figure 1). The area over which seepage occurs (A_s) will be somewhat larger than the area susceptible to evaporation (A_e). This is especially common in older ponds where the freewater pool covers only one part of the surface but saturated tailings extend across the whole area at some distance under the surface. Based on calculations shown in reference 1 and the authors' familiarity with a large number of tailing ponds, the area of seepage of the model pond is assumed to be twice the area of evaporation. Equation (1) can now be written:

$$Q_{ppt} + Q_{mill} = Q_{entr} + Q_{unit\ evap}\ (A_e) + Q_{unit\ seep}\ (A_s) \tag{2}$$

$Q_{unit\ evap}$ is 1.50×10^4 m³/yr-ha. Substituting known quantities and substituting for A_s:

$$2.79 \times 10^5\ m^3/yr + 4.60 \times 10^5\ m^3/yr = 1.64 \times 10^5\ m^3/yr + 1.50 \times 10^4\ m^3/yr\text{-}ha\ (A_e)$$

$$+ Q_{unit\ seep}\ (2A_e)$$

or:

$$A_e = \frac{5.75 \times 10^5\ m^3/yr}{1.50 \times 10^4\ m^3/yr\text{-}ha + 2\ (Q_{unit\ seep})} \tag{3}$$

$Q_{unit\ seep}$ can be calculated from Darcy's Law:

$$Q = KA\frac{H}{L} \tag{4}$$

where:

Q = Discharge, m³/sec, through a porous media having a cross-sectional area A, m².

K= Hydraulic conductivity, m/sec. This parameter is difficult to quantify and may have a wide range of values. Homogeneity must be assumed, but anisotropy can be dealt with by using horizontal (K_h) and vertical K_v) hydraulic conductivities of 10^{-2} and 10^{-5} cm/sec, respectively. The hydraulic conductivity of unsaturated soil varies with moisture content.[3,4] It is beyond the scope of this paper to assess a year-by-year rate of infiltration due to an advancing saturation front; therefore, for simplicity it is assumed that the average hydraulic conductivity in the range of moisture contents likely to be encountered during the operation of a tailing pond area is about 10% of the saturated hydraulic conductivity. Consequently, for the purpose of calculating the seepage using equation (4), the vertical unsaturated hydraulic conductivity equals 10^{-6} cm/sec.

H= Head loss, dimensionless. It is assumed that the saturated tailing and overlying free-water level averages 4 m deep over the life of the project; therefore, the change in head over the 25 m of unsaturated thickness of subsoil below the tailing pond area is 29 m.

L= Length of flowpath, meters. Only the 25-m thickness of unsaturated subsoil immediately below the tailing pond area is considered. No flow resistance is attributed to the tailing in these calculations.

Substituting into equation (4) and converting units:

$$Q_{unit\ seep} = (10^{-6} cm/sec)\ (1\ ha)\ (\frac{29\ m}{25\ m}) \tag{5}$$

$$= 3.65 \times 10^3\ m^3/yr\text{-}ha$$

Substituting the values of $Q_{unit\ seep}$ into equation (3):

$$A_e = \frac{.75 \times 10^5\ m^3/yr}{1.50 \times 10^4\ m^3/yr\text{-}ha + 2\ (0.365 \times 10^4\ m^3/yr\text{-}ha)} \tag{6}$$

$$= 25.8\ hectares$$

The total evaporation on 25.8 hectares is 3.87×10^5 m³ /yr. The total seepage is 1.88×10^5 m³/yr from 51.6 hectares.

In summary, the water budget for the tailing pond area (shown on Figure 1) can be written as:

$$Q_{ppt} + Q_{mill} = Q_{entr} + Q_{evap} + Q_{seep}$$

$$2.79 \times 10^5\ m^3/yr + 4.60 \times 10^5\ m^3/yr = 1.64 \times 10^5\ m^3/yr + 3.87 \times 10^5\ m^3/yr + 1.88 \times 10^5 m^3/yr$$

Not taken into account in these calculations are complex phenomena such as the possible reduction in seepage because of the buildup of slimes and chemical reactions associated with seeping acidic solutions. Nevertheless, the size of existing uranium tailing ponds seems to confirm the assumptions and calculations used in this analysis.

After mill operations cease gravitational water begins to drain out of the tailing. If the mill operates for 15 years, there would be a total of 1.47×10^6 m^3 of gravitational water. It is assumed that this water drains out at a constant rate of 1.88×10^5 m^3/yr (Q_{seep} from above). This drainage would therefore last 7.88 years (or approximately 8 yrs) after mill operations cease. Therefore, seepage lasts for a total of about 23 years. Decreased head and decreased permeability due to decreased saturation during postoperational drainage would have the effect of gradually slowing the rate of seepage; however for simplicity it is assumed that gravitation water seeps at a constant rate.

2. Seepage Water Velocity in the Subsoil

The seepage water travels in two subsoil environments: (a) unsaturated subsoil directly beneath the tailing pond area, and (b) saturated subsoil below the water table. At the start of operations, it is assumed that the water table (25 m below the surface) will be nearly level, sloping at about 2.5 m/km (0.0025) toward the north. With high seepage rates into subsoils of low permeability, the water table could rise under the tailing pond area, forming a ground water "mound." The shape of this mound can be predicted by various mathematical techniques[5,6] and it can be shown that, because of the permeable subsoil, the rising water-table mound will not reach the bottom of the tailing during the 15-year life of the mill, and will subside after seepage ceases.

a. Velocity of Seepage Water in Unsaturated Zone

As a first approximation the Darcy Law velocity (V_d) in the 25 m of unsaturated subsoil equals the vertical partially-saturated hydraulic conductivity ($K = 10^{-6}$ cm/sec = 0.315 m/yr; see part 1 above) times the hydraulic gradient (H/L):

$$V_d = K H/L \tag{7}$$
$$= 0.315 \text{ m/yr } (29 \text{ m}/25 \text{ m})$$
$$= 0.365 \text{ m/yr}$$

The velocity of the seepage water itself (V_s) is equal to V_d divided by the effective porosity (N_e). Based on typical values of specific yield of silty sandy alluvium[2] N_e is assumed to be 10%. Thus:

$$V_s = V_d/N_e \tag{8}$$
$$= (0.365 \text{ m/yr})/0.10$$
$$= 3.65 \text{ m/yr}$$

It follows that seepage water first reaches the water table in 25 m/3.65 m/yr = 6.85 = approximately 7 years. It is assumed that seepage water is not stored in the 25 m unsaturated zone; i.e., postoperational moisture content and preoperational moisture are equal. Tailing seepage lasts for 23 years and hence ground water contamination occurs continuously from the 7th to the 30th year after mill operations start.

b. Velocity of Seepage Water in Saturated Zone

In this model it is proposed that there are two flow regimes that determine the velocity and dispostion of seepage water in the saturated zone: (a) radial forced flow, and (b) aquifer flow. The radial forced flow is determined by the flow pattern of the seepage water, typically expanding downward and outward from the place of origin. The aquifer flow, or regional downgradient flow rate of natural ground water, is determined by the subsoil permeability and natural hydraulic gradient.

The radially expanding seepage water forms a bulb that displaces some of the natural ground water. When the bulb reaches certain dimensions, the seepage-ground water mixture is swept downgradient. It is assumed that the bulb expands until its subsoil cross-sectional area normal to the regional ground water gradient is equal to the cross-sectional area carrying an equivalent natural ground water discharge. When this is achieved, the natural discharge in the subsoil can accommodate the seepage input, and the seepage water moves down-gradient.

There is both theoretical and field confirmation of the model described above. Examples of ground water contamination flowing radially and then down-gradient in the manner used in this model have been found at a thorium tailing area near Chicago[7], at the Idaho National Engineering Laboratory, formerly the National Reactor Test Station[8], at a uranium tailing pile in Wyoming[9] at a chromium waste dump in Michigan[10], and at other places.[11,12]

The natural ground water discharge through the subsoil deposit equals the saturated horizontal permeability ($K_h = 10^{-2}$ cm/sec = 3150 m/yr) times the slope of the water table (it is assumed H/L = 0.0025) times the subsoil cross-sectional area (A). The cross-sectional area through which a natural ground water discharge of 1.88×10^5 m^3/yr occurs can be solved as follows:

$$Q = K_h \times A \times (H/L) \tag{9}$$

$$1.88 \times 10^5 \text{ m}^3/\text{yr} = (3150 \text{ m/yr}) \text{ (A) } (.0025)$$

$$A = 2.4 \times 10^4 \text{ m}^2$$

The actual shape of the cross-sectional area will be influenced by the seepage area (A_s = 51.6 ha) and the subsoil anisotropy. The geometry of the seepage bulb in the saturated zone should take into account the K_h/K_v ratio for the subsoil, which is 1000. A flow net using isotropic media constructed with vertical exaggeration = $\sqrt{K_h/K_v} = \sqrt{1000/1}$ = 31.6 will correct this anistropy[2,13]. Evenly spaced, radially expanding flow lines are drawn, taking into consideration the seepage area (A_s = 51.6 ha), until a bulb cross section equal to 2.4×10^4 m^2 is reached (see reference 1 for figures and more complete discussion). The volume of saturated subsoil contained in the bulb would be 3.3×10^7 m^3. As explained above, it is assumed that when the expanding seepage bulb reaches this volume it will be swept downgradient.
The rate of seepage is 1.88×10^5 m^3/yr, which would produce a water volume of 4.3×10^6 m^3 in the 23 years during which seepage occurs. Since the subsoil effective porosity (N_e) is 10%, the volume of subsoil saturated in 23 years will be 4.3×10^7 m^3. This is greater than the volume of the seepage bulb described above. Therefore after (23) $3.3 \times 10^7/4.3 \times 10^7 \cong 17$ years, radial outward movement of the seepage bulb ceases, and the seepage begins to be carried downgradient.

In summary, seepage water reaches the water table 7 years after mill operations start. The tailing seeps for 23 years. Seepage enters the ground water from the 7th to 30th year, with radial flow occurring from the 7th to 24th year and downgradient movement from the 24th year to the 30th. The maximum lateral horizontal extent of the radial bulb of seepage water would be about 1000 m from the center of the tailing pond and the contaminants would be confined to the upper 20 m of the zone of saturation.

3. Downgradient Movement and Dispersion of Seepage Water

Dispersion is a general term which describes the mixing and spread of fluid. Dispersion in ground water can occur by diffusion, hydrodynamic processes due to aquifer anisotropies, and chemical processes.

Ion exchange is a well-known chemical phenomenon whereby dissolved high-valence solute cations replace other cations. Robinson[14], who wrote a review of the principles of ion-exchange processes and their roles in the disposal of high-level radioactive wastes, states that clay minerals (e.g., montmorillonite and vermiculite) and zeolites have high exchange capacities, and points out that high-valence cations with high atomic weight have great replacing pwoer.

The "distribution coefficient," or "adsorption coefficient" (K_d) is a laboratory determination of the amount of solute left on a soil sample after it has been mixed and allowed to reach equilibrium with the soil[15]

$$K_d = \bar{C}/C \tag{10}$$

where $\bar{C}$ = concentration sorbed per gram of soil

              c = equilibrium concentration in external waste solution.

If $K_d = 0$, there is no sorption whatsoever; this is commonly the case with anionic solutes such as sulfate and chloride, and is the reason why these anions are good ground water tracers.

There is no accepted theory of solute movement encompassing all of the above phenomena of dispersal that can be used to predict the time rate of change in concentration (i.e., dilution) of a solute. A one-dimensional laboratory study of sorption, determined by a soil column or by shaking a solution with soil, can be extrapolated to predict the amount of sorption likely to occur in the field. The velocity of the ion transport in the field can be predicted using the formula developed by Hajek[15]:

$$V_i = \frac{V}{1 + \frac{K_d B_d}{\theta}} \tag{11}$$

where V_i = ion velocity

            V = solution velocity

          K_d = distribution coefficient

         B_d = bulk density

         θ = volumetric moisture content

Convection, hydrodynamic dispersion, and diffusion are not accounted for in equation 11; the seeping water is assumed to move straight down in a column. Gaussian-curve formulas have been developed which show that the solute concentration at distance x is very weak at first, then increases rapidly before a final slow approach to the injection concentration[16].

A survey of the literature shows that there are many specific instances of ground water contamination, and there are many general theories and mathematical models used to explain the movement and dispersion of toxic elements or tracers in ground water. Gureghian[17] presents an areal finite-element model in this symposium proceedings. This paper utilizes only basic mathematical and hydrogeologic assumptions and the construction of a simple analytical model to determine the extent of ground water contamination.

From the calculations above it was shown that after seepage water drifts downgradient with the natural ground water after the 24th year after mill operations start. Longitudinal dispersion is assumed to begin with the downgradient movement. The following simple mathematical model is used to determine the rate of longitudinal dispersal, and the dilution and chemical changes accompanying dispersion.

The average downgradient seepage velocity (V_s) of the slug of contaminants can be calculated from equation 8:

$$V_s = V_d/N_e$$

$$= K_h \, (H/L)/N_e$$

$$= (3150 \text{ m/yr}) \, (0.0025) \, (0.1)$$

$$= 79 \text{ m/yr}$$

If no longitudinal dispersion occurs, the seepage water bulb would move downgradient as a slug at a velocity of 79 m/yr. Assuming a "longitudinal dispersivity" (a_I) of 5 m, the "longitudinal dispersion coefficient" (K_L) is:

Table 1. **Chemistry of tailing pond liquid.**

	Column A	Column B	Column C	Columr
	Mill Effluent Liquid Waste Concentration (mg/L unless otherwise specified)	Column A times 130 %	USPHS-USEPA Max. Perm. Concentration (mg/L unless otherwise specified)	Ratio Columr Columr
Al	2,000	2,600	(no limit)	--
As	0.2	0.26	0.05	5.2
Ca	500	650	200	3.3
Cd	0.2	0.26	0.01	2.6
Cl	300	390	250	1.6
Cu	50	65	1	65
F	5	65	1.4-2.4	~ 3.4
Fe	1,000	1,300	0.3	4,300
Hg	0.07	0.09	0.005	18
Mo	100	130	(no limit)	--
Mn	500	650	0.05	13,000
Na	500	650	200	3.3
NH_4	500	650	(no limit)	--
Pb	7	9	0.05	180
Se	20	26	0.01	2,000
SO_4	30,000	39,000	250	156
V	0.1	0.13	0.1	1.3
Zn	80	104	5	21
TDS	35,000	45,500	500	91
pH (units)	2	1.9	6-9	--
U-nat (pCi/L)	5,400	7,020	550	13
Ra-226 (pCi/L)	400	520	5	104
Th-230 (pCi/L)	150,000	195,000	2,000	98
Pb-210 (pCi/L)	400	520	100	5.2
Po-210 (pCi/L)	400	520	700	0.9
Bi-210 (pCi/L)	400	520	400	1.3

$$K_L = V_s \times a_I \qquad (12)$$

$$= (79 \text{ m/yr}) \ (5 \text{ m})$$

$$= 395 \text{ m}^2/\text{yr}$$

Times of arrival of seepage water at various distances are shown by "break-through curve" concentration plots. Because the reverse S-shaped curves are essentially cumulative normal (Gaussian) curves, the standard deviation (σ) is defined as the spread between the 50% and the 16% or 84% value. The longitudinal dispersion coefficient is related to the standard deviation and time (t) as follows[11]:

$$\sigma = (2 \ K_L t)^{1/2} \qquad (13)$$

Tailing pond seepage water breakthrough curves for locations directly downgradient from the seepage bulb were constructed (see reference 1 for details). Figure 2 shows the resulting histograms for contaminant concentration at 2, 8, and 30 km downgradient. It is assumed that no lateral or vertical dispersion occurs beyond the 2000-m-wide bulb and 20-m-thick zone described above.

Table I shows the expected concentrations of dissolved substances contained in the tailing pond water at the model mill (Column A), and the maximum per-missible concentrations (MPC) for drinking water established by the U.S. Public Health Service, as modified by the U.S. Environmental Protection Agency in 1975 (Column C). It was shown above that the evaporation rate ($Q_{evap} = 3.87 \times 10^5 \text{m}^3/\text{yr}$) is greater than the precipitation rate ($Q_{ppt} = 2.79 \times 10^5 \text{ m}^3/\text{yr}$). Therefore, there is a deficit of 1.08×10^5 m^3/yr of pure water, and the concentrations of dis-solved substances in the mill discharge ($Q_{mill} = 4.60 \times 10^5$ m^3/yr) become increased to 4.60/4.50-1.08 = 130% of the residual tailing pond liquid (see Column B). Column D is the ratio of Column B to Column C, and indicates that the water is acidic, and that concentrations or iron, manganese, sulfate, selenium, radium, thorium, lead, and other trace metals are high. In addition to the contaminants commonly found in uranium tailing pond water listed in Table I there may be other trace metals in a specific ore which could be present in the tailing pond liquid, such as boron, barium, or chromium.

The fates of the major contaminants are analyzed below:

· pH--The volume of subsoil necessary to react with the neutralize seeping sulfuric acid can be predicted. The general equation is:

$$H_2SO_4 + CaCO_3 \rightarrow CaSO_4 + H_2O + CO_2 \qquad (14)$$

Water with a pH of 2 has by definition a concentration of 0.01 moles per liter (0.01 equivalents/L) of hydrogen ions. Assuming the subsoil has a dry density of 2 g/cc and contains 1% CaCO$_3$, there would be 0.02 g/cc CaCO$_3$. This is equal to 2×10^{-4} moles/cc = 4.0×10^{-4} equivalents/cc = 0.4 equivalents/liter. It follows that the ratio of equivalents of hydrogen ions per liter to calcium carbonate ions per liter is 0.01/0.4 = 1/40. Therefore 1 m^3 of subsoil could neutralize about 40 m^3 of pH 2 seepage water, if complete mixing could be obtained. Because the effective porosity (N_e) of the subsoil is 10%, 1 m^3 of subsoil at the model site could provide space for 0.1 m^3 of acidic seepage. In other words, if all the CaCO$_3$ is utilized, the subsoil has 400 times the capacity necessary to neutralize the advancing acid seepage water. It is not known how much of the available CaCO$_3$ will actually react with and contribute to the neutralization process; coal mine drainage studies have shown that precipitation crusts may form around limestone fragments. Nevertheless the 400 x neutralizing capacity present in the subsoil seems adequate.

· Iron--Although Column D of Table I shows that the theoretical expected con-centration of iron in the tailings liquid is 4300 times greater than the MPC for drinking water. Most is expected to precipitate out of solution as iron oxides in the tailing pond area and subsoil as neutralization of the seepage water occurs. Although the contamination of ground water by iron may not represent a serious hazard to man, calculations made to predict the rate of spread of iron can be used

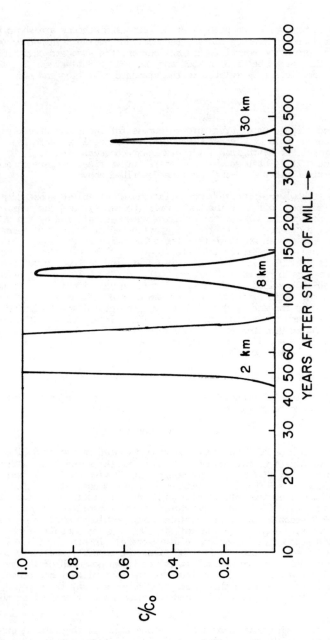

Figure 2. Contaminant concentrations in ground water 2, 8, and 30 kilometers downgradient from edge of impoundment.

for other contaminants as well.

As the seepage water is neutralized, the solubility of dissolved iron decreases dramatically. In a neutral solution, the solubility of ferric ions is so low that most will be precipitated, and the concentration of ferrous ions in ground water is probably limited by the solubility of ferrous carbonate, but still ranges between 1 and 10 mg/L if the pH is between 6.0 and 8.0[2]. If the seepage is neutralized (as discussed above for pH), then the highest concentration of total ferrous and ferric iron will be about 10 mg/L. This is, however, still about 30 times the MPC for drinking water.

If no further adsorption or cation exchange occurs, the changes in concentration of the 10 mg/L iron solution as it migrates downgradient can be determined from Figure 2. For example, at a hypothetical well located 2 km downgradient from the edge of the impoundment, contaminated water begins to arrive at about 45 years after mill operations begin, reaching a maximum $C/C_0 = 1$ (10 mg/L Fe) from 50 to 70 years, then declining until 80 years when $C/C_0 = 0$ (background concentration).

· Manganese--The geochemistry of manganese is similar to that of iron. Both metals are precipitated as hydrous oxides in response to neutralization of acidic water. In natural waters the concentration of manganese is typically less than one-half that of iron[18]. Above it was assumed that the iron concentration would be 10 mg/L after neutralization; therefore, as a rough approximation it is assumed here that the concentration of manganese (800 mg/L in tailings liquid at pH = 2) will be reduced to near 5 mg/L after neutralization. This is still 100 times the MPC for drinking water.

Based on the curves shown in Figure 2 and assuming no further precipitation or ion exchange en route, it can be shown that maximum manganese concentrations are:

   --5 mg/L from 50-70 years at 2 km downgradient

   --4.75 mg/L peak at 124 years at 8 km downgradient

   -- < 3.25 mg/L peak at 400 years at 30 km downgradient

· Sulfate--As shown in Column B of Table I, the residual tailings pond water will have 39,000 mg/L sulfate, which is 156 times greater than the 250 mg/L MPC for drinking water. As the tailings water seeps into the subsoil, it was calculated above that all of the sulfuric acid (H_2SO_4) will react with limestone ($CaCO_3$) or caliche in the subsoil to produce anhydrite ($CaSO_4$) and carbonic acid (H_2CO_3) (see equation 14). The carbonic acid will decompose to water and carbon dioxide. These reactions are reversible, and it can be shown that the maximum dissolved sulfate that can occur at equilibrium is governed by the solubility of anhydrite at pH 7 and will be about 2,000 mg/L. As the sulfuric acid is neutralized, calcium and (depending on subsoil mineralogy) sodium and potassium ions will be added so that the solution will have equal milliequivalents of cations and anions. (If only calcium ions were added, the concentration of calcium would be about 1040 mg/L, compared with the 200 mg/L MPC for drinking water).

In essence, calcium and sulfate are expected to reach the water table in quantities about five to ten times greater than the recommended limits for drinking water. These contaminants will flow downgradient as discussed above for iron, and will probably be further reduced by precipitation and ion exchange en route. Neither of these contaminants poses a serious health hazard.

· Selenium--Although selenium is a rare element, it is commonly associated with uranium deposits. A concentration of 32 ppm selenium is present in the residual tailing pond water at a pH of 2. This toxic element has caused poisoning of livestock in numerous areas in western United States. The geochemistry of selenium is poorly understood, but it is known that selenium can form an anion (selenate) similar to sulfate that is not subject to cation exchange. It can be assumed, therefore, that in the worst case no reduction of selenium concentration will occur due to changing pH or ion exchange. Therefore, based on Figure 2, the following selenium concentrations may be expected.

--26 mg/L from 50-70 years at 2 km downgradient

--25 mg/L peak at 125 years at 8 km downgradient

-- < 17 mg/L peak at 400 years at 30 km downgradient

· Radium and Thorium--As shown in Table I, the residual tailing pond water will contain 520 pCi/L radium and 195,000 pCi/L thorium. These are, respectively, 104 and 98 times greater than drinking water MPC.

A conservative value for the distribution coefficient of radium in nearly neutral water is 10 mL/g.

Modifying equation (11):

$$V_{Ra} = \frac{V}{1 + K_d \lambda \over N_e} \qquad (15)$$

where
V = water seepage velocity = 3.65 m/year (from above)

λ = dry density of subsoil = 2.12 g/cc

N_e = effective porosity = 10% = 0.10

It is shown by solving equation (15) that the radium will move at 1/215 the velocity of seepage water, or 0.017 m/year, so it would move 0.39 m in 23 years, the life of seepage from the pond. Thus, no ground water contamination by radium is expected. Thorium has a very high distribution coefficient[9,19], and would be expected to be fixed within a few centimeters of the tailing pond bottom.

· Other Possible Contaminants--It is shown on Table I that other ions are present in the tailing pond liquid in concentrations greater than the MPC. These include arsenic, cadmium, copper, mercury, lead, and zinc. Concentrations of aluminum and ammonia also are high. It is assumed that as the pH increases from 2 to neutrality, as discussed above, most of the trace metals will precipitate out. Further, trace metal cations are generally subject to ion exchange and will substitute for sodium, calcium, or potassium, particularly in clayey portions of the subsoil. Those cations with the higher oxidation state (such as $zinc^{+2}$, $lead^{+2}$, etc.) tend to exchange more readily than those with lower valence. However, some ions, such as arsenic, selenium, molybdenum, cyanide, and chromium, can behave as anions as well as cations, and as such should be considered especially dangerous contaminants.

The geochemistry of many trace metals and their effects on human health are generally poorly understood. In this study it is assumed that the neutralization of the seeping water will allow for the precipitation of aluminum, cadmium, copper, mercury, lead, and zinc so that seepage reaching ground water will have concentrations near the MPC values. Arsenic may not be affected by pH and can be expected to follow the concentration curves shown in Figure 2, where C_0 = 0.26 mg/L.

Ammonia and possibly other gases, such as hydrogen sulfide, are not expected to remain in the seepage solution.

Following the cessation of mill operations, seepage from the tailing pond area will be substantially reduced because there will be no more tailing water discharged from the mill.

CONCLUSION

The conclusion of the analysis herein is that no contamination of ground water with radioactive material will be caused by seepage from the typical unlined mill tailing pond, but that contamination from sulfate, iron, manganese, selenium, and possibly other trace elements will occur. Hydrogeologically the assumed mill site is favorable for the disposal of radionuclides because of the semi-arid climate, the large distance (25 m) to the water table.

Support of these predictions comes from examination of existing mills and from other studies. In a recent study[20] using a computer model and laboratory distribution coefficient it was predicted that radium would be reduced to 10% of its original concentration at 3 m below the bottom of a seeping uranium tailing pond after 20 years.

Field studies of sorption of radium and thorium, which would support laboratory values of the distribution coefficient of radionuclides and the analysis presented herein, are meager. Among the better is a study of soil and ground water contamination from a large abandoned thorium waste pile in West Chicago, Illinois[17], where preliminary findings are that dissolved anions such as sulfate have migrated through till and limestone for 50 years and have contaminated the ground water up to 1400 m away, but the dissolved radionuclides (radium and thorium) have not traveled more 100 m. Other field studies documenting the movement of radium and thorium below tailings ponds show even less spread of radionuclides than the West Chicago study. One report on the hydrogeology of five uranium mills in the Grants, New Mexico, area showed that vast quantities of radionuclides were seeping or being injected into ground water, but only one of 72 water wells tested within about 5 km of the tailing ponds showed any radioactivity above background[20]. Other ongoing research by Ford, Bacon and Davis, Utah, Inc., for the U.S. Dept. of Energy has included the drilling of test holes around the 22 inactive tailing piles in the western states. The preliminary conclusions of these studies are that remnant thorium and radium concentrations in the soil beneath the piles typically drop off to background within a meter below the bottom of the tailing piles.

REFERENCES CITED

(1) NRC. In preparation. Generic environmental statement for uranium milling. U.S. Nuclear Regulatory Commission, NUREG - _____.

(2) Davis, S. N., and R. J. M. DeWeist. 1966. Hydrogeology. John Wiley & Sons, New York, 463 pp.

(3) Elzeftway, A., and R. S. Mansell. 1975. Hydraulic conductivity calculations for unsaturated steady-state and transient-state flow in sand. Proc. Soil Science Soc. of Am., Vol. 39, pp. 599-603.

(4) Mualem, Y. 1978. Hydraulic conductivity of unsaturated porous media: Generalized macroscopic approach. Water Resources Research, Vol. 14, No. 2, pp. 325-334.

(5) Walton, W. C., 1970. Groundwater resource evaluation. McGraw-Hill Book Co., New York, 664 pp.

(6) Brock, R. R. 1976. Hydrodynamics of perched mounds. Hydraulics Div., Am. Soc. Civil Engr., 102 (H78), pp. 1083-1100.

(7) ANL. 1977. Environmental assessment related to Kerr-McGee Chemical Corporation--West Chicago: Characterization of geohydrology and subsurface chemistry. Division of Environmental Impacts Studies, Argonne National Laboratory, 66 pp. and Appendices, July 1977.

(8) Barraclough, J. T., et al. 1966. Hydrology of the National Reactor Testing Station Idaho. U.S. Geological Survey, Open-File Report TID 4500.

(9) D'Appolonia. 1977. Report 3, Environmental effect of present and proposed tailings disposal practices, Split Rock Mill, Jeffrey City, Wyoming. Project No. RM 77-419 D'Appolonia Consulting Engineers, for Western Nuclear, Inc., Vols. I and II, October 1977.

(10) Deutsch, M. 1972. Incidents of chromium contamination of groundwater in Michigan, in: Wayne A. Pettyjohn (editor), Water quality in a stressed environment, Burgsess Publ. Co., Minneapolis, pp. 149-159.

(11) Fried, J. J. 1975. Groundwater pollution, theory, methodology, modelling, and practical rules, in: Ven Te Chow (editor), Developments in water science, 4, Elsevier Scientific Publishing Co., New York, 330 pp.

(12) Gerahty and Miller. 1975. Groundwater pollution problems in the North-western United States. Geraghty and Miller, Inc., for Environmental Protection Agency, PB-242 860, 361 pp., May, 1975.

(13) Cedergren, H. R. 1967. Seepage, drainage, and flow nets, John Wiley & Sons, New York, 489 pp.

(14) Robinson, B. P. 1962. Ion-exchange minerals and disposal of radioactive-wastes--a survey of literature. U.S. Geological Survey Water-Supply Paper 1616, 132 pp.

(15) Hajek, B. F. 1969. Chemical interactions of wastewater in a soil environment. Water Pollution Control Federation, Vol. 41, No. 10, pp. 1775-1786.

(16) Routson, R. C. 1974. An evaluation of predictive methods of determining the effects of solid-liquid phase interactions on the movement of solutes in soils. Battelle Pacific Northwest Laboratories, BNWL-1196, Richland, Washington.

(17) Gureghian, A. B., N. J. Beskid, and G. J. Marmer. 1978. Predictive capabilities of a finite element model in the ground water transport of radionuclides. Seminar on management, stabilization and environmental impact of uranium mill tailings. Organization for Economic Cooperation and Development, Paris, France. Albuquerque, New Mexico, July 24-28, 1978.

(18) Hem, J. D. 1968. Aluminum species in water. in: R. A. Baker, Trace Inorganics in Water, Am. Chem. Soc., Advances in Chemistry Series 73, Washington, D. C., pp. 98-114.

(19) Sears, M. B., R. E. Blanco, R. C. Dahlman, G. S. Hill, A. D. Ryan, and J. P. Witherspoon. 1975. Correlation of radioactive waste treatment costs and the environmental impact of waste effluents in the nuclear fuel cycle for use in establishing 'as low as practicable' guides-milling of uranium ores. Oak Ridge National Laboratory, ORNL-TM-4903, Vol. 1, May 1975.

(20) Kaufmann, R. F., G. G. Eadie, and C. R. Russell. 1976. Effects of uranium mining and milling on ground water in the Grants Mineral Belt, New Mexico, Ground Water, Vol. 14, No. 5, pp. 296-308.

DISCUSSION

J. MONTGOMERY, United States

Do you have any opinion or have you become involved or in contact with the problem that Cutter Corporation has had in their Canyon City Mill ?

P.H. RAHN, United States

No. Sorry. Once in awhile you hear about a case where there is some high radium coming out. I wish people would document these and publish reports. There is very few hard facts.

N. FOURCADE, France

I am a little bit surprised by the high concentration of radium in the water that you apply in the United States, because in France this is a little bit lower. You may wonder why in fact this standard is adopted, because in France in many of our spring and well waters we have radium concentrations which are often higher than 5 picocuries per liter. So, I wonder how you can apply a limit of 3 per liter. We have done statistics on mineral water and we have reached values higher than 5 picocuries per liter, for a quantity of springs greater than 25 and this is considerable, I think. What do you think about this ?

P.H. RAHN, United States

Well, the question was about radium being 5 picocuries liter drinking water limit. I think maybe during the Panel Session it would be a good time to explore this. It is really a health physics thing and not so much ground water. But, you know, these limits change and there used to be people who would go to radium springs thinking they were getting a cure for their ills and I think we have to take these limits with a little grain of salt. What is the limit today may change tomorrow.

N. FOURCADE, France

The problem for us is that we could never apply this standard of 3 picocuries per liter in drinking water, we could not do it. We would have to purify our natural water.

LEAD ISOTOPES AS INDICATORS OF ENVIRONMENTAL CONTAMINATION FROM THE URANIUM MINING AND MILLING INDUSTRY IN THE GRANTS MINERAL BELT, NEW MEXICO

David B. Curtis and A. J. Gancarz
Los Alamos Scientific Laboratory
Los Alamos, New Mexico, USA

The unique isotopic composition of lead from uranium ores can be useful in studying the impact of ore processing effluents on the environment. "Common" lead on the earth's surface is composed of 1.4% ^{204}Pb, 24.1% ^{206}Pb, 22.1% ^{207}Pb, and 52.4% ^{208}Pb. In contrast, lead associated with young uranium ores may contains as much as 95% ^{206}Pb. These extreme differences provide the means to quantitatively evaluate the amount of lead introduced into the environment from the mining and milling of uranium ores by measuring variations of the isotopic composition of lead in environmental samples.

We will discuss the use of Pb isotopes as diagnostic tools in studying the hydrologic transport of materials from U ore dressing plants in the Grants Mineral Belt, New Mexico, USA. Preliminary measurements on effluents intimately associated with processing wastes are consistent with a simple model in which "radiogenic" lead from the ores is mixed with "common" lead from the uncontaminated environments.

UTILISATION DES ISOTOPES DU PLOMB COMME TRACEURS DE LA CONTAMINATION DE L'ENVIRONNEMENT DUE AUX ACTIVITES D'EXTRACTION ET DE TRAITEMENT DU MINERAI D'URANIUM DANS LA ZONE MINIERE DE GRANTS (NOUVEAU-MEXIQUE)

RESUME

La composition isotopique unique du plomb dérivé des minerais d'uranium peut présenter de l'intérêt pour l'étude des incidences, sur l'environnement, des effluents de traitement du minerai. Le plomb "commun" se trouvant à la surface de la terre est composé de 1,4 % de Pb-204, de 24,1 % de Pb-206, de 22,1 % de Pb-207 et de 52,4 % de Pb-208. Par contre, le plomb associé à des minerais jeunes d'uranium peut contenir jusqu'à 95 % de Pb-206. Ces différences extrêmes permettent d'évaluer quantitativement le plomb introduit dans l'environnement par suite de l'extraction et du traitement du minerai d'uranium en mesurant les variations de la composition isotopique du plomb sur des échantillons prélevés dans l'environnement.

Cette communication traite de l'emploi des isotopes du plomb comme moyens de diagnostic dans l'étude du transport hydrologique de matériaux provenant d'installations de traitement de l'uranium situées dans la zone minière de Grants, Nouveau-Mexique (Etats-Unis). Les mesures préliminaires effectuées sur des effluents intimement associés aux déchets de traitement concordent avec un modèle simple dans lequel du plomb "radiogénique" provenant du minerai est mélangé à du plomb "commun" provenant d'un environnement exempt de contamination.

Work done under the auspices of the U. S. Department of Energy

INTRODUCTION

Processing of uranium ore presents waste disposal problems not greatly different from other ore dressing procedures except for the unique composition of the materials being handled. Generally the processing involves crushing and grinding the raw materials to an appropriate size, selectively leaching uranium from the main mass, and further processing the pregnant solution to obtain a high grade uranium product. Leaching is usually done with sulfuric acid or sodium carbonate solutions, although other techniques are used in the United States on a small scale. (1) Disposal of solid and liquid wastes constitutes a major problem since these materials contain elevated levels of radioactive nuclides as well as non-radioactive materials. Tailings streams typically contain potential contaminants introduced as part of the ore dressing processes such as sulfuric acid, sulfates, carbonates, chlorides, nitrates, ammonia, lime, magnesia, caustic, potassium permanganate, copper sulfate, manganese dioxide, cyanide, polyacrylamides, several alkylphosphates, tertiary amines, alcohols, kerosene, and fuel oils. (1) In addition, the effluent may contain high concentrations of constituents from the ore such as iron, copper, vanadium, molybdenum, arsenic, lead, fluorine, selenium, and up to 70% of the radioactive materials initially in the uranium deposits. (1) Failure to effectively contain these wastes obviously represents a significant hazard to the environment and the well being of the local population.

Typically, solid tailings containment is accomplished by a retention dam constructed of local mine wastes or previous tailings materials. Piles of these solid wastes can reach appreciable size; some of them in the Grants region are comparable to the prominent natural land forms. Every ton of ore processed produces one to five tons of liquid wastes. Disposal of these liquids is accomplished by evaporation, seepage into the underlying alluvium or release into local rivers and streams. In the Western United States, a typical acid leach plant would require pond areas of several hundred acres to evaporate the liquids generated. (1) There are plants with very large evaporating ponds, although it is estimated that in many cases seepage losses may account for as much as 80% of the liquid loss from waste disposal ponds. It is clear that this seepage of liquid wastes and leaching of solid wastes by invading waters represent a potential source of contamination of local aquifers and water supplies.

This paper will discuss the principles of a technique, based upon lead isotopic systematics, which addresses some aspects of the transport of materials from these waste disposal areas. Because of the unique composition of the ore bodies, the lead isotopes are likely to be diagnostic in studying the movement of material from these waste disposal areas.

LEAD ISOTOPES IN NATURE

The chart of the nuclides indicates that natural lead is composed of 1.4% ^{204}Pb, 24.1% of ^{206}Pb, 22.1% of ^{207}Pb, and 52.4% ^{208}Pb having an atomic weight of 207.2. (2) However, it was demonstrated many years ago that lead from U ores had atomic weights less than this value and those from Th-rich materials were heavier. This perturbation is the result of the radioactive decay of isotopes of uranium and thorium. Uranium-238 decays through a series of radioactive progeny to produce the stable isotope ^{206}Pb, ^{235}U decays to ^{207}Pb and ^{232}Th eventually produces ^{208}Pb. The light isotope ^{204}Pb is not the end product of any known radioactive decay chain, but results from the processes of nucleosynthesis, which produced the elements. It is obvious that "natural" lead can have variable isotopic composition depending upon the relative abundances of U, Th, and Pb and the length of time this melange of elements has been in close association.

"Common" lead, i.e., that with isotopic composition given on the chart of the nuclides is pervasive in the crust of the earth. It reflects the isotopic composition at the time the earth was formed plus the evolution of radiogenic lead in an environment of constant average composition with respect to U, Th, and Pb. Variations from "common" lead occur when these elements are fractionated, as in the formation of ores. Such variations can be quite large. For instance, consider the formation of lead in an average uranium ore from the Grants region. Such an ore was deposited $\sim 10^8$ years ago and presently contains 0.68% uranium. (3) Application of the familiar law of radioactive decay indicates that this ore would contain 90 ppm of lead produced *in situ* by the decay of U isotopes. This lead has a $(^{207}Pb/^{206}Pb)$ ratio of 5×10^{-2} and will hereafter be referred to as radiogenic lead. The average measured concentration in these ores is 80 ppm (3) and the $(^{207}Pb/^{206}Pb)$ ratio is 8.7 $\times 10^{-2}$, as opposed to the 8.3×10^{-1} (4) found in "common" lead. The close correspondence between the isotopic composition of pure radiogenic lead and the lead from these ores indicates that the majority of lead was produced *in situ* by the decay of uranium.

Table I presents the average isotopic composition of lead in uranium ores from the Colorado Plateau (5) and the isotopic composition of "common" lead. (4) The differences are significant and provide a means of distinguishing between lead originating in uranium ores or ore residue and lead normally found in the natural environment.

TABLE I

LEAD ISOTOPE RATIOS IN SOME NATURAL MATERIALS

	"Common"[1]	Uranium Ores[2]
$^{204}Pb/^{206}Pb$	$(5.35 \pm 0.09) \times 10^{-2}$	$(3.9 \pm 3.2) \times 10^{-3}$
$^{207}Pb/^{206}Pb$	$(8.34 \pm 0.1) \times 10^{-1}$	$(8.7 \pm 4.6) \times 10^{-2}$
$^{208}Pb/^{206}Pb$	2.06 ± 0.04	$(9.0 \pm 12) \times 10^{-2}$

[1]Stacey and Kramers (1975)
[2]Miller and Kulp (1969)

LEAD ISOTOPES AS ENVIRONMENTAL TRACERS

Isotopic labeling involves the introduction, of material that is isotopically different but chemically identical to that normally found in the system. Mixing of the isotopically different species provides a means to study the dynamic behavior of that constituent in the system. Isotopic tracers may be radioactive, which can be detected by counting techniques, or they may be non-radioactive, which requires more difficult techniques for detection. Isotopic tracers offer an advantage over chemical ones in that the measured results are normalized relative to chemically identical species, i.e., results may be determined in terms of the specific activity of a radioactive isotope or the isotopic ratio of nonradioactive isotopes. Variations of these quantities directly reflect the mixing of isotopically distinct entities *independent of* the absolute concentration of the chemical species in the system. Changes in the isotopic composition are dependent upon chemical and physical processes only as they influence the mixing of labeled and unlabeled species.

Stable Isotopes

Consider as an illustration, a uranium mill tailings pile where liquids "percolate" through the tailings and seep into the underlying soil to the surrounding environment. The liquid may be of any origin; leaching agents which eventually become liquid wastes, rainwater, or groundwater which flows into the pile. Initially any such liquid is likely to contain lead that is isotopically normal. Such lead represents the endpoint labeled as "common" in Fig. 1. When this liquid impacts the tailings pile, the lead isotopes in the liquid begin to exchange with isotopes in the wastes and the isotopic composition of lead in the liquid moves down the line toward the "radiogenic" endpoint as shown in Fig. 1. The shift of the isotopic composition toward this endpoint depends upon the relative abundance of lead with the unique isotopic compositions and the efficiency with which they are mixed (either chemically or physically). *It does not depend upon a change in the absolute*

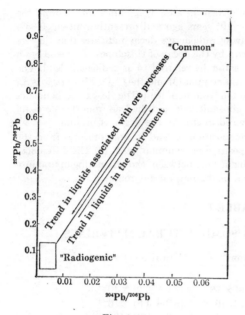

Fig. 1.

Isotopic mixing diagram showing change of isotopic composition resulting from mixing of "radiogenic" lead with "common" lead. Measured isotopic composition in uranium ores from the Colorado Plateau are defined by the box near the "radiogenic" endpoint.

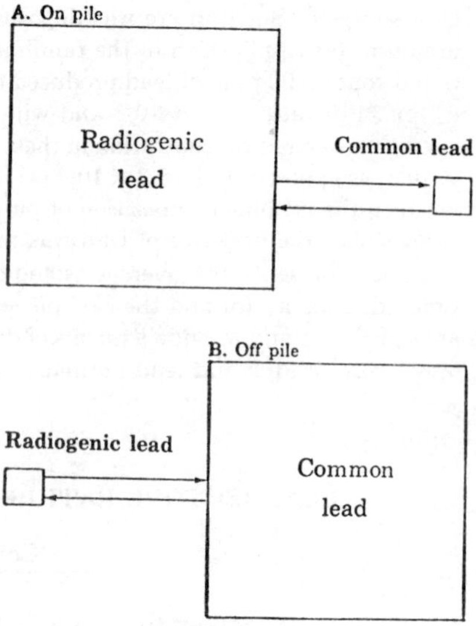

Fig. 2.

A diagram illustrating the differences in the size of the reservoirs of "radiogenic" lead and "common" lead on the wastes from a uranium ore processing plant (A) and in the environment near the waste pile (B).

concentration of lead in any phase. It is likely that the situation on the pile will be like that represented in Fig. 2A. Radiogenic lead is probably an enormous reservoir relative to "common" lead. Further, since the tailings have been ground and leached specifically to promote the exchange of materials, it is possible that exchange of lead between phases is rather fascile. Consequently, the isotopic composition of lead in liquids impacting the tailings piles is likely to be strongly shifted toward the "radiogenic" endpoint of the mixing diagram in Fig. 1.

Once the liquid leaves the pile, the situation is reversed as shown in Fig. 2B. Now "common" lead in soils and uncontaminated waters is a large reservoir relative to radiogenic lead leached from the pile. The isotopic composition of lead in the waters will be shifted toward the "common" endpoint as shown in Fig. 1. As before, the extent of the shift will depend upon the relative magnitude of the abundances of the two types of lead and the facility with which they exchange.

The mixing of end members with unique isotopic compositions is a simple isotope dilution problem. A mixing ratio (ρ) can be calculated from measurements of isotopic ratios, assuming the end members are well defined. The equation for calculating this ratio is

$$\rho = \frac{(^{207}\text{Pb}/^{206}\text{Pb})_C - (^{207}\text{Pb}/^{206}\text{Pb})_M}{(^{207}\text{Pb}/^{206}\text{Pb})_M - (^{207}\text{Pb}/^{206}\text{Pb})_R}$$

The subscripts C, M, and R refer to common lead, the ratio measured in the sample and radiogenic lead respectively. The mixing ratio is just the quantity of radiogenic ^{206}Pb in the sample relative to ^{206}Pb from lead with "common" isotopic abundances. The fraction of total lead in any sample which originated in the uranium ores (radiogenic lead) may be easily calculated from the mixing ratio.

The previous example is merely an illustration. Such simple models are unlikely to represent the complex interactions that actually exist when the natural environment is impacted by human activities. However, the power of the technique is that it does not rely on any assumptions regarding the chemical and physical behavior of Pb in the system. The only significant assumption in the example involves the isotopic composition of lead in the constituents. Lead from the ore or ore residue from the Grants mineral belt has a unique isotopic composition compared to lead found in the uncontaminated environment. Mass spectrometric techniques that can measure these compositional differences with a precision of a few hundredths of a percent, provide the means to unambiguously identify small proportions of this source of anthropogenic lead in any environmental media. The ability to make such observations provides a tool to study lead contamination from waste associated with the uranium mining and milling industry. It is possible that lead isotope ratios may also have a broader application as general indicators of the encroachment of the anthropogenic wastes on their immediate environment.

MEASUREMENTS OF LEAD ISOTOPIC RATIOS

Because of the interest in lead isotopic ratios as geochronometers, mass spectrometric techniques have been developed to measure highly precise isotopic ratios in as little as a nanogram of lead. The major problem is to chemically separate lead from elements that interfere with the mass spectrometric measurements. Lead introduced by reagents and glassware during separations has the isotopic composition of "common" lead. So the effects of this contaminant will be a shift toward the "common" lead endpoint in Fig. 1. Since lead is a trace element in most natural materials, separations often involve as little as a few nanograms of the element. Extraordinary precautions are necessary to assure that the chemistry is done free of lead contaminants.

The spectrometer separates isotopes of lead and focuses them seqentially on a detector which produces a signal proportional to the abundance of the isotope. No attempt is made to relate the intensity of the output signal to the absolute abundance. Instead, the intensity from each isotope is measured relative to a reference isotope. In our case, all measurements are taken relative to ^{206}Pb and reported as $^{x}Pb/^{206}Pb$ ratios. Figure 3 is a strip chart output from the mass spectrometer showing the intensity of lead isotopes in "common" lead. As indicated in Table I, ^{208}Pb is nearly twice as abundant as ^{206}Pb, ^{207}Pb is slightly less abundant than ^{206}Pb, and ^{204}Pb is 20 times less abundant than ^{206}Pb. Figure 4 is the mass spectrometer output of lead isolated from liquid solutions residing on top of solid wastes from a currently active acid leach processing plant. Contrast this with Fig. 3; ^{208}Pb is $\sim$3 times less abundant than ^{206}Pb, ^{207}Pb is $\sim$5 times depleted relative to ^{206}Pb and the minor isotope ^{204}Pb is underabundant by a factor of 100 relative to ^{206}Pb. It is clear from the mass spectrum that this liquid contains a large proportion of radiogenic lead.

APPLICATIONS OF LEAD ISOTOPE TECHNIQUES

As previously indicated, clean separation of lead from matrix material poses the major obstacle to implementing the technique. Most of our efforts have been directed toward the development of such separation procedures. Although this development work is not complete, preliminary results unmistakably identify the presence of radiogenic lead in samples associated with mill tailings piles. Table II presents these preliminary data. Sample 1 is solution taken from the top of a waste disposal pile at a currently active acid leach processing plant, the mass spectra of lead isotopes from this sample were presented in Fig. 4. Sample 2 represents the lead isotopic composition in a sample of solution taken from a well on

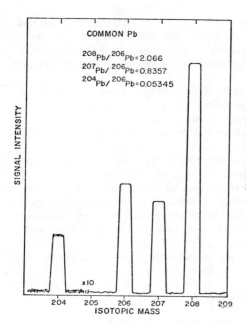

Fig. 3.
The mass spectrum of "common" lead showing the relative abundances of the four isotopes as found in most materials in the earth's crust.

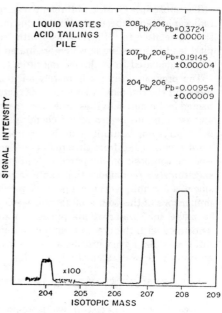

Fig. 4.
The mass spectrum of lead extracted from solutions from the top of the waste disposal pile at a currently active acid leach processing plant. The relative prominence of the signal from ^{206}Pb compared to "common" lead is indicative of a large fraction of radiogenic lead in this sample.

TABLE II

LEAD ISOTOPE RATIOS IN LIQUIDS ASSOCIATED WITH URANIUM ORE WASTES

	Sample 1	Sample 2	Sample 3
$^{204}Pb/^{206}Pb$	$(9.54 \pm 0.01) \times 10^{-3}$	$(2.277 \pm 0.009) \times 10^{-2}$	$(5.04 \pm 0.01) \times 10^{-2}$
$^{207}Pb/^{206}Pb$	$(1.9145 \pm 0.0004) \times 10^{-1}$	$(3.895 \pm 0.009) \times 10^{-1}$	$(7.900 \pm 0.005) \times 10^{-1}$
$^{208}Pb/^{206}Pb$	$(3.724 \pm 0.001) \times 10^{-1}$	$(8.937 \pm 0.01) \times 10^{-1}$	1.9332 ± 0.0007

top of a waste pile associated with a carbonate leach facility. This pile has not been used for 15 yr. Sample 3 is water taken from a well about 5 m from the inactive carbonate waste pile. The well is believed to be hydrologically downgradient from the wastes.

Figure 5 is a plot of the data on a lead isotope mixing diagram exactly like that presented for illustrative purposes as Fig. 1. The close fit to the mixing line indicates that our original assumption was correct. Samples associated with uranium wastes do indeed appear to be simple mixtures of radiogenic lead from the ores and "common" lead. It must be emphasized that this is preliminary data! Quantities of lead introduced into the samples during the separation procedures have not been assessed. Such contamination would shift the points up the mixing line toward the "common" endpoint. As a result, the percent of radiogenic lead given in Fig. 5 must be considered only as a lower limit on the actual

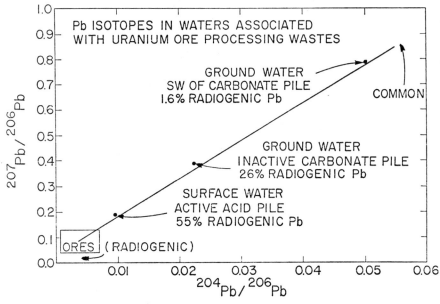

Fig. 5.

Preliminary data showing the measured isotopic composition on or near uranium ore tailings piles. The data is superimposed on an isotopic mixing diagram identical to Fig. 1. The shift away from the "common" endpoint is unambiguous evidence of the presence of radiogenic lead indicating different degrees of contamination of the solutions by lead from the ore residue.

proportion of radiogenic lead in the sample. Nevertheless, these results are extremely important. They demonstrate that the radiogenic lead signature is strongly present in solutions directly associated with the wastes. If these solutions seep from the pile, the lead isotopic ratios are quantitative indicators of the impact. Such seepage is indicated by the data from solutions taken adjacent to the inactive carbonate pile. Lead isotope ratios unmistakably show the presence of radiogenic lead, albeit small, in these liquids.

Preliminary data indicates that lead isotopes are a viable means of studying element migration from wastes associated with the uranium mining and milling industry. They can be used to trace the extent of migration and to study the fundamental physical and chemical processes associated with the movement of materials from anthropogenic sources.

REFERENCES

1. Merritt, Robert C. (1971), The Extractive Metallurgy of Uranium, Colorado School of Mines Research Institute.

2. Holden, Norman E. and F. William Walker (1972), Chart of the Nuclides, Knolls Atomic Power Laboratory, General Electric Co., Schenectady, NY 12345.

3. Squyres, Jon Benjamin (1970), Uranium Deposits of the Grants Region, N.M., Ph.D. thesis, Stanford University.

4. Stacey, J. S. and J. D. Kramers (1975), Approximation of Terrestrial Lead Isotope Evolution by a Two Stage Model, *Earth and Planetary Science Letters*, **26**, 207-221.

5. Miller, Donald S. and Laurence Kulp (1969), Isotopic Evidence on the Origin of Colorado Plateau Uranium Ores, *Geochim et Cosmochim Acta*, **33**, 1247-1294.

DISCUSSION

M.C. O'RIORDAN, United Kingdom

        No doubt you are expecting to be asked what advantages
this method has over lead-210 as a tracer, perhaps you could summarize
those for us.

D.B. CURTIS, United States

        Well, it is the same advantage. I think it is likely to
enable us to give a transport rate type of information, but only in
association with what we have already measured here. The ability to
to normalize the lead isotope or look at the lead isotope ratios
takes us away from, completely divorces us from, problems associated
with the chemistry of the material. We are not looking for changes
in chemical concentration. So, if you looked at the lead-210 and you
looked at the absolute concentration of lead-210, it is going to drop
down into the background of the natural materials and be no longer
indicative of anything. However, if you normalize the lead-210 to
the radiogenic lead-206, you are no longer dependent upon the absolute concentration of the lead. We have a normalization, the ability
to normalize an isotope, which is chemically identical and we have
taken away all assumptions with regard to the chemistry of the material.

W.E. KISIELESKI, United States

        Pursuing the lead-210, there has always been the attempt
to establish some bioassay technique to represent body burden ; and,
of course, lead-210 has been used somewhat controversely. I wonder,
if you feel your techniques might lend themselves to this sort of a
bioassay test ?

D.B. CURTIS, United States

        What is the interest ? Are you interested in the absolute
abundance of lead ?

W.E. KISIELESKI, United States

        No, I am thinking of urinary or blood level concentrations
reflecting total body burden or establishing some risk relationship.

D.B. CURTIS, United States

        You are still looking at abundances of lead in the urine
or whatever. It may have an advantage in that it is analytically
more sensitive than other lead methods. It has the isotopics. The
advantage of these is establishing a source of the material that you
are interested in.

S.A. BAMBERG, United States

        The question comes up as the movement of the lead through
the environment. Lead is one of the lease mobile of the elements,
and you mentioned this briefly. Would you describe again how you
plan to determine the rate of movement and the mixing of lead as it
moves through the soil : I guess at this point, as aqueous media.

D.B. CURTIS, United States

We really are not to that stage yet. The least we can
expect out of such a study is insight into the migration of lead
"per se". Whether it has applicability as a general tracer of
movement of material off the sources remains to be seen for this
very reason. Lead may be so immobile that it just does not move,
whereas selenium which seems to be a likely candidate may be quite
mobile and be moving. But the disadvantage of selenium, is that of
most other elements, is that they do not have the isotopic signature,
so the ability to unambiguously identify of the source is not there
for the other chemical species.

J.D. SHREVE Jr., United States

It seems that if you are in an area where there are uranium
deposits, this migration must have been going on for a long, long
time. I thought at first that maybe your techniques eliminated the
need for background measurements. It really does not, does it ?

D.B. CURTIS, United States

Indeed it does not. In fact, one of the first things we
have to do is to establish some background measurements. My answer
to that is that if we find the evidence of isotopic lead from young
uranium ores and in leads and waters, we have discovered a really
wonderful exploration technique.

D. JACOBS, United States

Turning to the previous question. I would like to make a
comment on that in that the talks that were given here this morning
concerning the migration of radionucleides, particularly radium or
the other solutes in ground water. Usually, models are good for
predicting general behavior, but when you get into the field, you
find that the heterogenity of the situation allows seeps to occur
that are very non-homogenous and you find leakage. I think that the
big advantage of the lead isotope method would be that if you do
find a seep high in lead content then you can relate that back to
where it came from rather than it just being a chemical anomaly.

D.B. CURTIS, United States

Right. And I think Dr. Rahn provided me with an interesting
opening talk ; but some of the piles that he showed where the seepage
was would be a great place to go and try to apply some of our techniques and see if we can understand what is happening.

D.R. WILES, Canada

I would be interested in knowing how you intend to handle
the question of varying backgrounds if, let's say, you compare two
lead samples from an area in which there is a lot of natural lead
and an area where there is not a lot of natural lead, you find they
will give you quite different signatures for identical seepages.

D.B. CURTIS, United States

That is correct.

D.R. WILES, Canada

Is there any easy way to handle that ?

D.B. CURTIS, United States

No, I am not sure I caught the essence of your question, but I think one of the advantages of the technique is that it is likely that the background isotopic composition is going to be incredibly narrowly defined, so that we do not have the problem which arises with so many measurements of absolute chemical abundance. When you have such a large fluctuation in background, distinguishing the signal from the noise becomes a real problem. I think that the background is likely to be very tight for the isotopic composition.

D.R. WILES, Canada

This is quite true, of course, but the problem is going to come when you try to find out how much of the signal to noise ratio really comes from the background, and how much of it is differing amounts of seepage.

D.B. CURTIS, United States

The only way I can think of is to try to go to places where we believe there to be no possibility of encroachment by the uranium ores or by the subsequent residues and begin looking at the lead isotopic composition and other materials, and see what kind of variation and infer from that what the isotopic background is. I do not know how else to do it.

D.R. WILES, Canada

You would be susceptible, I suspect, to interference from a nearby major highway.

D.B. CURTIS, United States

No, I do not think so because that is going to be common lead.

D.R. WILES, Canada

But you do not know how much of it ?

D.B. CURTIS, United States

No. How much of it is another problem which I have not really addressed here. We will begin to look for correlation between the isotopic signature and the absolute abundance of lead or selenium or whatever else is interested. The isotopes provide the signature. If we could begin to see the correlations between the signature and the various other possible contaminents, then I think we have a very direct evidence of what is going on. But the absolute abundance of lead does not really come into the question with regards to the isotopic systematics.

J.D. SHREVE Jr., United States

What is the cost of analysis ?

D.B. CURTIS, United States

This is a question that always comes up. It is expensive. It is a difficult analysis, and it is a very delicate analysis. We estimate a mandate for results. It is not a technique which one is going to apply without some careful consideration of what one wants to learn from making the measurements. But I think that the power of the tool is such that is warrants that careful consideration.

PREDICTIVE CAPABILITIES OF A TWO-DIMENSIONAL MODEL IN THE GROUND WATER TRANSPORT OF RADIONUCLIDES

A. B. Gureghian, N. J. Beskid, and G. J. Marmer

Division of Environmental Impact Studies
Argonne National Laboratory
Argonne, Illinois 60439 USA

ABSTRACT

The discharge of low-level radioactive waste into tailings ponds is a potential source of ground water contamination. The estimation of the radiological hazards related to the ground water transport of radionuclides from tailings retention systems depends on reasonably accurate estimates of the movement of both water and solute. A two-dimensional mathematical model having predictive capability for ground water flow and solute transport has been developed. The flow equation has been solved under steady-state conditions and the mass transport equation under transient conditions. The simultaneous solution of both equations is achieved through the finite element technique using isoparametric elements, based on the Galerkin formulation. However, in contrast to the flow equation solution, the weighting functions used in the solution of the mass transport equation have a non-symmetric form. The predictive capability of the model is demonstrated using an idealized case based on analyses of field data obtained from the sites of operating uranium mills. The pH of the solution, which regulates the variation of the distribution coefficient (K_d) in a particular site, appears to be the most important factor in the assessment of the rate of migration of the elements considered herein.

POSSIBILITES DE PREVISION D'UN MODELE BIDIMENSIONNEL EU EGARD AU TRANSPORT DE RADIONUCLEIDES DANS LA NAPPE PHREATIQUE

RESUME

L'évacuation des déchets de faible activité dans des bassins prévus à cet effet est une source potentielle de contamination de la nappe phréatique. L'évaluation des risques radiologiques liés au transport de radionucléides dans la nappe phréatique à partir de systèmes destinés à retenir les résidus dépend d'une estimation raisonnablement exacte du mouvement de l'eau et du soluté. On a mis au point un modèle mathématique bidimensionnel permettant de prévoir le mouvement de l'eau et du soluté dans la nappe phréatique. L'équation d'écoulement des fluides a été résolue en régime stationnaire et l'équation de transport de masse, en régime transitoire. La solution simultanée de ces deux équations s'obtient par la méthode des éléments finis au moyen d'éléments isoparamétriques suivant la formule de Galerkin. Cependant, par opposition à la solution de l'équation d'écoulement des fluides, les fonctions de pondération utilisées pour la solution de l'équation de transport de masse ont une forme non symétrique. On démontre les possibilités de prévision du modèle en utilisant un cas optimisé fondé sur l'analyse des données recueillies sur le site d'installations de traitement de l'uranium en service. Le pH de la solution, qui détermine la variation du coefficient de distribution (K_d) sur un site particulier, semble être le principal facteur intervenant dans l'évaluation de la vitesse de migration des éléments considérés en l'occurrence.

Introduction

In recent years, radioactive waste generation from the development of
nuclear industry has posed a potential threat to the environment; in particular,
to the quality of ground water resources. Mathematical modeling, one of the few
alternatives for predicting the extent and impact of ground water contamination,
is a method for designing mitigative or preventative programs. The equations
that are used in these models are well established. They govern the movement of
the water and the reacting solutes, which may undergo decay and adsorption in
porous media.

In this paper, predictive limits of migration and relative concentrations
for Uranium (U), Radium (Ra-226), and Arsenic (As) from tailings pond seepage of
a hypothetical uranium processing mill are presented. In attempts to simulate
the movement of ground water, the following assumptions are made: the unsaturated
region of the aquifer is of relatively negligible thickness and, hence, ignored;
the equipotentials are normal to the aquifer bed, assumed to be impervious; the
flow is steady, and rainfall and evaporation are neglected. In attempts to simulate the solute movement, the pH of the soil water solution is not constant,
variation of the distribution coefficient K_d with concentration is ignored, the
density of solution is assumed to be constant and negligible background concentration for each of the various elements is allowed to prevail in the aquifer prior
to dumping operations.

The steady-state flow pattern was obtained after solving the Dupuit-
Forchheimer equation using the classical Galerkin-type finite element approach.
The mass transport equation was solved by the finite element approach using
asymmetric weighting functions, which take into account the numerical oscillations
frequently encountered in the approximate solutions of mass transport equations.

The value of the various soil and aquifer parameters utilized in the solution of the steady-state flow pattern and mass transport equations is representative of a typical uranium mill operating in the semi-arid western United States.

Solute Equation

The movement of a dissolved substance, i.e. metal, tracer or radioactive
material, which may undergo decay through a sorbing anisotropic unconfined
aquifer, is governed by the mass transport conservation requirement for the
solute and the equation of motion for the fluid. The conservation equation for
a solute in a three-dimensional saturated flow system [1] may be written as:

$$L(C) = \theta\frac{\partial C}{\partial t} + \rho\frac{\partial S}{\partial t} - \nabla \cdot (\theta D \cdot \nabla C - C\bar{v}) + \lambda\theta C + \lambda\rho S = 0 , \qquad (1)$$

where L is the defined differential operator
  C is concentration of the solute in solution (ML^{-3})
  S is concentration of the solute in the adsorbed phase $(M°)$
  ρ is soil bulk density (ML^{-3})
  θ is volumetric moisture content $(L°)$
  λ is first order rate constant for decay (T^{-1})
  t is time (T)
  $\bar{v}$ is specific discharge or Darcy velocity (LT^{-1})
  D is hydrodynamic dispersion tensor (L^2T^{-1}).

Equation (1) implies that the solution density is constant. It is assumed equal
to the one of the solvent (in our case water) such that the distribution of the
solute will not influence the flow pattern.

For an aerial two-dimensional situation such as the one considered, the
corresponding equation may be obtained by integrating Eq. (1) over a vertical line:

$$\int_{Z_1}^{Z_2} L(C) = 0 , \qquad (2)$$

where Z_1 is an elevation of the aquifer base measured from a chosen datum
Z_2 is free surface elevation as measured from a chosen datum.

Applying Leibnitz's rule* for differentiation of an integral to each term of Eq. (1) and assuming that:

$$\tilde{C} = C(x,y,Z_1) = C(x,y,Z_2)$$

where C is the solution concentration averaged over the aquifer thickness and corresponds to $\tilde{C}$, one obtains:

$$L(C) = h\theta\frac{\partial C}{\partial t} + h\rho\frac{\partial S}{\partial t} - \nabla \cdot (\theta D^* \cdot \nabla C - C\bar{q}) + \lambda h\theta C + \lambda h\rho S + F_z\Big|_{Z_2} - F_z\Big|_{Z_1} = 0 \quad (3)$$

where ∇ is gradient operator $\frac{\partial}{\partial x}, \frac{\partial}{\partial y}$

h is saturated thickness of the aquifer (L)
D^* is effective hydrodynamic dispersion tensor over the aquifer thickness (L^3T^{-1})
$\bar{q}$ is discharge per unit width of the aquifer, normal to the direction of the flow (L^2T^{-1})
F_z is mass flux across the upper or lower boundaries of the aquifer defined as:

$$F_z = -\theta D\frac{\partial C}{\partial z} + v_z C \ . \quad (4)$$

If the aquifer at base Z_1 is assumed impermeable, implying no mass transfer through that particular boundary, the last term of Eq. (3) becomes zero.

Only aquifer recharge resulting from seepage of radioactive waste from a disposal pond is considered in this work. Since the present mathematical approach is not intended to measure the aquifer's unsaturated region, which could retard the time of arrival of the solute front at the location of the free surface, it is assumed that the base of the pond is in immediate contact with the free surface. At a localized point (x_i, y_i, Z_2, t) on the interface of the source and the free surface, and by imposing continuity requirements on the fluxes, F_z may be expressed by:

$$F_z\Big|_{Z_2} = QC^*\Big|_{Z_2+\epsilon} = -\theta D\frac{\partial C}{\partial z} + v_z C\Big|_{Z_2-\epsilon} \ , \quad (5)$$

where ϵ is the interface thickness (L), assumed to be an infinitesimal positive quantity,
C^* is the concentration of the fluid source (ML^{-3}), and
Q corresponds to a virtual seepage rate (LT^{-1}) from the pond generally expressed by:

$$Q = \sum_{i=1}^{N} Q_w(x_i,y_i,t) \ \delta(x - x_i)(y - y_i) \ ,$$

where Q_w is the volumetric discharge (L^3T^{-1}),
N is the number of source points, and
$\delta(x - x_i)(y - y_i)$ is the Dirac delta function.

---

*Leibnitz's rule for differentiation of an integral I with respect to a parameter α

$$I = \int_{Z_1(\alpha)}^{Z_2(\alpha)} f(z, \alpha)dz, \text{ namely, } \frac{dI}{d\alpha} = f(z, \alpha)\frac{dz}{d\alpha}\Big|_{Z_2} - f(z, \alpha)\frac{dz}{d\alpha}\Big|_{Z_1} + \int_{Z_1}^{Z_2} f_\alpha'(z, \alpha)dZ$$

Flow Equation

Based on Dupuit's assumption [2] that in most ground water flow the slope of the phreatic surface is small, the flow across a unit width of an aquifer of thickness (h), according to Darcy's law, may be expressed by:

$$\bar{q} = -T \cdot \nabla H \qquad (6a)$$

with

$$T = Kh \qquad (6b)$$

and

$$h = H - Z \, , \qquad (6c)$$

where K_{ij} is hydraulic conductivity tensor (LT^{-1})
H is height of the free surface from a chosen datum (L)
Z is height of the aquifer base from a chosen datum (L)
T_{ij} is aquifer transmissivity tensor $(L^2 T^{-1})$.

With the fluid in the aquifer assumed incompressible, the shape of the free surface under steady-state conditions may be obtained from the solution of the Dupuit-Forchheimer equation written as:

$$L(H) = \nabla \cdot T \cdot \nabla H + Q = 0 \, . \qquad (7)$$

The solution of Eq. (3) requires the knowledge of two additional parameters; the effective hydrodynamic dispersion D^* and the adsorption rate $\partial S/\partial t$.

A survey on the subject of dispersion in saturated porous media is presented by Shamir and Harleman [3]. Experimental and theoretical studies by Saffman [4], Scheidegger [5], Bear and Bachmat [6], suggest that the coefficient of hydrodynamic dispersion D may be defined as the sum of the coefficient of mechanical dispersion $\bar{D}$, and of molecular diffusion $\bar{D}_d$.

$$D_{ij} = \bar{D}_{ij} + \bar{D}_d \tau_{ij} \, . \qquad (8)$$

Following Bear [1], the general form for D^* may be written as:

$$D^*_{ij} = \lambda_t |V| \delta_{ij} + (\lambda_\ell - \lambda_t) V_i V_j |V| + \bar{D}_d \tau \delta_{ij} \, . \qquad (9)$$

where V_i is the ith component of the average apparent flow per unit width of the aquifer written as:

$$V_i = q_i / \theta \, , \qquad (10)$$

and

$$|\bar{V}| = \left(V_i^2 + V_j^2 \right)^{1/2} \, , \qquad (11)$$

where λ_ℓ is longitudinal dispersivity of the medium (L)
λ_τ is lateral dispersivity of the medium (L)
δ_{ij} is Kronecker delta, i.e., $\{(\delta_{ij} = 0, \ (i \neq j), \text{ and } \delta_{ij} = 1, \ (i = j)\}$
τ is tortuosity (dimensionless)
$\bar{D}_d$ is an effective molecular diffusion coefficient $(L^3 T^{-1})$.

Although this mathematical model is capable of handling two types of adsorption isotherms under equilibrium condition; linear and nonlinear as described by Freundlich or Langmuir relations*; in the section dealing with the predictive capabilities of this model, only the linear form written as:

$$S = K_d C \qquad (12)$$

will be considered. K_d in Eq. (12) is the distribution coefficient $(L^3 M^{-1})$.

---

*Freundlich: $S = kC^n$, k and n are empirical constants.

 Langmuir: $S = a \left\{ C / \left(\frac{1}{b} + C \right) \right\}$, a and b are empirical constants.

Substituting Eqs. (5), (7), and (12) in Eq. (3), one may obtain:

$$L(C) = h\theta R_f \frac{\partial C}{\partial t} - \nabla \cdot (\theta D^* \cdot \nabla C) + \bar{q} \cdot \nabla C + \lambda h \theta R_f C - Q(C^* - C) = 0 , \tag{13}$$

where R_f is the retardation factor defined as:

$$R_f = 1 + K_d \rho/\theta . \tag{14}$$

Boundary Conditions

The flow domain R in Figure 1 is bounded by the curve:

$$\xi = \xi_1 + \xi_2 .$$

where ξ_1 is AC and BD, an impermeable boundary
ξ_2 is AB and CD, respectively, referring to the upstream and downstream
boundaries of the aquifer; the latter assumed to be in contact with a
river
R_p is the area of a disposal pond.

a) Flow Equation:

$$H = H_i \qquad\qquad \text{on AB} \tag{15a}$$

$$H = H_o \qquad\qquad \text{on CD} \tag{15b}$$

$$\nabla H \cdot \bar{n} = 0 \qquad\qquad \text{on AC and BD} \tag{15c}$$

where $\bar{n}$ is the direction of the outward flow normal to the boundary at a given
point. Eq. (15a), Eq. (15b), and Eq. (15c), respectively, refer to Dirichlet and
Neuman conditions [7].

b) Solute Equation:

In the simulation of radionuclide migration from the site of a disposal
pond, the upstream boundary is considered impermeable since chances of solute
migration upstream of the pond location from the ground water mound tends to
diminish with increasing distance. Thus:

$$\nabla C \cdot \bar{n} = 0 \qquad\qquad \text{on AB, AC and BD, } t > 0 . \tag{16a}$$

A mixed-type boundary condition [7] had to be considered for the boundary DC
in contact with the river:

$$(-\theta D^* \cdot \nabla C + qC) \cdot \bar{n} = qC(t') \cdot \bar{n} \quad \text{on CD, } t > 0 , \tag{16b}$$

where $C(t')$ is the concentration of solute in water discharging into the river at
time $t' = t - \Delta t$.

At the nodes located at the site of the pond Dirichlet-type boundary conditions are written as:

$$C(x,y,t) = C^*(t) \qquad\qquad \text{on } R_p, \ t > 0 , \tag{16c}$$

were imposed.

The decision to use Eq. (16c) was based solely on the numerical consideration
that stable solutions often are desired. This type of boundary condition initially
violates the continuity of flux [Eq. (5)] across the interface between the pond
and the free surface, up to time t_1 when the concentration at the pond does come
close enough to C^*.

Experience has shown that t_1 is not significant when simulation time is
generally expressed in terms of decades. Considering that conservative predictions are always of great concern, this approach seems to be viable.

The initial conditions are:

$$C(x,y,0) = C_b \qquad \text{on R} \tag{17}$$

where C_b is a background concentration prevailing in the flow region R prior to onset of discharge from the pond.

Finite Element Formulation

The spatial discretization of Eq. (7) and Eq. (13) in the solution domain R is obtained through a Galerkin type finite element approach [8]. The unknown functions of H and C throughout the solution domain R are approximated, respectively, by:

$$H(x,y) = \sum_{i=1}^{\Omega} N_i(x,y)H_i \tag{18}$$

$$C(x,y,t) = \sum_{i=1}^{\Omega} N_i(x,y)C_i(t) \tag{19}$$

where N_i are the basis functions, defined piecewise (element by element) satisfying the boundary conditions
H_i and C_i are the unknown parameters
Ω is the total number of nodes.

In the summation process, the appropriate function for each particular point in space must be used. An additional approximation for the components of T is considered [9]. Thus:

$$T_{xx} = \sum_{I=1}^{\Omega_K} N_I(x,y)T_{xxI} \tag{20}$$

A similar expression may be obtained for T_{yy}; the subscript k refers to the number of relevant nodes in a particular element e. Using the above, an approximation of q_x, is obtained from Eq. (6a) written as:

$$\hat{q}_x = -\sum_{I=1}^{\Omega_K} N_I(x,y)T_{xxI}\frac{\partial H}{\partial x} . \tag{21}$$

In a manner similar to Oden's conjugate function theory [10], q_x at each node of the flow region is computed through an orthogonalization applied to Eq. (21). This is written as:

$$\int_R N_i L(\hat{q}_x) \, dR = 0 . \tag{22}$$

where $L(\hat{q}_x)$ is:

$$L(\hat{q}_x) = \hat{q}_x + \sum_{I=1}^{\Omega_K} N_I(x,y)T_{xxI}\frac{\partial H}{\partial x} \tag{23}$$

and using a trial solution for q_x expressed in series form as:

$$\hat{q}_x = \sum_{i=1}^{\Omega} N_i(x,y)q_{xi} \tag{24}$$

yields a relation which matrix form may be written as:

$$[D]\{q_x\} + \{P\} = 0 \tag{25}$$

where:

$$D_{ij} = \int_{R_e} N_i \sum_j N_j q_{xj} \, dR \qquad (26a)$$

and:

$$P_i = \int_{R_e} N_i \sum_{I=1}^{\Omega_\kappa} N_I(x,y) T_{xxI} \sum \frac{\partial N_j}{\partial x} H_j \, dR \qquad (26b)$$

A similar expression may be obtained for q_y. The computation of q_x and q_y requires prior knowledge of H at the nodes. Having computed the component of $\vec{q}$ at each node in the system, the corresponding values of D* are easily obtained from Eq. (9). An approximate solution of Eq. (7) and Eq. (13) is sought through an orthogonalization process [11] written as:

$$\int_R W_i L(\phi) \, dR \qquad i = 1, 2, \ldots, \Omega \ , \qquad (27)$$

where W_i is a linearly dependent weighting function
ϕ is the unknown function.

When solving the flow equation, Eq. (7), the weighting functions are chosen to be the same as the basis functions, which is the conventional Galerkin finite element method. However, when solving the solute transport equation, Eq. (13), at the exception of the last term, which is dealt in the same way as the flow equation; the upstream weighting finite element was applied [12], in which the weighting functions would correspond to asymmetric basis functions.

The weighted and integrated residuals resulting from substituting Eq. (7) and Eq. (13), respectively, with the approximations for the variable and their coefficients; Eq. (18), Eq. (20), Eq. (19), Eq. (24), in Eq. (27); and applying Green's theorem [8] yields two sets of algebraic equations, which in matrix form are written as:

$$[M]\{H\} + \{G\} = 0 \qquad (28)$$

$$[A]\{C\} + [B]\frac{dC}{dt} + \{F\} = 0 \ , \qquad (29)$$

where

$$M_{\kappa\ell}^e = \int_{R_e} T_{ij} \frac{\partial N_\kappa}{\partial x_i} \frac{\partial N_\ell}{\partial x_j} \, dR \qquad (30a)$$

$$G_\kappa^e = -\int_{R_e} N_\kappa Q \, dR \qquad (30b)$$

Vector $\{G\}$ will be dependent on the boundary conditions as a result of the Dirichlet nodes given by Eq. (15a) and Eq. (15b).

$$A_{\kappa\ell}^e = \int_{R_e} \left[\theta D_{ij}^* \frac{\partial W_\kappa}{\partial x_i} \frac{\partial N_\ell}{\partial x_j} + q_i W_\kappa \frac{\partial N_\ell}{\partial x_i} + \lambda h \theta R_f N_\kappa N_\ell + N_\kappa Q N_\ell \right] dR \qquad (31a)$$

$$B_{\kappa\ell}^e = \int_{R_e} h \theta R_f W_\kappa N_\ell \, dR \qquad (31b)$$

$$F_\kappa^e = -\int_\zeta W_\kappa \theta D_{ij}^* \frac{\partial C}{\partial x_j} 1_i \, d\zeta - \int_{R_e} N_\kappa Q C^* \, dR \qquad (31c)$$

where l_i is the direction cosine of the normal to the boundary at a particular model.

To take care of a mixed-type boundary condition as given in Eq. (16b), the second term of Eq. (31a) must be modified to:

$$-\int_{R_e} q_i \frac{\partial W_\kappa}{\partial x_i} N_\ell \, dR$$

whereby the first term of Eq. (31c) becomes:

$$\int_\zeta W_\kappa qC(t) \cdot \bar{n} \, d\varsigma$$

Note that Eq. (16c) will adequately substitute for the last terms in Eqs. (31a) and (31c), which will be ignored in the computation of Eq. (29). Approximations for the components of D^* and q similar to Eq. (20) are also considered in Eq. (31a).

Element Formulation

The isoparametric quadrilateral element was used to discretize the flow region R [7]. The bilinear polynomial basis function for the ith node, as given by Ergatoudis et al. [13], may be written in terms of the local normalized coordinates ε and η as:

$$N_i = \frac{1}{4} (1 + \varepsilon_0)(1 + \eta_0) \tag{32}$$

with

$$\varepsilon_0 = \varepsilon\varepsilon_i \qquad \eta_0 = \eta\eta_i \tag{33}$$

and

$$\varepsilon_i = \pm 1 \qquad \eta_i = \pm 1$$

Details regarding the relationship between cartesian coordinates and the new coordinates, and the numerical mapping may be found in Zienkiewicz [7].

The asymmetric weighting functions proposed first by Christie et al. [14] and applied subsequently to field problems by Heinrich et al. [12] and Huyakorn and Pinder [15] may be written as:

$$W_i = W_i(\varepsilon) \cdot W_i(\eta) \tag{34}$$

with

$$W_i(\varepsilon) = N_i(\varepsilon) + \alpha_{ij} F_{ij}(\varepsilon) \tag{35a}$$

$$W_i(\eta) = N_i(\eta) + \beta_{ik} F_{ij}(\eta) \tag{35b}$$

where $F_{ij}(\varepsilon)$ and $F_{ij}(\eta)$ are quadratic functions of coordinates x and y
ij and ik are the sides of the element common to node i
α and β are scalar quantities and are function of the average velocity V and dispersion coefficient D' along the side ij and ik.

The damping factors α_1, β_1, α_2, β_2 associated with sides 1-2, 2-3, 4-3, 1-4 of an element (Fig. 2) may be calculated from a relation given by Christie et al. where the optimal value of say α_{ij} is written as:

$$\alpha_{ij} = \coth \gamma/2 - 2/\gamma \tag{36}$$

with

$$\gamma = VL'/D'$$

where L' is the length of side ij.

Whereas V is easily computed, the computation of D' would generally involve a transformation of axes [16], when the direction of the sides of the elements in the system is not parallel to the principal axes. After some mathematical manipulation one may obtain:

$$D'_{ij} = \frac{1}{2} \left\{ \cos^2 \theta^* \left(D_{xx_i} + D_{xx_j} \right) + 2 \cos \theta^* \sin \theta^* \left(D_{xy_i} + D_{xy_j} \right) + \right.$$

$$\left. \sin^2 \theta^* \left(D_{yy_i} + D_{yy_j} \right) \right\} \tag{37}$$

where $\bar{D}_{ij}$ is the average value of D' along a side ij, and
θ^* is the angle of the direction of element side with the x axis.

Solution Scheme

a) Flow Equation

The matrix Eq. (28) with its boundary conditions Eq. (15) generates a system of nonlinear first order differential equations with the nonlinear coefficient T. An initial guess of H is required to evaluate [M], the natural elevation of the free surface in the absence of vertical recharge, and a first approximation is obtained after linearizing the resulting set of equations. This process is repeated until the values of H in two successive iterations are found not to alter appreciably. (2% variation was chosen as the convergence criteria.)

b) Solute Equation

The matrix Eq. (29) with its boundary conditions Eq. (16) generates a system of linear differential equations of first order, unless the adsorption is one of a nonlinear type. When the type derivative is replaced by a weighted finite difference approximation [17], the general matrix equation Eq. (29) becomes:

$$\theta[A]\{C\}_{t+\Delta t} + (1 - \theta)[A]\{C\}_t + [B] \left(\{C\}_{t+\Delta t} - \{C\}_t \right) / \Delta t + \{F\} = 0 \tag{38}$$

where θ is a scalar quantity.

The Crank-Nicolson method of solution is adopted and this formulation is obtained after setting $\theta = 0.5$ in Eq. (38).

The flow in the aquifer, assumed under steady-state conditions, follows that matrices [A] and [B] are time independent; hence, these need to be evaluated only once, unless adsorption becomes nonlinear.

Referring to matrix [B], this does not generally appear in spatial discretization when finite differences are used and when it corresponds to a unit matrix. Therefore, by suitable scaling, [B] may be reduced to a finite difference equivalent if this is assumed to be lumped [18]. More explicitly, [B] will be lumped once all of its rows are summed up and a diagonal matrix is obtained as written:

$$B^e_{ii} = \Gamma \int_{\Gamma^e} h\theta R_f \, W_i N_i \, d\Gamma \Big/ \sum_{i=1}^{n^e} \int_{\Gamma^e} W_i N_i \, d\Gamma \tag{39}$$

where n^e is the number of nodes within a typical element e
Γ is the area of the element. Note that all other nondiagonal elements of B are zero.

The set of linear equations is solved in-core or out-of-core with an algorithm utilizing equilibration and partial pivoting [19]. The time step initially is small and allowed to increase by a factor of 1.2 up to the limit set for the particular problem (in our case 2.5 months).

The accuracy of our numerical scheme was satisfactorily tested against an analytic solution given by Cleary [20] and applied to a field problem [20,21].

Table I. System Parameters

Symbol	Values	Units
ρ	1.9	g/cm^3
θ	0.4	
λ_l	5.0	m
λ_t	1.0	m
T	0.6	
D_d	0.03	m^2/day
K_{xx} K_{yy}	4.0	m/day
Q_w	3780	m^3/day
H_i	200	m
H_o	188	m
AB,CD	1000	m
AC,BD	2500	m
Rp	2500	m^2
x coordinates of Rp	475	
	525	m
y coordinates of Rp	475	
	525	m

Table II. Decay Constants
of Various Elements

Element	λ(month-1)
U	1.3×10^{-11}
Ra-226	3.6×10^{-5}
As	0

Table III.   Distribution Coefficients K_d (cm^3/g) of
Various Elements as a Function of pH

Element	pH			
	2.0	3.5	5.0	7.0
U	50.0	75.0	150.0	450.0
Ra-226	0.0	7.0	35.0	100.0
As	0.0	10.0	60.0	300.0

Table IV. Concentrations of Various Elements

Element	Source	Background	Units
U	16.9×10^{-3}	6.8×10^{-6}	$\mu Ci/ml$
Ra-226	4.8×10^{-6}	2.0×10^{-9}	$\mu Ci/ml$
As	1.44	1.0×10^{-5}	mg/L

Results and Discussion

For the obtention of the finite element solution of Eqs. (7) and (13), the flow region R, as shown in Figure 1, was divided into isoparametric elements of rectangular shape (190 elements and 220 nodes). As mentioned earlier, the flow equation was solved independently of the mass transport equation. Values of the parameters entering Eqs. (7) and (13) are given in Tables I through IV.

Flow Pattern

When solving the flow equation, the free surface initially was assumed to have a slope of 4.8/1000. With a daily seepage rate from the pond to correspond to 3780 m^3, which is fairly representative of the average daily output of a uranium mill, the resulting flow pattern computed after Eq. (25) is shown at the top of the second page of figures. A distortion of the flow pattern may be observed in the region upstream of the pond and its vicinity resulting from the ground water mound generated by pond seepage. Streamlines seem to be heading towards the upstream boundary of the aquifer for some distance before altering their course in the direction of the river and the regional ground water gradient.

Ground Water pH

A spatial variation of the soil water pH is assumed to prevail in the aquifer at the onset of the mill operations and maintained as such during the whole simulation period. The soil water at the pond location is assumed to be highly acidic with pH = 2 gradually becoming neutral to pH = 7 beyond a radius of 250 m from the center of the pond.

During operations, the rise in the pH value of the solution seeping out from the pond may be attributed to the large inflow of fresh water at pH = 7 entering the aquifer at 46,150 m^3/day, which corresponds to about 12.2 part of ground water to one part tailings water, in addition to other geochemical processes such as precipitation and adsorption. The distribution coefficient K_d, bearing a strong dependence to the solution pH [22,23] for a given concentration, can no longer be considered constant. Values for K_d for U, Ra-226, and As at different pH's are shown in Table II. With the values for the decay constants and selected concentrations of the various elements considered (Tables III and IV), the simulation of the movement of these species for a period of 50 years was then carried out.

U

Isopleths for the relative concentration of 0.5×10^{-3} are shown in the second set of figures. For this case, the rate of migration of the radionuclide seems to be quite appreciable for the initial period of 10 years after which it decreases; this is a reflection of the increasing pH in the aquifer. The relative concentrations of 0.9, 0.2, 0.01, and 0.0005 are shown. This summarizes some of the results and shows, as expected, the rate of migration for the high value of relative concentration (0.9) is negligible.

Ra-226

Isopleths for the relative concentration of 0.1 and 0.005 are shown in the third set of figures. The same trend in the rate of migration as for U is observed. The isopleths for the relative concentration of 0.8, 0.3, 0.1, and 0.005 after a period of 50 years are shown.

For a 10 year period, after the beginning of operations, the isopleth for a relative concentration of 0.1 has the same trend as in the previous case (U). However, following that period, the rate of advance of the solute front is more appreciable than in the case of U; this may be attributed to a lower value of K_d. The relative concentration of 0.005 for isopleths demonstrates the rate of migration decreases with time. Note that when the isopleth corresponding to a period of 10 years lies in the same zone of pH as the one at 30 years, the rate of migration tends to decrease, becoming almost negligible beyond that mark.

As

Isopleths for the relative concentrations of 0.1 and 0.005 are shown in the fourth set of figures. The trend in migration rate observed is similar to those of U and Ra-226. The isopleths for the relative concentrations of 0.8, 0.3, 0.1, and 0.005 after a migration period of 50 years are also shown.

For the first 10 year period after initiation of pond operations, the isopleth for a relative concentration of 0.1 has the same trends as U. The 30 year isopleth also has a similar trend. But the 50 year isopleth follows the uranium trend rather than the Ra-226 trend. Apparently, arsenic, like uranium migration, is retarded relative to radium migration in the later stages. However, all migration rates as derived from the model have the strong dependence on pH; this is primary to K_d determinations used in the model.

Conclusions

In evaluating the results, the most important finding was that the pH of the solution, which regulates the variation of distribution coefficient (K_d) in a particular site, appears to be the most important factor in the assessment of the rate of migration of the elements considered. There are two other important factors that should be remembered. First, the examples given were based on hypothetical situation. As a result, these results should not be generalized to represent all situations. Rather, each case must be developed based on information for the system of interest. Second, it should be stressed that this program does not currently include chemical reactions among solutes and desorption; a phenomenon that has received little attention by investigators. It is anticipated, however, that once experimental data is available on these reactions, it could be incorporated into the model.

In general, the program gives solutions to the flow and solute equations, which predict the spread of ground water contaminants. These predictions, given the current experimental limitations, seem very reasonable.

The program was run on an IBM 370/175 computer for a simulation covering 50 years with one constituent. The program ran for about three minutes and required 275k of storage.

References Cited

1. Bear, J.: "Dynamics of Fluids in Porous Media," American Elesevier Publishing Co., New York, N.Y. (1972).

2. Dupuit, J.: "Etudes Theoriques et Pratiques sur le Mouvement des Eaux," Ed. No. 2, Dunod, Paris (1863).

3. Shamir, U.Y., and Harleman, D.R.F.: "Numerical and Analytic Solutions of Dispersion Problems in Homogeneous and Layered Aquifers," M.I.T., Dept. Civ. Eng., Hydrogen Lab., Cambridge, Mass., Rep. No. 89, pp. 206 (1966).

4. Saffman, P.G.: "Dispersion Due To Molecular Diffusion and Macroscopic Mixing Inflow Through a Network of Capillaries," J. Fluid Mech., Vol. 7, pp. 194-208 (1960).

5. Scheidegger, A.E.: "General Theory of Dispersion in Porous Media," J. of Geophys. Res., Vol. 66(10), pp. 3273-3284 (1961).

6. Bear, J., and Bachmat, Y.: "A Generalized Theory on Hydrodynamic Dispersion in Porous Medium," Int. Assoc. of Sci. Hydro., Pub. No. 72, Symposium of Haifa, pp. 7-16 (1967).

7. Zienkiewicz, O.C.: "The Finite Element Method in Engineering Science," McGraw-Hill, New York, N.Y., pp. 521 (1971).

8. Kantorovich, L.V. and Krylov, V.I.: "Approximate Method of Higher Analysis," Noordhoff, Netherlands, pp. 681 (1958).

9. Pinder, G.F., Frind, E.O., and Papadopoulos, S.S.: "Functional Coefficients in the Analysis of Ground Water Flow," Water Resources Res., Vol. 9(1), pp. 222-226 (1973).

10. Gallagher, R.H.: "Finite Element Analysis Fundamentals," Prentice-Hall, Inc., Englewood Cliffs, New Jersey (1975).

11. Finlayson, B.A.: "Method of Weighted Residuals and Variational Principles with Application in Fluid Mechanics, Heat, and Mass Transfer," Academic Press, New York, N.Y., pp. 412 (1972).

12. Heinrich, J.C., Huyakorn, P.S., Zienkiewicz, O.C., and Mitchell, A.R.: "An Upwind Finite Element Scheme for Two-Dimensional Convective Transport Equations," Int. J. Num. Meth. Eng._, Vol. 10, pp. 1389-1396 (1976).

13. Ergatoudis, J., Irons, B.M., and Zienkiewicz, O.C.: "Curved Isoparametric, Quadrilateral Elements for Finite Element Analysis," Int. J. Solids Structures, Vol. 4, pp. 31-42 (1968).

14. Christie, I., Griffiths, D.F., Mitchell, A.R., and Zienkiewicz, O.C.: "Finite Element Scheme for Two-Dimensional Convective Transport Equations," Int. J. Num. Meth. Eng._, Vol. 11, pp. 131-143 (1977).

15. Huyakorn, P.S., and Pinder, G.F.: "A Pressure Enthalpy Finite Element Model for Simulating Hydrothermal Reservoirs," Paper presented at the 2nd Int. Symp. on Computer Methods for Partial Differential Equations, June 22-24, Lehigh University, Bethlehem, Penn. (1977).

16. Nye, J.F.: "Physical Properties of Crystals," Oxford University Press, pp. 322 (1957).

17. Mitchell, A.R.: "Computational Methods in Partial Differential Equations," John Wiley, New York, N.Y., pp. 17-99 (1971).

18. Clough, R.W.: "Analysis of Structural Vibrations and Dynamic Response," Recent Advances in Matrix Methods of Structural Analysis and Design, R.H. Gallagher, Y. Yamada, and J.T. Oden (eds.), University of Alabama Press, Huntsville, Ala. (1971).

19. Martin, R.S., and Wilkinson, J.H.: "Solution of Symmetric and Unsymmetric Band Equations and the Calculations of Eigenvectors of Band Matrices," Numerische Mathematik, Vol. 9(4), pp. 279-301 (1967).

20. Cleary, R.W.: Final Report 208 Project, Nassau Suffolk Regional Planning Board, Hauppauge, New York, N.Y. (1978).

21. Gureghian, A.B., and Cleary, R.W.: "Three-Dimensional Ground Water Pollution Modeling," Paper presented at the American Geophysical Union Conference in San Francisco, Calif. (1976).

22. Rancon, D.: "The Underground Behavior of Uranium and Thorium Rejected by the Nuclear Industry," IAES-SM, pp. 333-346, Vienna (1978).

23. Rojkova, E.V. et al.: "The Role of Absorption of Uranium Concentration in Sedimentary Rock," IIe Conf. Pacific Utilization of Atomic Energy, Actes Conf., Geneva (1958), ONU, New York, N.Y. (1958), 160 Version Anglaise, Vol. 2, p. 420.

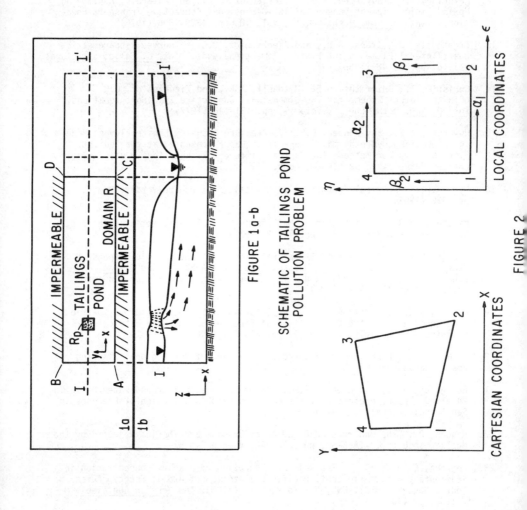

FIGURE 1a-b

SCHEMATIC OF TAILINGS POND
POLLUTION PROBLEM

LOCAL COORDINATES

CARTESIAN COORDINATES

FIGURE 2

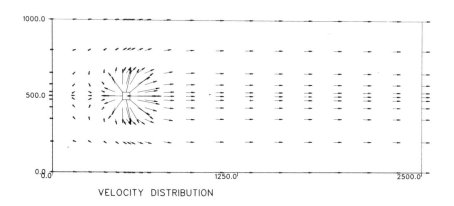

VELOCITY DISTRIBUTION

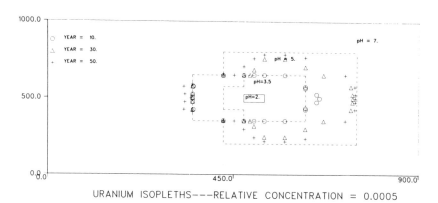

URANIUM ISOPLETHS---RELATIVE CONCENTRATION = 0.0005

URANIUM ISOPLETHS---50 YEAR CONTOURS

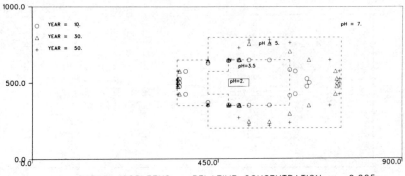

RADIUM ISOPLETHS---RELATIVE CONCENTRATION = 0.005

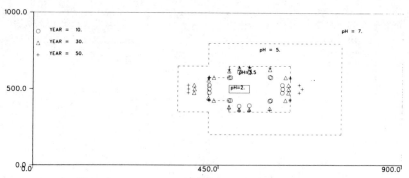

RADIUM ISOPLETHS---RELATIVE CONCENTRATION = 0.1

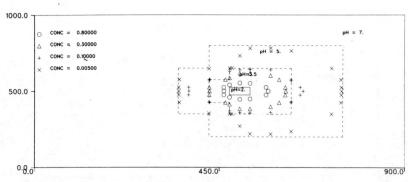

RADIUM ISOPLETHS---50 YEAR CONTOURS

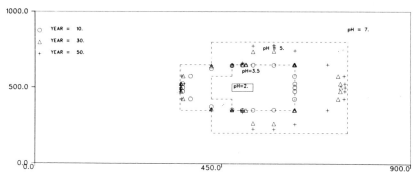

ARSENIC ISOPLETHS---RELATIVE CONCENTRATION = 0.005

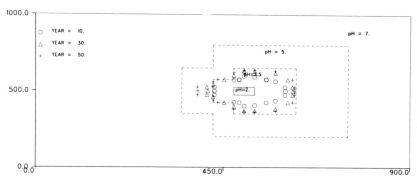

ARSENIC ISOPLETHS---RELATIVE CONCENTRATION = 0.1

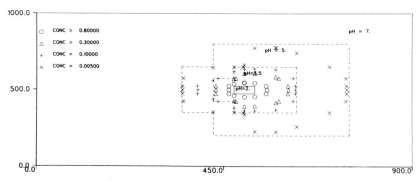

ARSENIC ISOPLETHS---50 YEAR CONTOURS

DISCUSSION

D.M. LEVINS, Australia

I think one of the real problems in doing this even though the analysis is quite sophisticated, is the actual input data and particularly the distribution coefficients. Now you quote some figures here, and I suspect that some of the distribution coefficients are also a function of other factors. For example, in the case of radium, it may well be a function of the sulphate concentration. I wonder if you could comment on how accurate you regard this sort of data that you are using, and do you think there is a need for better data ?

A.B. GUREGHIAN, United States

Well, I fully agree with you that the input parameters will have a strong bearing on the final results. Although, some of these parameters such as the bulk values of dispersivities or hydraulic conductivities may be determined through some well known experimental techniques, however, the distribution coefficient which is function of the soil structure (composition) the pH of the soil matrix and solution as well as other chemical reactions is generally difficult to estimate in-situ therefore, one is left with no other choice but to rely on laboratory experiments performed on soil cores extracted from the site. I must admit that the evaluation of the distribution coefficient determined by this method is not always representative of the field situation. Thus, until we shall be able to describe the complete adsorption process in a more elaborate manner through a mathematical expression embodying all the parameters liable to affect this, one would have to rely on the value of this parameter generated by the method known to us and eventually improve its accuracy after a model calibration through data obtained after a monitoring period.

M.H. MOMENI, United States

In spite of the lack of the precise knowledge on the coefficients, one cannot do without a model, as such. In a field expedition, without a model to dictate the choice of monitoring locations, it could be a complete failure. Example of it is, if it takes 50 years for a front to move to 50 ft. or 100 ft., digging a well for monitoring 200 ft. away will not provide any answer. So choice of parameters, either on the conservative side or very liberal side, could provide the first data base for going to the field for a successful condition. Would you care to comment on it ?

A.B. GUREGHIAN, United States

Here again, I can no more agree with you. One has certainly to implement initially the available mathematical model with a first approximation of the parameters involved and generally the choice of selecting these will be on the conservative side. The movement of the radionuclides or other elements in groundwater being generally slow an elaborate monitoring programme of the site will adequately enable one to reassess the value of these parameters.

L.A.I. BAECKSTRÖM, Sweden

I may be indulging myself in a question which follows what Dr. Levins posed a little earlier. The distribution coefficients you have quoted for uranium or arsenic, do you measure them separately or correlate the aspect of pH. Arsenic, for example, in the presence of iron or calcium or phosphate of arsenic is going to complicate your numbers. So what sort of reliance do we have on the calculated numbers ?

A.B. GUREGHIAN, United States

There again, I did briefly mention in my talk that the data that I used were just collected from the literature and from data submitted by the mill operators as I have not done any private investigations, on site or experimentally, to generate those numbers. However, I deem that the values we have used are very conservative in a way. Therefore, to answer your question, the actual values of distribution coefficients could be higher by perhaps an order of magnitude.

D.B. CURTIS, United States

One thing that I would try to emphasize in my talk, which was brought out by Dr. Levins, is that the importance in understanding the fundamentals of what is going on, or trying to understand the fundamentals of the physics is that the chemistry cannot be ignored in solving these problems and the mathematical models are only as good as our understanding of the physics and the chemistry, and yet there seems to be a reluctance to fund this kind of work, but I think that the importance of it is so obvious.

J. HOWIESON, Canada

One of the big advantages as I see it, of having a mathematical model, is that you can do sensitivity analysis and determine that way what the important characteristics of the problem are. Have you undertaken any such sensitivity analysis, and can you give us any information on it ?

A.B. GUREGHIAN, United States

We normally perform sensitivity analysis but an analysis of this kind did not seem appropriate in this case, since the mathematical model could not be tested with any actual parameters obtained from field data.

M.C. O'RIORDAN, United Kingdom

Do you think that you have an unfair advantage over meteriologists because of your time scales ? It is struck me that when you were talking about the time scales involved in your models, you had a tremendous advantage over the poor meteriologist.

A.B. GUREGHIAN, United States

As a matter of fact, I am at a distinct disadvantage compared to the meteorologist, since an inaccurate prediction by him will be probably be forgotten the next day, whereas, I will have to live with mine for centuries.

MILL TAILINGS DISPOSAL AND ENVIRONMENTAL MONITORING
AT
THE NINGYO-TOGE URANIUM PROCESSING PILOT PLANT

I. Iwata, Y. Kitahara, S. Takenaka, Y. Kurokawa
Power Reactor and Nuclear Fuel Development Corporation
Minato-ku, Tokyo (Japan)

ABSTRACT

The tailings from the uranium processing pilot plant with a maximum ore processing capacity of 50 t/d are transferred to a tailings dam. The overflow from the dam is chemically treated and through settling ponds, sand filters to be discharged into a river. The concentrations of U, ^{226}Ra, pH, S.S., COD, Fe, Mn, Cl and F were monitored periodically and they were all below the control values. The results of monitoring on the river bed and rice paddy soil showed no signs of accumulation of U and ^{226}Ra in it.

EVACUATION DES RESIDUS DE TRAITEMENT DU MINERAI ET
SURVEILLANCE DE L'ENVIRONNEMENT A L'USINE PILOTE
DE TRAITEMENT DE L'URANIUM DE NINGYO-TOGE

RESUME

Les résidus de l'usine pilote de traitement du minerai d'uranium, ayant une capacité maximale de 50 tonnes par jour, sont transférés dans un bassin de boues résiduaires. Le trop plein de ce bassin est soumis à un traitement chimique, puis passe par des bassins de décantation, des filtres à sable pour être rejeté dans un cours d'eau. Les concentrations d'uranium, de Ra-226, le pH, la DCO, les matières solides en suspension, la teneur en Fe, Mn, Cl et F font l'objet d'une surveillance périodique et sont tous inférieurs aux valeurs témoins. Les résultats de la surveillance au niveau du lit du cours d'eau et du sol des rizières n'ont pas révélé de traces d'accumulation d'uranium et de Ra-226 dans ces milieux.

1. Introduction

The Ningyo-toge Mine is located on the watershed of the Chugoku mountain range in the southwest part of the mainland of Japan. The Yoshii River originating from the Mine has a total length of about 138 km and flows into the Seto Inland Sea. (Fig. 1)

Since the discovery of uranium ore deposits in the Ningyo-toge Mine in 1955, the uranium mining and milling experiments have been continued.

The uranium deposit is of the Tertiary sedimentary type and consists of several ore bodies. Total ore reserves of the uranium deposits are estimated at approximately 2,500 tons of U_3O_8 if the cut-off grade is assumed to be 0.1% U. The pilot plant (with a maximum ore processing capacity of 50 t/d) was built in 1964 and since then experiments at the plant have been made on the PNC process. The PNC process is a wet process by which UF_4 is made directly from the uranium ore and thus obtained UF_4 is converted to UF_6.

The open pit mining and heap leaching test have been started in 1978 and the operation will be started on the PNC process demonstration plant (with a processing capacity of 300 t-UF_6/y) in 1979.

The solid and liquid wastes from the pilot plant are transferred to the tailings dam. The overflow from the dam is chemically treated and passed through settling ponds and sand filters to be discharged into a river.

The concentrations of U, ^{226}Ra, pH, S.S., COD, Fe, Mn, Cl, and F are monitored periodically and they are all below the control values. The results of monitoring on the river bed and rice paddy soil show no signs of accumulation of U and ^{226}Ra in it.

2. Uranium Ore Process and Liquid Waste Treatment

The uranium ore process (the PNC process) which is now

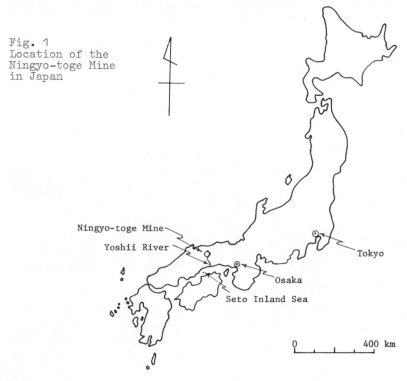

Fig. 1
Location of the
Ningyo-toge Mine
in Japan

Ningyo-toge Mine

Yoshii River

Tokyo

Osaka

Seto Inland Sea

0   400 km

being tested is schematically shown in Fig. 2. The following is the description of the process from the ore to UF_6 and the disposal of the wastes discharged from the various steps in the process.

The ore is all crushed by the crusher to minus 100 mm and mixed with water and screened by means of a drum washer with trommel and then the ore over 20 mm in size is crushed to smaller sizes by means of impact crusher and then to minus 3 mm by a rod mill. The 50 to 60% solids pulp is transferred to the leaching drum, where uranium is leached out with sulfuric acid. Thereafter, the pulp is flown to the classifier, by which it is separated into sand and slime. The slime is separated into the leaching solution and waste spigot by the thickener, using COD. The waste spigot in a slurry form and the sand from the classifier are continuously flown to the neutralization tank so that the sand and slime are neutralized with lime in a milky form and then transferred through a pipeline to the tailings dam. The amount of the liquid which is neutralized is approximately 1.2 m^3/t-ore (60 m^3/d).

The overflow from the thickner is delivered to the solvent extraction process after passing through the filter press to remove a small quantity of solids. Uranium is extracted from the leaching solution with a solvent containing tri-n-octylamine in a kerosene diluent. Then the uranium in the solvent is converted with HCL, and UO_2Cl_2 is stripped with water. The liquid waste produced during the solvent extraction process is approximately 2.3 m^3/t-ore (115 m^3/d). The liquid waste, as the residues in the solid-liquid separation process, is also neutralized with lime and transferred to the tailings dam.

UO_2Cl_2 is reduced to UCl_4 by electrolytic reduction. The electrolytic reduction process uses cooling water but discharges no liquid waste. When HF is added to the UCl_4 solution in the hydrofluorination tank, green salt is precipitated. The precipitate is filtered and dried to obtain $UF_4 \cdot nH_2O$ in particles ranging from 50 to 150μ in size. The hydrofluorination precipitation is a continuous process and the overflow from the hydrofluorination tank contains 0.07 to 0.09 g-U/l, 1 to 2 g-F/l, and 56 to 58 g-Cl/l. The overflow is mixed with the filtration and washing liquid waste, then the mixtures are distilled to remove F and Cl, thereby to recover uranium. The recovered uranium is recycled to the leaching solution. The removed F and Cl are neutralized with lime and passed through the filter and also F is kept in storage in the form of CaF_2 cakes. The reason for this is that CaF_2 has a relatively high solubility of approximately 10 ppm, and therefore if it is deposited in the tailings dam it would dissolve in rainwater and flow out of the dam. This being so, only the filtrate is transferred to the tailings dam. The hydrofluorination process discharges liquid waste of approximately 0.015 m^3/t-ore (0.75 m^3/d).

The $UF_4 \cdot nH_2O$ is fed to the dehydration process, where the crystallized water is removed from the grains to obtain unhydrous UF_4 in the dehydrating tower of the fluid bed type. The obtained UF_4 is reacted with F_2 gas in the reaction tower of the fluid bed type to obtain UF_6. Since the dehydration and conversion processes are dry operations, they produce almost no liquid waste.

The amounts of the required acid and neutralizing lime are about 50 kg/t-ore, respectively.

Table I presents the pH and concentrations of U, ^{226}Ra and F in the liquid waste before and after neutralized, in the dam overflow and in the effluent. The ^{226}Ra concentration in the liquid waste before it is neutralized is 20 to 80 times higher than the control value but the concentration of ^{226}Ra in the water of the river to which the liquid waste is discharged has decreased to 1/10 to 1/100 of the control value owing to the tailings dam and chemical treatment process.

3. Tailings Dam

The tailings dam at the Ningyo-toge Mine is a gravity dam about 65 m wide and 10 m high. It has total area of about 30,000 m^2

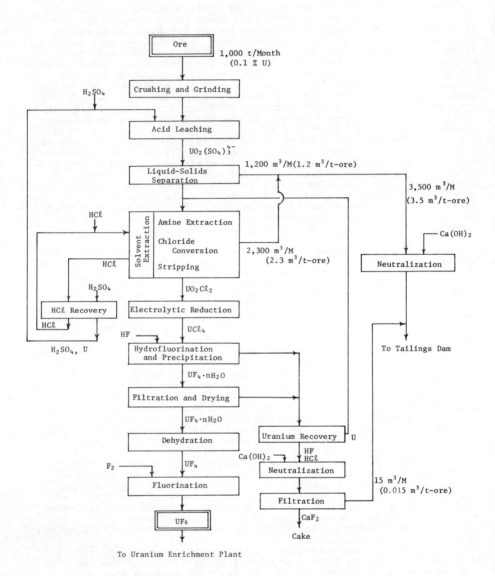

Fig. 2 Flowsheet of the PNC process at the Ningyo-toge
pilot plant

Table I. Liquid waste treatment

		pH	U	^{226}Ra	F	Volume of liquid (per month)
Before neutralization	Solids-liquid separated, solvent extract liquid waste	1.2-2.0	1.3 ~ 2.3 x 10^4 pCi/l	200 ~ 800 pCi/l	–	3,500 m^3
	Fluorinated and precipitated liquid waste	< 1	2.3 ~ 3.0 x 10^4 pCi/l	–	1 – 2g/l	20 m^3
After neutralization	(Dam overflow)	7 – 8	50 ~ 100 pCi/l	70 ~ 500 pCi/l	0.1-0.5mg/l	10,000 m^3 (Rainwater flows in)
After waste water treatment	(Discharged water)	6.3-7	10 ~ 100 pCi/l	0.1 ~ 1 pCi/l	0.1-0.5mg/l	10,000 m^3
Control value		5.8-8.6	600 pCi/l	10 pCi/l	15 mg/l	–

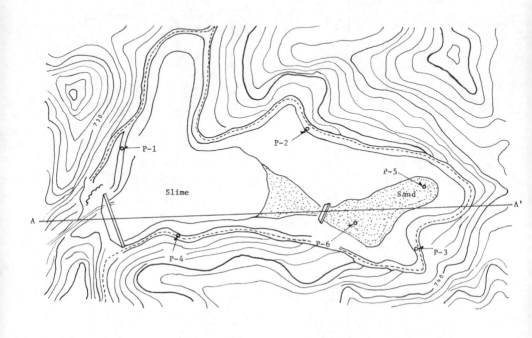

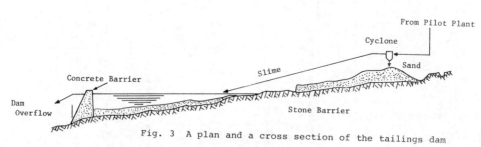

Fig. 3 A plan and a cross section of the tailings dam

and is capable of holding about 60,000 tons of tailings (about 36,000 tons of sand and about 24,000 tons of slime) and also of about 10,000 m³ of water. A plan and a cross section of the tailings dam are shown in Fig. 3.

The neutralized tailings from the pilot plant are separated into sand and slime by means of the cyclone located on the upstream side of the dam, and the sand is deposited on the upstream side while the slime on the downstream side of the dam. Table II shows the deposition of tailings up to the spring of 1978 and the U and ^{226}Ra concentrations in them. The water in the dam is very clear and flows over the upper edge of the dam. There is the necessity for Ra removing facility because the ^{226}Ra concentration in the clear top water is several tens as high as the control value as presented in Table I. In order to prevent rainwater flowing into the dam from the surrounding hills, the waterways are provided on the hillsides.

4. Radium Removing Facilities

In order to remove radium from the dam overflow, there are provided the liquid waste treatment facilities for effecting chemical and settling treatment as shown in Fig. 4. The dam overflow is added a barium chloride at the rate of 0.05 kg/m³-dam overflow in the agitating tank (1 m³) so that it reacts with the sulfate radical which is present in small quantities in the liquid to form very small particles of $BaSO_4$ with a low solubility. Similarly the radium in the liquid waste is also precipitated in the form of sulfate together with the small particles of $BaSO_4$. Fig. 5 shows the relation between the amount of barium chloride added and the ^{226}Ra concentration in the dam overflow. As seen from this graph, the addition of $BaCl_2$ in excess of 0.02 kg/m³-dam overflow, the concentration of ^{226}Ra in the dam overflow decreases below the control value.

However, a problem is posed by the fact that the above-stated particles are so small in diameter as 1 to 2μ that their sedimentation velocity is so low as 0.2 mm/min. Since it is technically difficult to filter such small particles, first $BaCl_2$ is added and then bentonite (0.1 kg/m³-dam overflow) to cause the particles to unite together to form large flocks. Then a precipitant (0.01 kg/m³-dam overflow) is added to increase the sedimentation velocity to 2.4 mm/min to facilitate the sedimentation in the settling.

There are provided five settling ponds (total area: 2,200 m²) and 10 sand filters (total area: 930 m²). The settling pond overflow is discharged into a river after being filtered by the sand filters. There are two such treatment systems running parallel to each and normally only one of them is used. Each system has a effluent treating capacity of 1 m³/min. At the time of heavy rainfall, both systems are operated to provided a total capacity of 6 m³/min.

5. Environmental Monitoring

Around the Ningyo-toge Mine, the external radiation does rate, airborne activity, uranium and radium contents in the drinking water, river water, river bed and rice paddy soil, and the ordinary pollutants in the river water have been monitored periodically. Table III shows the measurements of the external radiation does rate and radium concentration at a height of 1 m from the dam surface. The external radiation dose rate in the area of the sand sediments on the upstream side of the tailings dam was found to be 200 μR/hr while that of the soil in the surroundings of the dam was about 10 to 90 μR/hr.

Table IV presents the results of monitoring on the concentrations of ^{226}Ra, U and ordinary pollutants in the effluent and river waters during a period from 1972 through 1977. The volume of river water flow at about 10 km downstream from the waste water outlet into the river was about 1.3 x 10⁶ m³/month.

Table II. Amounts of tailings deposits, and U, ^{226}Ra contents

| | Tailings storing capacity (ton) | Accumulated tailings deposits (ton) (As of spring of 1978) | Contents | |
			^{226}Ra	U
Sand	36,000	15,000	2.0×10^2 pCi/g	$3.3 \sim 6.7$ pCi/g
Slime	24,000	10,000	2.7×10^2 pCi/g	$26.7 \sim 33.3$ pCi/g
Total	60,000	25,000	−	−

Table III. External radiation dose rate and radon concentration at 1 m above the dam surface

Measurement point*	External radiation dose rate (μR/hr)	Radon concentration (pCi/1)	Remarks
P-1	90	< 1	*Refer to Fig. 3
P-2	16	< 1	Weather: Fine
P-3	14	< 1	Temperature: 21°C
P-4	16	< 1	Wind direction: SW
P-5	200	< 1	Wind velocity: 3 - 5 m/sec.
P-6	200	< 1	

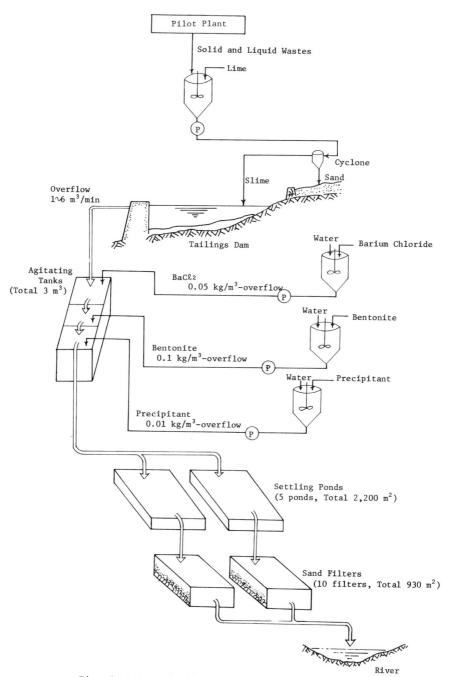

Fig. 4 Schematic diagram of the tailings disposal

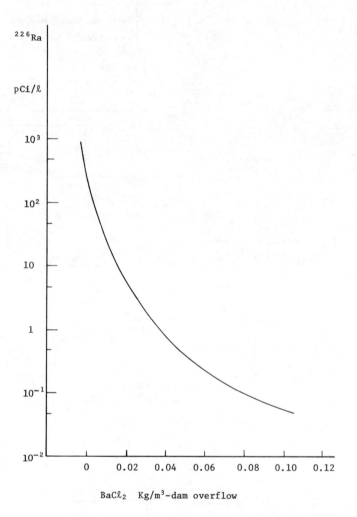

Fig. 5 Effect of BaCℓ₂ content on
        the removing of ^{226}Ra in the dam
        overflow

The ^{226}Ra concentration in the river bed soil about 1.3 km downstream is 3.4 to 17 pCi/g-dry soil, according to the monitoring data for the past six years. The ^{226}Ra concentration in the rice paddy soil about 1.7 km downstream is within the range of 1.1 to 7.4 pCi/g-dry soil. There are no signs of ^{226}Ra accumulation in both cases.

6. Postscript

Table V shows the construction and operation costs of the tailings treatment and disposal facilities. As seen from the table, the total construction cost of all those facilities amounted to $181,000.

The uranium and radium concentrations in the effluent have been controlled to be below the control values. In an effort to further decrease those concentrations in the effluent, a study is now in progress on the recovery of uranium from the dam overflow by the use of an ion exchange absorbent. As for the removal of radium, efforts are being made for the development of an efficient absorbent and improvement of filtration equipment.

The construction of the PNC process demonstration plant which will be about ten times as large as the existing one at the Ningyo-toge Mine is now planned to be completed in 1980. It will incorporate the most up-to-date techniques and equipment, thereby to further decrease the concentrations of radioactive materials in the effluent.

When the tailings dam is full, it will be covered with soil and planted with trees to ensure a higher degree of stability.

Table IV. Results of environmental monitoring
of discharged water and river water

(1972 - 1977)

Measuring point	Discharged water	Okutsu (about 10 km downstream)	Tsuyama (about 50 km downstream)	Control value
^{226}Ra (pCi/1)	0.06 - 1.2	0.04 - 0.93	0.13 - 0.48	10 pCi/1
U (pCi/1)	3.8 - 290	0.02 - 0.09	N.D. - 0.06	600 pCi/1
pH	5.8 - 7.7	6.6 - 7.4	7.4 - 7.7	5.8 - 8.6
S.S. (mg/1)	<1.0 - 10.7	<1.0 - 15.2	<1.0 - 22.4	40 mg/1
COD (mg/1)	0.5 - 5.4	0.2 - 2.6	1.0 - 3.5	15 mg/1
Fe (mg/1)	0.0 - 0.35	0.0 - 0.6	0.02 - 0.14	10 mg/1
Mn (mg/1)	0.18 - 0.55	N.D. - 0.02	N.D. - 0.09	10 mg/1
Cl (mg/1)	36 - 525	1.8 - 9.2	5.0 - 10.3	-
F (mg/1)	0.1 - 1.3	-	-	15 mg/1

Table V. Construction and operation costs
of tailings treatment facilities

(¥230/$)

	Construction cost	Operation cost	
(1) Neutralization and transporting facilities	$32,000	$3.4/ton-ore	
		Operators	29%
		Reagents	58%
		Electric power	8%
		Maintenance	5%
(2) Tailings dam	$124,000	−	
(3) Radium removing facility	$25,000	$0.18/m^3-dam overflow	
		Operators	39%
		Reagents	33%
		Electric power	6%
		Maintenance	22%
Total	$181,000	−	

JOINT PANEL

ON

OCCUPATIONAL AND ENVIRONMENTAL RESEARCH

FOR URANIUM PRODUCTION

P. Hamel, G. Zahary

Paper presented by

J. Howieson
Energy, Mines and Resources Canada
Canada

INTRODUCTION

The first stage of the nuclear power cycle, uranium production, is the least systematically studied and documented. Many of the operating procedures in current use have developed from practical experience gained in full scale operations and have neither been written down nor fully analyzed. The potentially harmful side effects of some of these operations are now more fully appreciated than they have been in the past.

Since 1974 dramatic changes have taken place. In a period of three years a national uranium policy was defined, a royal commission report on the health and safety of mine workers was prepared, and environmental assessment hearings were initiated in two provinces. Public awareness of the issues related to the development of nuclear power and, to some extent, of uranium supply has expanded greatly. This increased awareness has focussed attention on a wide variety of problems and has itself contributed to the resolution of many which are essentially of an administrative character, including posting of environmental monitoring data, re-evaluation of the roles of Atomic Energy Control Board, labour unions, mines inspectorate, etc. Fundamental changes in the tradit: responsibility-authority structure of the industry have been made and are in the process of implementation. This fuller appreciation of the industry will not, of course, resolve problems which are essentially of a scientific or technological character.

In August 1976 a meeting was held under the auspices of CANMET, Department of Energy, Mines and Resources, to discuss how a more comprehensive program of research might be organized. Representatives of various sectors of government, industry and labour attended and one outcome of this meeting was the Joint Panel on Occupational and Environmental Research for Uranium Production.

The first report on the Joint Panel's operations is prepared to report on progress being made in meeting the Panel's objectives. The Panel is seen to be a practical means of identifying, discussing and disseminating scientific information and we are encouraged to develop it further in the future.

PURPOSE

Occupational health and environmental effects associated with the production of uranium are of concern to labour, industry, government and the public. To alleviate this concern the knowledge base for the industry must be expanded.

At present there is no single organization with the authority, expertise or research resources to perform this task independently. Hence, there is a need for cooperation between all organizations active in research and special studies related to the effects of uranium production.

This agreement sets out the principles and operating policies of the Joint Panel on Occupational and Environmental Research for Uranium Production.

OBJECTIVES

1. To produce an annual inventory of relevant research being conducted by the supporting members.

2. To exchange information on projected and completed research.

3. To identify research needs and to stimulate research activity.

PARTIES TO THE AGREEMENT

Supporting Members of the Joint Panel are those who have an active and continuing involvement in research on the environmental or health effects of uranium production.

Associate Members are those who have responsibilities or authority in the same area but are not directly engaged in research.

GUIDING PRINCIPLES

1. Participation on the Joint Panel will in no way constrain members from pursuing the objectives and meeting the obligations of their respective organizations.

2. Supporting Members are prepared to contribute to the development of a comprehensive program of research compatible with their existing roles and responsibilities.

3. Supporting Members recognize the obligation to make information widely available by:

> describing on-going research in a common format;
>
> informing the Joint Panel of progress on specific projects through interim reports;
>
> reviewing reports prepared by other Supporting Members and providing comment;
>
> providing Associate Members with information on the work of the Joint Panel and giving serious consideration to their research needs.

4. Associate Members are prepared to identify the need for new knowledge and research in a common format, and to inform the Joint Panel of this need at an early stage.

5. All members recognize the obligation to make final research results available to the public. Distribution of interim progress reports remains at the discretion of the organizations responsible for the work.

6. Participation on the Joint Panel by others is welcomed. Membership is open at any time to any organization that is prepared to accept the principles outlined here and the operating procedures that follow.

OPERATING PROCEDURES

1. The Joint Panel will meet semi-annually in June and December.

2. The June meeting will be devoted to a discussion of the distribution of reports on completed work and examination of the need for additional research.

3. The December meeting will be devoted to an exchange of information on projected research and to the preparation of an annual report which will be available to the public.

4. Projected research will be described in a common format, such as the attached "Research Project" form, and reported to the Joint Panel annually.

5. Research needs will be described in a common format, such as the attached "Research Problem" form, and reported to the Joint Panel at least annually.

6. The Joint Panel will review research plans and needs in a systematic manner, as illustrated by the attached logic diagram, "Consolidated Program of Research".

7. Additional procedures will be introduced as necessary to ensure satisfactory communication between Supporting Members, for example, scheduling of special meetings and seminars, quarterly reports on on-going work, etc.

8. The Chairman of the Joint Panel will serve for a term of three years.

9. Department of Energy, Mines and Resources, Ottawa, will provide the necessary secretariat and nominate the first chairman.

RESEARCH PROJECT

No.

TITLE

	CODE
	REF.

SCALE

	TOTAL PROJECT		CURRENT YEAR	
	EST. COST	ELAPSED TIME	MAN YEARS	BUDGET
	$	YEAR OF	PROF. TECH.	$

OBJECTIVE

PURPOSE OF WORK — RELEVANCE - JUSTIFICATION - RATIONALE

WORK PLAN — TASKS - ACTIVITIES - DECISIONS | MILESTONE DATES

CURRENT STATUS

SPONSOR

CONTACT	PHONE	DATE

RESEARCH PROBLEM

No.

TITLE

CODE	
REF.	

PROBLEM

OBJECTIVE

CRITICAL FACTORS

RESOURCES — TECHNICAL — HUMAN

WORK STATEMENT

KEY DECISIONS — EVENTS — ACTIVITIES

SCALE

	ELAPSED TIME	MAN YEARS	SPECIAL COST	ESTIMATED COST
	YEARS	PROF. _ TECH.	$	$

OUTPUT

NATURE OF · FIRST USER — DISSEMINATION

ORIGINATOR

CONTACT	PHONE
	DATE

CONSOLIDATED PROGRAM OF RESEARCH

Occupational and Environmental Research for Uranium Production

	OCCUPATIONAL	ENVIRONMENTAL

DEFINE
    Conditions
    of Exposure		

IDENTIFY
    Adverse
    Effects		

RELATE
    Exposure
    with Effect		

ASSESS Risk		
CONTROL Hazard		

SUPPORTING MEMBERS

Atomic Energy Control Board

Denison Mines Limited

Department of Energy, Mines and Resources

Department of Fisheries and Environment

Department of National Health and Welfare

Eldorado Nuclear Limited

Ministry of Environment for Ontario

Ministry of Labour for Ontario

Rio Algom Limited

United Steelworkers of America

ASSOCIATE MEMBERS

Agnew Lake Mines Limited

AMOK Limited

Department of Consumer Affairs and Environment, Newfoundland and

Department of Indian and Northern Affairs, Ottawa   Labrador

Department of Labour, Saskatchewan

Department of Mines and Energy, Newfoundland and Labrador

Department of Mines and Petroleum Resources, British Columbia

Labour Canada

Madawaska Mines Limited

Mines Accident Prevention Association of Ontario

Ministry of Natural Resources, Ontario

CURRENT RESEARCH BY ORGANIZATION

Atomic Energy Control Board (AECB)

1. Sputum Cytology and CEA in Uranium Workers.
2. Field Test of MOD Working Level Dosimeter.
3. Field Test 'H & H Custom Work' Personal Alpha Dosimeter.
4. Collaboration EMR 5.
5. Radiation Exposure and Lung Cancer Mortality Spectra of Miners.

Denison Mines Limited (DEN)

1. Field Evaluation of a Powered Respirator.
2. Occupational Exposure Indices to Air-Borne Dust.
3. Collaboration EMR 5.
4. Investigation of Accuracy in Radiation Exposure (WLM) Estimate Data.
5. Effect of Filter Pore-Size on the Dissolved Radium-226 Acceptability Criteria.
6. Feasibility Study for Pyrite Removal from Mill Tailings.
7. Pyrite Distribution in Mine Backfill Products.
8. Sulphuric Acid Elution.
9. Serpent River Radium Deposition.

Eldorado Nuclear Limited (ENL)

1. Originator of AECB 1.
2. Revegetation of Beaverlodge Tailings.
3. Methods of Recovering Uranium from Mine Water.
4. Quality of Effluents and Escape of Elements in Addition to Radium-226.

Energy, Mines and Resources (EMR)

1. Effect of a Number of Variables on Three Types of Dust Samplers.
2. Dust Sources and Control.
3. Comparison of Gravimetric Dust Samplers.
4. Infrared Determination of Quartz Dust.
5. Evaluation of CEA-STEPPA Alpha Dosimeter.
6. Radon Daughter Measurement Techniques.
7. Radon Sources and Control.
8. Dust-Radon Daughter Dosimeter.
9. Effect of Hydrocarbon Inhalation on Lung Function.
10. Diesel Emissions.
11. Diesel Emissions Controls.
12. Surface Treatment of Tailings.
13. Weathering of Tailings.
14. Effluent Control in Tailings Basin.

15. Vegetation of Tailings.
16. Use of Tailings as Backfill in Uranium Mines.
17. Stability of Barium/Radium Sulphate Sludges.
18. Miners Work Suit and Helmet.
19. Continuous Monitoring.
20. Effluent Control in Mill.
21. Ventilation Strategies for Uranium Mines.
22. Industrial Hygiene Survey of the Uranium Industry.
23. Noise Exposure in Mines.
24. Mine Ventilation Models.
25. An Interdisciplinary Research and Public Information Project.

Health and Welfare Canada (HWC)

1. Occupational Radiation Exposure Evaluation of Uranium Miners.
2. ^{210}Pb Measurements in Uranium Miners.
3. Collaboration AECB 5.

Ministry of Environment for Ontario (MOE)

1. Effect of Depressed pH on Fish.

Ministry of Labour (MOL)

1. Lung Cancer in Ontario Uranium Miners.
2. Sputum Cytology in Uranium Miners.

Rio Algom Limited (RIO)

1. Experimental Work on EMR 3.
2. Collaboration EMR 5.
3. Collaboration EMR 12.
4. Collaboration EMR 15.

United Steelworkers of America (USW)

1. Health Effects of Diesel Equipment in Confined Spaces.
2. Collaboration HWC 2.

PROJECT DESCRIPTIONS

1. Sputum Cytology and CEA in Uranium Workers   AECB 1, ENI

    To study uranium workers by means of sputum cytology
    and carcinoembryonic antigen (CEA) assay with a view
    to detecting at a potentially reversible or curable
    stage changes that may progress to invasive lung
    cancer; to determine the relative carcinogenicity of
    cigarette smoking and exposure to uranium.

2. Field Test of MOD Working Level Dosimeter   AECB 2

    To develop an arrangement for wearing TLD dosimeters
    that is convenient and acceptable to miners and to
    correlate dosimeter results with conventional radon
    daughter measurements. Report circulated.

3. Field Test 'H & H Custom Work'
    Personal Alpha Dosimeter   AECB 3

    To test in the mine environment the performance
    of a personal alpha dosimeter of Canadian
    manufacture.

4. Radiation Exposure and Lung Cancer Mortality
    Spectra of Miners   AECB 5, HW

    To obtain a spectrum of radiation exposure (WLM)
    and lung cancer deaths as function of months exposed
    to radiation and as a function of the rates of
    exposure from a varitey of mining sites across Canada.

5. Field Evaluation of a Powered Respirator   DEN 1

    Evaluate the S.M.R.E. airstream dust helmet under
    field conditions and establish exposure protection
    criteria.

6. Occupational Exposure Indices to Air-Borne Dust   DEN 2

    Establish dust exposure indices for various occupa-
    tional groups by full-shift gravimetric sampling
    and X-ray diffraction analysis for quartz.

7. An Investigation of Accuracy in
    Radiation Exposure (WLM) Estimate Data   DEN 4

    Establish confidence limits for present method
    of measurement and calculation of a miners cumula-
    tive exposure to air-borne radon daughters and
    determine the optimum sampling frequency.

8.   Effect of Filter Pore-Size on the Dissolved
      Radium 226 Acceptability Criteria in Ontario
      Water Quality Objective   DEN 5

          Investigate the effect of filter pore-size on
          dissolved Radium 226 activity level and quantify
          the possible variations that might occur using
          3.0μ or the 1.2μ Millipore filters.

9.   A Feasibility Study For Pyrite
      Removal From Mill Tailings   DEN 6

          To study the effects and economics of pyrite
          removal from mill tailings with a view to
          producing pyrite-free tailings.

10.   Pyrite Distribution in Mine Backfill Products   DEN 7

          To determine the distribution of the pyrite,
          uranium, radium and thorium in the sand and slime
          fractions of the proposed sand backfill plants.

11.   Sulphuric Acid Elution   DEN 8

          To determine the effects of using sulphuric
          acid elution in the ion exchange process with a
          view to reducing nitrogen compounds in mill
          effluents.

12.   Serpent River Radium Deposition   DEN 9

          To determine the extent, if any, and location of
          Barium-Radium sulphate deposits along the Serpent
          River.

13.   Revegetation of Beaverlodge Tailings   ENL 2

          To determine the feasibility of using vegetation
          to stabilize alkaline uranium tailings in the
          northern climate.

14.   Methods of Recovering Uranium from Mine Water   ENL 3

          To investigate methods of recovering uranium
          from mine-water and assess the effect on effluent
          quality of mine-water additions to the tailings.

15.   Quality of Effluents from Present Treatment
      Procedures and Escape of Elements in
      Addition to Ra^{226} ENL 4

          To specify the quality of present effluents with
          particular reference to the stability of $Ba.RaSO_4$
          precipitates and develop alternate methods for
          effluent treatment.

16. Effect of a Number of Variables
    on Three Types of Dust Samples   EMR 1

        A laboratory study to determine the effects of
        wind speed and direction, sampler inlet velocity,
        cyclone orientation etc. on the quality of dust
        collected by 3 types of mass samplers. Completed.

17. Dust Sources and Control   EMR 2

        To determine dust produced at a number of under-
        ground operations e.g. drilling and slushing, using
        gravimetric methods; and to assess the effectiveness
        of water sprays.

18. Comparison of Gravimetric Dust Samplers   EMR 3   RIO

        To compare, under mining conditions, the dust collection
        characteristics of three samplers differing with
        respect to method of size-selection i.e. impactor,
        cyclone, horizontal elutriator. Experimental work
        completed.

19. Infrared Determination of Quartz Dust   EMR 4

        To develop the infrared method of analyzing for
        quartz in dust samples and to compare results
        with those obtained by X-ray diffraction methods.

20. Evaluation of CEA-STEPPA Alpha Dosimeter   EMR 5 AECB
                                                      DEN 3 RIO 2
        To field test a track-etch dosimeter which will
        reliably measure accumulated exposure of miners
        to air-borne radioactivity.

21. Radon Daughter Measurement Techniques   EMR 6

        To obtain a direct reading instrument for radon
        daughter concentration and build a solid state
        alpha particle spectrometer (Argonne National
        Laboratory design); to build a multi-head counter
        for counting up to 10 Lucas cells simultaneously;
        and to develop a more efficient technique for
        analyzing low-level radon sources.

22. Radon Sources and Control   EMR 7

        To determine the relation between radon emanation
        rate and properties of tailings backfill e.g. radium
        content, size distribution, moisture content etc.;
        to evaluate ventilation in filled stopes and
        effectiveness of various coatings.

23.   Dust-Radon Daughter Dosimeter   EMR 8

       To extend the existing personal dust sampler
       (CAMPEDS) into a combined dust-radiation dosimeter
       by incorporating a thermoluminescent sensor.

24.   Effect of Hydrocarbon
       Inhalation on Lung Function   EMR 9

       To review the medical evidence relating loss of
       pulmonary function to inhalation of hydrocarbon
       contaminants.

25.   Diesel Emissions   EMR 10

       To develop strategies encompassing machine design,
       engine selection, operating conditions and use of
       emission control ancillaries to minimize exhaust
       pollutants.

26.   Diesel Emission Controls   EMR 11

       To investigate methods of reducing levels of
       diesel soot in underground operations. The
       efficiencies of two proprietary wet scrubbers
       are being tested as well as the feasibility of
       collecting soot by filtration.

27.   Surface Treatment of Tailings   EMR 12, RIO 3

       To rank four types of surface treatment i.e. wood
       products, glacial till, vegetation and water
       flooding on the basis of quality and quantity of
       seepage water. Effluent from five test pits is
       to be monitored to 1981.

28.   Weathering of Tailings   EMR 13

       To develop economical and safe means of inhibiting
       biological oxidation of pyrite in mine tailings.
       Laboratory studies of biological control mechanisms,
       mine-mill organics, chemical biocides are underway.

29.   Effluent Control in Tailings Basin   EMR 14

       To establish the most practicable means of controlling
       acid, toxic metals, dissolved salts and radioisotopes
       from tailings. Effluent from an isolated inactive
       tailing basin is being monitored to identify critical
       mechanisms and develop an effluent model.

30.   Vegetation of Tailings   EMR 15 RIO

         To develop and demonstrate the best practicable
         methods for the vegetation of uranium tailings.
         Existing test plots are being monitored and a
         larger test plot (6 acres) will be established
         to test the CANMET vegetation "recipe".

31.   Use of Tailings as Backfill
       in Uranium Mines   EMR 16

         To evaluate the feasibility of using tailings as
         backfill in Canadian Uranium Mines e.g. stabilization
         of ground, degradation of the underground environment
         etc., and to identify the knowledge gaps. Completed.

32.   Stability of Barium-Radium Sulphate Sludges   EMR 17

         A laboratory study to assess the stability of the
         sludges over a fixed range of temperature, pH and
         dissolved sulphate in the overlying liquid.
         Completed.

33.   Miners Work Suit and Helmet   EMR 18

         To design, fabricate and test under field
         conditions a miners work suit, an integrated
         protective system consisting of ear muffs, improved
         lighting, powered filtered-air supply and battery.
         Completion by end 1977.

34.   Continuous Monitoring   EMR 19

         To develop and demonstrate a system capable of
         continuously measuring airborne contaminants and
         ventilation flow rates in mines, transmitting and
         displaying the information on surface. Completion
         March, 1978.

35.   Effluent Control in Mill   EMR 20

         To identify potential techniques for extracting
         pyrite and Radium-226 in the mill. An investigation
         is underway of locations in the mill where radium
         might be removed.

36.   Ventilation Strategies for Uranium Mines   EMR 21

         To review ventilation practice and design criteria
         and make recommendations on how the ventilation
         option might be used to improve environmental
         conditions in uranium mines. Completed.

37. Industrial Hygiene Survey of
    the Uranium Industry   EMR 22

        To conduct a preliminary industrial hygiene
        survey of the uranium industry (mining and milling)
        and make recommendations for further work required.

38. Noise Exposure in Mines   EMR 23

        To develop and evaluate simplified methods of
        specifying the full shift exposure of mine workers
        to noise e.g. one approach is to identify "noise-
        exposure-elements" which can be combined to
        calculate full shift exposure.

39. Mine Ventilation Models   EMR 24

        To identify critical features of existing
        uranium-mine ventilation systems, assess the
        potential of available computer models to
        represent them and to begin collecting data to
        validate the models.

40. An Interdisciplinary Research and
    Public Information Project   EMR 25

        To determine the feasibility of establishing
        an independent centre to carry out research
        and disseminate information on occupational and
        environmental health issues related to the uranium
        industry. Contract has been awarded to the
        Elliot Lake Centre.

41. Occupational Radiation Exposure
    Evaluation of Uranium Miners   HWC 1

        To establish in computer files the radiation
        dose records of uranium workers for use in
        exposure control and epidemiological records.

42. ^{210}Pb Measurements in Uranium Miners   HWC 2, USW 2

        To determine the feasibility of using the Whole
        Body Counter for measuring radioactivity in the
        bodies of uranium miners and nuclear fuel workers.

43. Effect of Depressed pH on Fish   MOE 1

        To determine levels of pH which impair reproduction,
        growth, and survival of Brook Trout under soft water
        conditions and correlate these changes with that
        of flag fish used as test species in toxicity
        laboratories.

44. Lung Cancer in Ontario Uranium Miners   MOL 1

     To study the long-term effects of radon exposure
     on Ontario uranium miners as reflected in mortality
     data. Study began in 1974.

45. Sputum Cytology in Uranium Miners   MOL 2

     To determine the feasibility of using sputum
     cytology as a method for early diagnosis of
     lung cancer.

46. Health Effects of Diesel Equipment in
     Confined Spaces   USW 1

     To identify the health hazards associated with
     the use of diesel equipment in underground mines.

NOTE:
     Department of Fisheries and Environment projects
     to be included in a subsequent report.

CONSOLIDATED PROGRAM OF RESEARCH

Occupational and Environmental Research for Uranium Production

	OCCUPATIONAL	ENVIRONMENTAL

DEFINE
 	
Conditions
of Exposure
	2 16 20* 31* 42
	3 17* 21 34
	6 18 22* 36*
	7 19 23 38
	          25 39
		8 12 27* 32
		    13 28
		    14* 29*
		    15*

IDENTIFY Adverse Effects	10* 24 31* 46 37	10*
RELATE Exposure with Effect	1 41 4 44 45	43
ASSESS Risk		

CONTROL
 	
Hazard
	5 17* 22* 33
	        26 36*
		9 11 27* 30
		    14* 29* 35
		    15*

Special Study   40   40

* Included in more than one Category

JOINT PANEL SUPPORTING MEMBERS

1.   Atomic Energy Control Board
      P.O. Box 1046
      Ottawa, Ontario, K1P 5S9

              Dr. H. Stocker   613-996-3783
              Associate Scientific Adviser
              Research and Coodination Directorate

2.   Denison Mines Limited
      P.O. Box B2600
      Elliot Lake, Ontario

              Mr. J.L. Chakravatti   705-848-2221
              Sr. Environmental Engineer

3.   Department of Energy, Mines and Resources
      Elliot Lake Laboratory
      CANMET
      P.O. Box 100
      Elliot Lake, Ontario, P5A 2J6

              Dr. D.G.F. Hedley   705-848-2236
              Coordinator Mining Research
              Minerals Research Program

4.   Department of Fisheries and Environment
      Fontaine Building
      Ottawa, Ontario, K1A OH3

              Dr. E.F. Roots   613-997-2393
              Science Advisor
              Office of the Science Advisor

5.   Health and Welfare Canada
      Tunney's Pasture
      Ottawa, Ontario, K1A OL2

              Dr. D.H. Niblett   613-992-7736
              Senior Medical Advisor, Room 130
              Environmental Health Centre

6.   Eldorado Nuclear Limited
      255 Albert Street
      Ottawa, Ontario

              Mr. C.F. Smith   613-238-5222
              Director,Environment Protection   Ext. 70
              and Health

7.   Ministry of Environment for Ontario
445 Albert Street, E.
Sault Ste. Marie, Ontario

        Mr. G.J. LaHaye   705-949-4640
        District Officer

8.   Ministry of Labour for Ontario
15 Overlea Blvd.
Toronto, Ontario

        Dr. J. Muller, M.D.   Chief   416-965-6375
        Environmental Health Studies Service
        Occupation Health Protection Branch

9.   Rio Algom Limited
P.O. Box 1500
Elliot Lake, Ontario

        Mr. E. Barnes   705-848-2234
        Research Superintendent
        Process and Development

10.   United Steelworkers of America
20 Alberta Road
Elliot Lake, Ontario

        Mr. H. Seguin   705-848-2226
        Staff Representative

JOINT PANEL EXECUTIVE

Chairman

        Mr. P.E. Hamel, Director   613-992-1226
        Research Coordination Directorate
        Atomic Energy Control Board
        P.O. Box 1046
        Ottawa, Ontario, K1P 5S9

Secretary

        Mr. G. Zahary   705-848-2236
        Elliot Lake Laboratory
        Mining Research Laboratories
        CANMET, E.M.R.
        P.O. Box 100
        Elliot Lake, Ontario   P5A 2J6

PUBLICATIONS - FILE

1. Yourt, R.   "Field Test of MOD Working Level Dosimeter",
Proc. of the Specialist Meeting, Elliot Lake, Nuclear
Energy Agency, 1976.

2. Washington, R.A.   "Proceedings of a Seminar on Radon and Radon
Daughter Measurements in Underground Uranium Mines,
Elliot Lake, Ontario, September 16-17, 1976",
MRP/MRL 77-27(TR).

3. Stewart, D.B., Dainty, E.D., and Mogan, J.P.
"Diesel Emissions with Respect to the Mine Environment",
CIM Bulletin, January 1976.

4. Stewart, D.B., d'Aoust, A., Dainty, E.D. and Mogan, J.P.
"Determination of Diesel Engine Parameters Under-
ground for a Load-Haul-Dump Vehicle", CIM Bulletin,
June 1976.

5. Moffett, D.   "Waste Disposal at Canadian Uranium Mines", MRP/MRL
76-43 and Canadian Mining Journal, January 1977.

6. Murray, D. and Moffett, D.   "Vegetating the Elliot Lake Uranium
Mine Tailings", MRP/MRL 76-114 and Journal of Soil and
Water Conservation.

7. Moffett, D.   "Environmental Aspects of Thorium", MRP/MRL 77-43
and Canadian Uranium Producers Meeting.

8. McCready, R.G.L.   "The Effects of Solvent Extraction Organics
on Thiobacillus Ferrooxidans", MRP/MRL 77-44 and
Canadian Uranium Producers Meeting.

9. Moffett, D.   "The Disposal of Solid Wastes and Liquid Effluents
from the Milling of Uranium Ores", CANMET Report 76-19.

Contract Research Reports available on "Open File" at the Mining
Research Laboratories in Ottawa and Elliot Lake.

10. University of Calgary.   "The Chemical and Physical Stability of
Barium/Radium Sludges under Varying Conditions",
June 1977.

11. James F. MacLaren Ltd.   "Industrial Hygiene Survey of the Uranium
Industry", July 1977.

12. Acres Consulting Services Ltd.   "Study on the Use of Tailings as
Backfill in Uranium Mines", July 1977.

13.   McGill University. "A Study on the Effect of a Number of
             Variables on Three Types of Dust Samplers", July 1977.

14.   Dames and Moore. "A Study of Ventilation Strategies for
             Uranium Mines", July 1977.

15.   Powers-Conspec Ltd. "Continuous Monitoring of the Underground
             Mine Environment".   (1st and 2nd year).

16.   Powers-Conspec Ltd. "Development of Sensors for Continuous
             Monitoring of Dust and Radon Daughters", July 1977.

CONCLUDING STATEMENT

The Joint Panel provides a unique mechanism for bringing together, on a voluntary basis, organizations with widely differing roles in the uranium production industry. This report is in line with its first objective of producing an inventory of research. Forty-six research projects have been identified as of the end of June, 1977. The second and third objectives will be addressed more fully in the coming months and the results included in the annual report.

The project descriptions have been kept brief. If additional information is required, the representative of the appropriate organization should be contacted. The Joint Panel as a whole has no authority over the research conducted by its members. A complete set of final reports will be maintained by the Secretariat and will be available on open file at its offices. If the reports are not otherwise available, arrangements can be made to borrow or in some cases to obtain copies from the Secretariat.

Session 3

MANAGEMENT AND STABILISATION

Chairman — Président
Dr. J.R. COADY
Canada

Séance 3

GESTION ET STABILISATION

WASTE MANAGEMENT AT DENISON MINES

E. LaRocque
R. Webber
Denison Mines Limited
Elliot Lake, Ontario, Canada

ABSTRACT

Milling operations began at Denison in May, 1957. Until
1960 tailings were discharged to local lakes and depressions
immediately south and west of the hydrometallurgical plant.
Since 1960, the tailings flow has been directed south into the
main containment area, (Long Lake).

To date some 20 million short tons of solids have been
deposited in the Long Lake basin. Further commitments for
the delivery of uranium concentrates to the year 2010 will
necessitate the improvement of present impervious dykes and
dams to contain some 155 million short tons of solids in this
area.

RESUME

Les opérations de traitement du minerai d'uranium à
Denison Mines débutèrent en Mai 1957. Depuis 1957, et jusqu'en
1960 les déchets solides provenant de l'usine de traitement furent
déversés dans les dépressions et lacs situés dans les environs
immédiats au sud et à l'ouest de l'usine. Depuis 1960, environ
20 millions de tonnes de déchets solides ont été dirigés vers
le sud dans le bassin du lac Long Lake. Les récents contracts
de livraison d'uranium concentré nécessiteront d'importantes
améliorations dans la construction de nouvelles digues pour
permettre l'addition de 155 millions de tonnes de déchets
solides dans ce même bassin, jusqu'à l'an 2010.

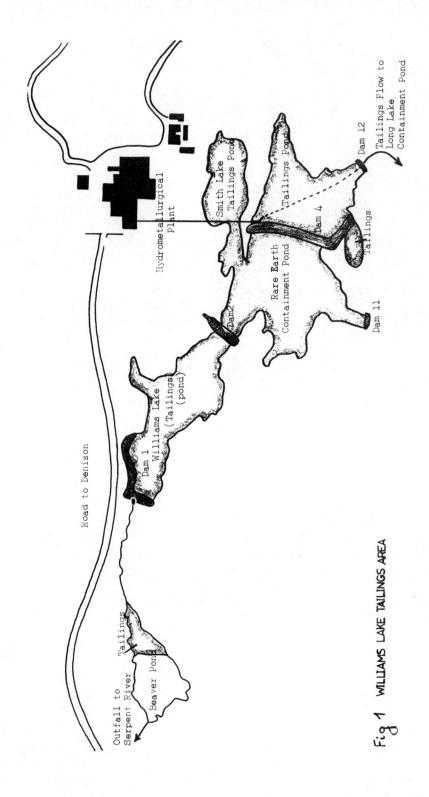

Fig 1 WILLIAMS LAKE TAILINGS AREA

Denison Mines is located in Northern Ontario about 10 miles north of the town of Elliot Lake approximately half way between the cities of Sudbury and Sault Ste. Marie.

The ore body is located in the Huronion sediments on the north limb of the Quirke Lake syncline. The hydrometallurgical plant began operations in May, 1957, and gradually reached its design capacity of 6,000 short tons per day by September, 1957.

In 1974, the capacity of the hydrometallurgical plant was increased to 7,100 tons per day and by 1982, a further expansion will increase the capacity to 12,000 or 15,000 tons per day.

Uranium is leached from the ore at elevated temperatures and high sulphuric acid concentrations. Liquid-solid separation combining acid wash thickening with rotary vacuum drum filtration is followed by clarification of the uranium-bearing pregnant solution, purification and concentration by strong base ion exchange and final precipitation with ammonia to produce an acceptable "yellowcake" for further refining.

The barren solids and ion exchange barren solution are neutralized with lime and pumped to the tailings area.

Williams Lake Tailings Area

From May, 1957, to 1959, the tailings slurry was pumped to an area of local lakes and depressions south of the hydrometallurgical plant (Smith Lake) Figure 1. As this area filled up, a lake to the west of this tailings area (Williams Lake) was used for tailings storage.

A dyke designated as Dam No. 1 was built to contain the solids and a decant structure constructed to allow the supernatant solution to flow through a drainage course west of the dam into the Serpent River.

After abandonment of this area in 1959, heavy spring run-off caused the migration of some 30,000 cubic yards of slimes through the decant structure into a beaver pond area below the dam. For some years, seepage from the dam combined with natural run-off cut a stream through these deposited tailings. The area has since been levelled and revegetated, and a hypalon-lined ditch diverts the fresh water around the tailings spill. Since September, 1975, sodium hydroxide has been added to this stream for pH control and barium chloride added to co-precipitate barium radium sulphate.

Long Lake Tailings Area

In 1959 permission was obtained to use Long Lake for tailings containment. This area has been used as the main Denison tailings area since that time and will be used for many years in the future (Figure 2).

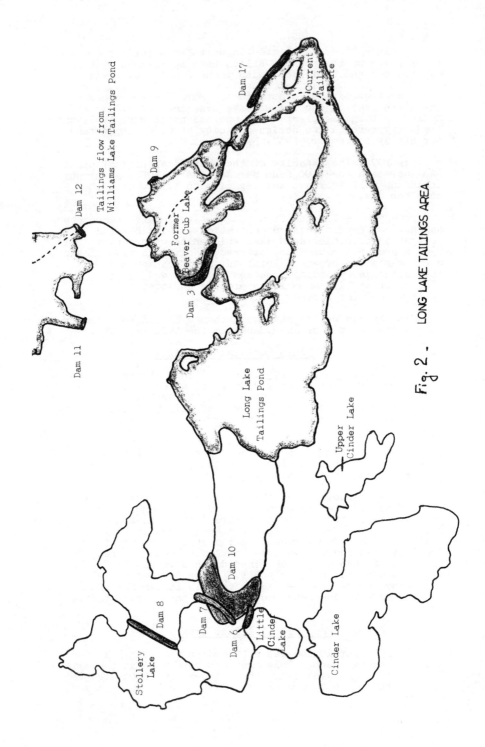

Fig. 2 - LONG LAKE TAILINGS AREA

To divert the free-flowing tailings from Williams Lake, a dam (Dam No.2) was constructed from cycloned tailings. To exercise control over this flow, a number of dykes were constructed. Dam No. 3 was raised to contain tailings in Bearcub Lake and ensure that the tailings flow would enter the east end of Long Lake. Another dam (Dam No. 4) was constructed with cycloned tailings to separate the solid tailings from the neutralized ion exchange barren solution. (Thorium and rare earth containment).

By 1960, substantial quantities of tailings were deposited in the Long Lake basin, and slimes were beginning to migrate toward the outlet of Long Lake located at the west end of the basin. The flow from the fresh water lake known as Cinder Lake was diverted by the construction of Dam No. 6. In 1966, the north side of the outlet of Long Lake was raised by Dam No. 7, isolating the tailings settling pond. However, by about 1970, the slimes in the west end of Long Lake had risen to the extent that the construction of a permanent dam, approximately 100' (30 metres) high was required to replace Dam Nos. 6 and 7.

Geotechnical engineers were commissioned in 1971 to undertake the design of what was designated Dam No. 10, Figure 3. At first a starter dyke was initiated using end dumped sand and gravel, and tailings (slimes) behind the starter dyke were removed, and later controlled by suction dredging operations. Further a 4' diameter (1.2 m) prestressed horizontal concrete pipe was placed on bedrock at the north abutment to act as part of the proposed decant system. This construction included a mass concrete thrust block/decant tower base located at the upstream end of the pipe. Sections of this pipe will be installed in stages as the tailings level rises and the dam is raised. An anti-vortex entry was installed on the top of the tower.

Seepage through Dam No. 10 is controlled in two ways:

(1)   The magnitude of flow and seepage gradients through the dam are minimized by the installation of a flexible plastic membrane. This membrane (hypalon) was installed in 1971 and consists of a monolithic sheet of flexible plastic which extends across the upstream shell of the dam.

(2)   The direction of the flow is controlled in that seepage is positively collected and channelled toward a location where it can be chemically treated; a properly bedded perforated pipe drain has been provided. To ensure that all seepage passing beneath the dam is collected, an interceptor zone, consisting of select drain material extends down from the collector pipe to the bedrock surface. The alignment of the toe collector system has been chosen to follow the optimum contours of bedrock surface underlying the dam. To prevent escape of seepage toward the fresh water lake (Cinder Lake), the outlet to that lake was dammed, raising the lake to a level which is about 2' above the highest portion of the collector pipe in that sector. The massive bedrock outcrop located downstream of the central portion of the dam acts as a barrier in this zone. Finally, some small impervious dykes have been built to ensure that all seepage is channelled toward the chemical treatment plant.

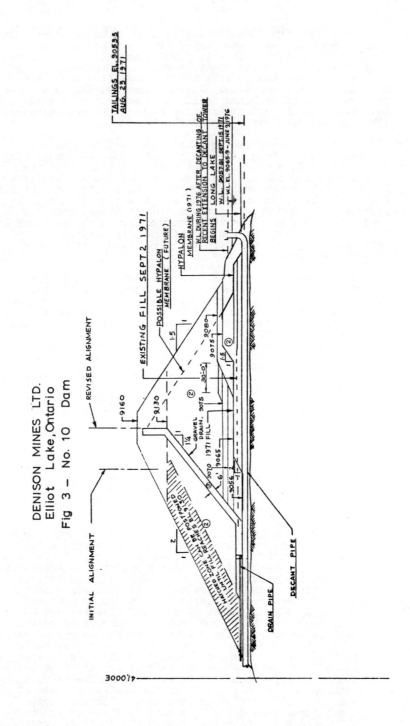

DENISON MINES LTD.
Elliot Lake, Ontario
Fig 3 — No. 10 Dam

Since 1971, (elevation 9065) raising of Dam No. 10 has continued on a staged basis, and raising of the body of the dam has been carried out and monitored by Denison Mines Limited. Compaction control and gradation analyses continued during construction periods (June to September). The initial schedule of tailings impoundment behind Dam No. 10 called for raising the headpond annually in 5' increments for the period of 1971 to 1981. However, the build-up of tailings behind Dam No. 10 has been much slower than originally envisaged, and consequently heightening of the decant tower was not necessary until May, 1976. Accordingly, in June and July, 1976, with the addition of the first tower segment, the headpond level was increased to the approximate elevation of 9070'.

The Long Lake valley is characterized by a relatively continuous rim of surrounding bedrock to the elevation of 9160'. To provide continuous containment around the periphery of the basin, several other dams of relatively modest size are required.

In 1972, subsurface investigations were made into designs for Dam Nos. 9, 16, and 17 which will close three saddles in the rim of the basin. The stage construction of these dams is underway.

Dam No. 17, the largest of the three, is situated in the south-east corner of the Long Lake basin. The predominant feature along the proposed dam alignment is a hill which divides the site into a northern and southern valley. The floor and the south wall of the northern valley are directly underlain by sound bedrock. Due to the relatively impervious nature of the dam foundations, the glacial till dam cores, and the tailings themselves, the seepage through the dam and their foundations should be negligible. Due to the topographic conditions downstream of Dam Nos. 9 and 17, any seepage which does occur can readily be collected.

Chemical Treatment

The supernatant liquid that is decanted from the west end of Long Lake is treated with a 20 % barium chloride solution at a rate of 22 mg. per litre to form a barium radium sulphate. The effluent flow varies from 2,000 Imperial gallons per minute during dry periods and winter to 15,000 Imperial gallons per minute during high run-off periods. The resulting precipitates are contained in a settling pond known as Stollery Lake which is located immediately downstream of the treatment facility.

This lake occupies approximately 50 acres (20 hectares) and contains about 95,000,000 gallons. The estimated retention time within Stollery Lake is 40 days. The flow from Stollery Lake passes through a culvert into the Serpent River. Daily monitoring of the outflow by Denison personnel is collected and a monthly effluent analysis report submitted to the federal and provincial governing agencies.

Since 1970, Denison has been meeting the Ministry of the Environment guidelines for dissolved radium 226 (Table 11).

TABLE 1   Denison Mines Limited
Elliot Lake Uranium Tailings

Constituents	% or Otherwise
Total S	2.37
Sulphate S	1.52
Sulphide S (by difference)	0.85
Total Fe	2.15
U_3O_8	0.003
Th	0.007
Rare Earths	0.05
Ti	0.15
Cu	0.01
P	0.02
Au	0.006 oz/ton
Ra226	290 pCi/gram

TABLE 11   Denison Mines Limited

Effluent Characteristics From the Long Lake Basin 1976

(Typical Analyses)

	Mill Effluents	Long Lake Effluents	Stollery Lake Effluents
Flow	140 1/s (1850 IGPM)	230 1/s (3050 IGPM)	230 1/s (3050 IGPM)
Dissolved Ra226	4500 pCi/1	350 - 1000 pCi/1	1.5 pCi/1
pH	9.5	8.0 - 9.5	8.0 - 9.5
Fe	-	0.5 mg/1	0.05 to 0.25 mg/1
TDS	4500 mg/1	3000 mg/1	2000 - 3000 mg/1
SO_4	2200 mg/1	1500 mg/1	1200 - 2000 mg/1
NH_3	130 mg/1 -N	50 mg/1 -N	35 - 70 mg/1 -N
NO_3	140 mg/1 -N	90 mg/1 -N	90
Trace Metals (Cu,Zn,Ni,Pb)	0.1 mg/1	0.1 mg/1	0.1 mg/1

Mine Water Treatment

Until 1960, mine water was pumped to surface and dis-charged into Quirke Lake. The pH of the mine water was about 2.8 and the pH in Quirke Lake dropped from 6.8 to 4.5. Several other mines situated around the lake disposed of mine water in a similar manner.

Between 1960 and 1962, efforts were made to neutralize the mine water underground with both dry lime and slaked lime. Gypsum build-up in the pipelines and handling problems made lime neutralization unattractive and since 1962, ammonia has been used to neutralize the mine water.

Ammonia is made up as a 3 % solution at the mill site and pumped to a storage tank underground. The pH of the water is automatically controlled and the solution is pumped to surface and used as primary solution in the grinding plant.

By December, 1979, stainless steel pumps and pipe lines will be installed underground. The mine water will be pumped to surface, neutralized with slaked lime and used as primary grinding solution.

At present, with the expansion partially completed, tonnage is averaging 7,100 tons/day (dry basis). The tailings slurry discharging at pH 9.0 is carried almost a full mile by pipeline and discharged at the east end of Long Lake. The present elevation of the tailings in Long Lake is 9075'; to date this represents 20 million tons of solids. The tailings Dam Nos. 9, 10 and 17 will be raised in stages as required to an elevation of 9160'. At this elevation an estimated 70 million tons of solids, (covering 640 acres) can be stored.

With the planned increase in tonnage, a 3,500 ton per day backfill plant will be constructed to aid in pillar recovery underground.

Future commitments for uranium concentrate to the year 2010 have initiated investigations into the possibility of increasing the tailings area to 9190' elevation*, an estimated extra 35 million tons of tailings. Assuming 9190' elevation will complete the tailings area, a cap will be installed over the entire area to a peak elevation of 9300'. This will close out the Long Lake containment area with an estimated total of 155 million tons of tailings solids.

* elevations are local datum only.

DISCUSSION

R.E. BOHM, United States

     Do you have any instrumentation in your dam, such as presometers ?

J.E. LaROCQUE, Canada

     Yes, we do. I could not get them on the slides but, we have them at regular intervals all along the surface of the dam.

R.E. BOHM, United States

     In the last phrase of your text, there you mentioned that 9190 would be your ultimate dam height and you would have a fill to 9300.

J.E. LaROCQUE, Canada

     A cap at 9300 and, again, if that is not acceptable, you will have to leave the dam itself to another 50 ft. by itself.

R.E. BOHM, United States

     But you are putting a 110 ft. cap on the top of your dam ?

J.E. LaROCQUE, Canada

     Well, that is what our geotechnical consultants say we should do. Yes, it will be quite a monument.

W.E. KISIELESKI, United States

     Have you made any measurements on the plant material or forage sometime after revegetation ?

J.E. LaROCQUE, Canada

     No, we have not. We have not done any in our material as yet.

D.C. McLEAN, United States

     Do not you have any potential for backfilling the mines with some of these tailings so that they all do not have to wind up on the surface ?

J.E. LaROCQUE, Canada

     We will be doing that at 35 tons a day in the future. You cannot put it all in the ground. You can only put half of it. It swells that much. You only backfill with half of your tons, actually.

D.C. McLEAN, United States

One more question on your backfill. What kind of a cut are you making there ?

J.E. LaROCQUE, Canada

A straight 50 % cut.

R.E. BOHM, United States

What particle size cut off are you making ?

J.E. LaROCQUE, Canada

Well, our regular material is about 45-50 % minus 200 mesh and 25 % plus 65 mesh and we are doing a straight 50 % and I think we will still end up with about 20 % minus 200 mesh in backfill material, actually. The size will grade up pretty well. People say that the material is fine and we have been putting it back in the tailings. Actually, it is still much coarser tailings than most mills in the world.

RETENTION OF RADIOACTIVE WASTES AT

AN OPERATING URANIUM MILL SITE

R.B. Dodds
Morton, Dodds & Partners Limited
Toronto, Ontario, Canada

E.I. Jurgens and C.A. Freitag
Kilborn Limited
Toronto, Ontario, Canada

ABSTRACT

A uranium mine in Eastern Ontario, Canada has recently reopened in a
predominantly tourist recreation area sensitive to hazardous waste
emissions. The tailings and waste water from the mine are ponded in
a bedrock basin filled with deep glacial outwash sands and gravels
(alluvium) at one end. The ponded tailings water is treated with
barium chloride, passed through a concrete settling basin and discharged into the alluvium. However, some tailings water and precipitation seeps through the tailings into the underlying deep alluvium
carrying dissolved Ra 226 into a nearby lake. This seepage is being
controlled by a grout curtain constructed of a slurry of clay,
bentonite and cement injected into the alluvium to the bedrock
surface.

RETENTION DE REBUT RADIOACTIF

DANS UN MOULIN D'URANIUM EN OPERATION

RESUME

Une mine d'uranium, dans l'est de l'Ontario, Canada, a dernièrement
recommencé sa production dans une région touristique. Les résidus et
les eaux rejetées de la mine sont retenus dans des alluvions (bassin
de roche de base, rempli de sable et de gravier provenant des gla-
ciers). Les eaux retenues dans un étang subissent un traitement au
chlorure de barium. Ensuite, elles s'acheminent vers un bassin de
décantation, construit en béton, pour finalement se déverser dans
les alluvions. Par contre, une quantité minime d'eau résiduaire et
de chlorure de barium s'infiltre à travers les résidus et traverse
les alluvions, entraînant dans son parcours une solution contenant
du Ra-226 vers les lacs avoisinants. Cette fuite est contrôlée par
un mur scellé, construit d'argile, de "bentonite" et de béton injec-
té dans les alluvions, jusqu'à la surface de la base rocheuse.

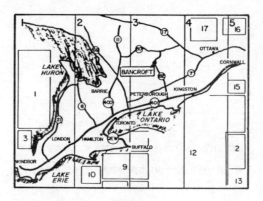

FIGURE 1

KEY PLAN

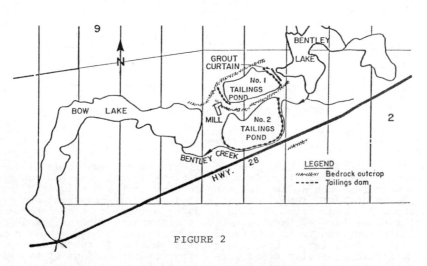

FIGURE 2

GENERAL LAYOUT OF THE TAILINGS DISPOSAL SITE

INTRODUCTION

This paper describes the tailings and waste water management of a Uranium Mine located near Bancroft, Ontario between Algonquin Park and Lake Ontario as shown on the Key Plan in Figure 1. The mine began operation in April, 1957 as Faraday Mines Limited and ceased operations in June, 1964. The mine was re-opened in the summer of 1976 as Madawaska Mines Limited.

The mine site is located in a predominantly tourist recreation area and is therefore extremely sensitive to hazardous waste emissions into the environment.

RADIOACTIVE WASTES

Wastes from ore processing in the present state-of-the-art of uranium mining include wash water, sands and waste solids and various process liquors. Customarily these liquid wastes are stored in tailings ponds or lagoons where the volume of waste is reduced by evaporation and seepage of the liquid portions.

Only uranium is recovered from the ore and the processed waste includes all of the radioactive daughters in varying amounts. Radium, at the moment, has not sufficient value to warrant recovery and more than 98% of the solid radium in the ore is discharged with the solid wastes.

Of particular concern is the amount of dissolved Ra226 that may seep into nearby natural bodies of water that are used for recreational purposes or contain biota. Currently the concentration of dissolved Ra226 that may be allowed to seep into the environment is controlled by the Provincial and Federal regulations.

The limits set by the Federal Government for dissolved Radium 226 are summarized below as picocuries per litre (pCi/l).

Objectives

Max. acceptable monthly arithmetic mean conc.	Max. acceptable conc. in a composite sample	Max. acceptable conc. in a grab sample
10.0 pCi/l	20.0 pCi/l	30.0 pCi/l

Frequency of Sampling

At least weekly if conc. is equal to or greater than	At least every two weeks if conc. is equal to or greater than	At least monthly if conc. is equal to or greater than
10.0 pCi/l	5.0 pCi/l	2.5 pCi/l

GEOLOGY OF SITE

The general layout of the tailings disposal site is shown in Figure 2. The surficial geology of the site principally consists of Precambrian rock knobby hills with a light soil cover and hardwood tree growth. The bedrock depressions have been filled with late glacial or post-glacial fluvial deposits sands and gravels to depths in excess of 31 meters (100 feet).

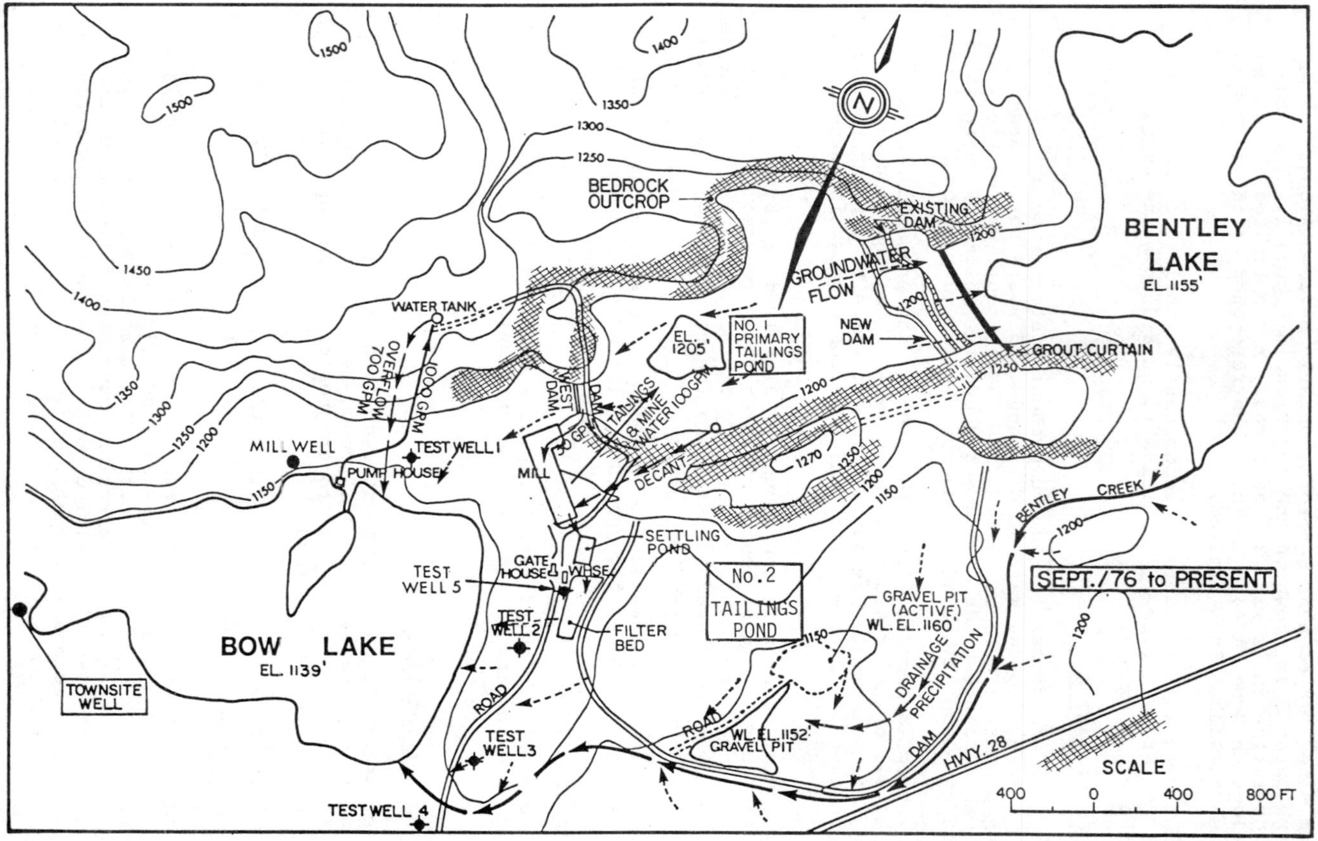

FIGURE 3

DETAILS OF TAILINGS DISPOSAL SYSTEM

Pressure tests carried out in shallow bedrock core holes on various parts of the site indicate that the bedrock is relatively tight. This is confirmed by the fact that approximately 45 km (26 miles) of subsurface mine works produce only 408 litres/min (90 imperial gallons per minute (IGPM)) of groundwater flow.

MINE OPERATIONS

The mine operation uses an acid leach process. Upon start-up in April, 1957 the mine waste water and tailings were initially deposited in a disposal area formed by dams made of mine wastes on three sides and rock outcrop on the fourth side, enclosing a depression filled with water and muskeg with a sand island in the centre. This area is presently referred to as Tailings Pond No. 2 as shown in Figure 3. The normal drainage from Bentley Lake to Bow Lake went through Bentley Creek which passes through this depression. The Creek was rerouted outside the tailings dam adjacent to Highway 28 and is presently in this location. The use of the area for tailings disposal was terminated in June 1962 with a maximum depth of tailings in the order of 12.2 metres (40 feet).

Thereafter the tailings were deposited in a second natural bedrock depression to the immediate north which is filled with a deep alluvium on the north-east side adjacent to Bentley Lake. This area is referred to as Tailings Pond No. 1. It was used until the closing of the mine in 1964 with the approximate depth of 6.1 meters (20 feet) of tailings.

Mine water was pumped directly into Tailings Pond No. 2 at a rate of 410 litres/min (90 IGPM) from 1967 until re-opening of the mine in 1976.

The mill began operating again in September 1976 and the current system of handling of the tailings and waste water is shown on a simplified flow diagram on Figure 3. The overflow from the tailings pond flows by gravity to the mill. A portion of this is diverted to the mill circuit, presently at approximately 230 litres/min (50 IGPM). The balance is treated with barium chloride at a rate of 50 milligrams per litre then flows into a concrete settling pond with about three days retention capacity. The overflow at an average rate of 950 litres/min (210 IGPM) from the settling pond flows into a filter bed consisting of a pit in the natural sand and gravel overburden approximately 2.7 meters (9 feet) deep with a surface area of 2540 square meters (27,300 square feet). A section through the tailings pond is shown on Figure 4. Seepage from Tailings Pond No. 1 is controlled at the west end by a concrete cut-off wall.

Seepage entrapped behind the wall is in the order of 137 to 227 litres per minute (30 to 50 IGPM) and is pumped into the mill for re-circulation. Seepage beneath the east end of Tailings Pond No. 1 takes place through the deep alluvial deposits towards Bentley Lake. The rate of seepage has been calculated to be in the order of 160 litres per minute (35 IGPM).

Although the bedrock was known to be shallower beneath the existing tailings dam at Bentley Lake, construction of a cut-off in this location would raise the groundwater levels beneath the toe of the new tailings dam and thereby reduce the stability. A suitable cross-section of dam to withstand the higher water levels would require much flatter slopes with an increase in cost of construction and a subsequent loss of storage capacity.

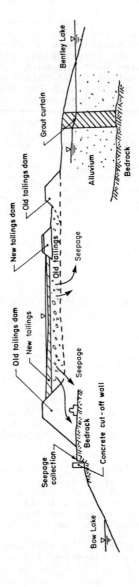

FIGURE 4

SECTION THROUGH TAILINGS POND NO. 1

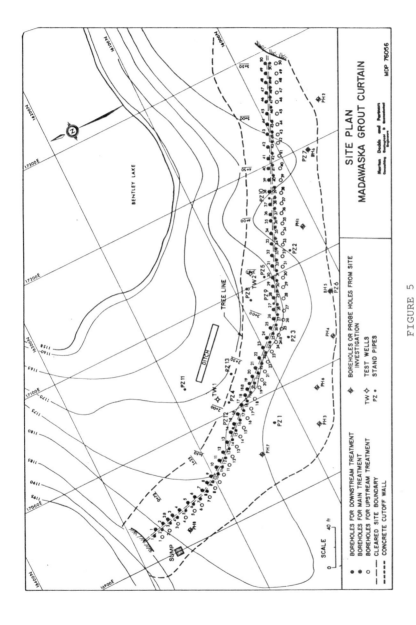

FIGURE 5

SITE PLAN OF GROUT CURTAIN

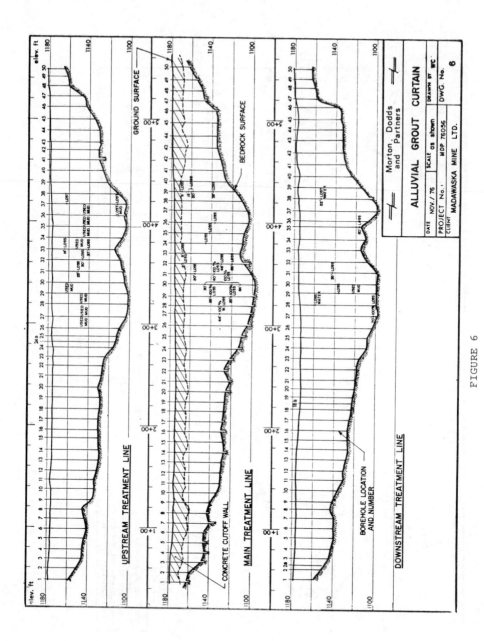

FIGURE 6

SUBSURFACE PROFILES ALONG GROUT CURTAIN

Essentially the construction process for the grout curtain consisted of installing slotted pipes through the alluvium down to and into the bedrock surface at a pre-determined spacing along three rows along the alignment of the curtain. A grout consisting of clay, water, cement, bentonite and occasionally calcium chloride was injected into the alluvium through the slotted pipes in the upstream and downstream row of holes. The layout of the slotted pipe locations and the subsurface profile determined from the installation are shown on Figures 5 and 6, respectively.

The centre line of holes were left untreated in order that access could be gained to the curtain at any future date to inject more grout or even a chemical grout if further cut-off of seepage is required. In addition, several test wells and a number of observation wells or piezometers were installed at the locations shown on Figure 5. These were used in pump draw-down tests conducted before and after construction of the curtain and are also used to obtain samples of the groundwater for analysis of dissolved Ra226. Several of the observation wells were constructed in sampling the grout curtain after completion.

The pump draw-down test conducted after completion of the grout curtain show that the alluvium has been successfully grouted and that the major part of the seepage through the upper part of the alluvium has been retarded. However, the results show that a thin zone adjacent to the bedrock surface still allows seepage of groundwater.

RADIUM 226 CONCENTRATION

During the period from 1966 to 1976 when the mine was closed, samples of the water were taken by Ontario Environment officials and tested for levels of concentration of Ra226. It is interesting to observe that levels of dissolved Ra226 in Bentley Creek near the gravel pit continued to reduce to a level of approximately 5 pCi/l despite the fact that from 1967 onward mine water was being pumped into this tailings area (Tailings Pond No. 2) at a rate of 410 litres/min (90 IGPM) with an estimated average dissolved Ra226 content of 35 pCi/l. This decline in Ra226 content is shown graphically on Figure 7.

Similar observations were recorded for Bow Lake where it crosses Highway 28.

In the current mine operation the overflow from the concrete settling basin at an average rate of 950 litres/min (210 IGPM) contains dissolved Ra226 varying from 12 pCi/l in the winter to 2 pCi/l in the summer.

Test wells constructed between the mill and nearby Bow Lake allow sampling of the groundwater for analysis of dissolved Ra226. The locations of the wells are shown on Figure 4. The results to date indicate that very little or no dissolved Ra226 is entering the groundwater or that the natural sands and gravels are filtering out most of the dissolved Ra226 entering from the mining operation.

The seepage through the west end of the tailings pond at a rate of 136 to 227 litres/min (30 to 50 IGPM) is trapped by a concrete cut-off wall and returned to the mill circuit (see Figure 4). The content of dissolved Ra226 has been measured at 50 to 70 pCi/l. The tailings effluent entering the pond contains in the order of 2000 pCi/l of Ra226. Limited dilution and filtering action of the old tailings and sand and gravel in the tailings dam therefore reduces the dissolved Ra226 content by a factor of 30 to 40.

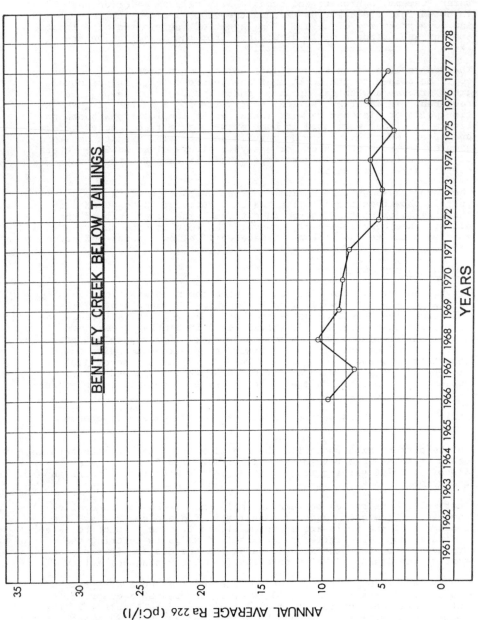

FIGURE 7 – RADIUM 226 CONCENTRATION IN BENTLEY CREEK BELOW TAILINGS

Calculations indicate that seepage takes place through the alluvium beneath the east tailings dam toward Bentley Lake at a rate of 160 litres/min (35 IGPM). This seepage consists of tailings effluent and precipitation infiltration through the new and old tailings and into the underlying natural sands and gravels. Although the grout curtain has not totally cut-off the seepage in this direction, it has apparently reduced the Ra226 content in the groundwater. The results of tests in water samples taken from observation wells upstream and downstream of the curtain and Bentley Lake are tabulated in Table I. The locations of the observation wells are shown on Figure 5. The measured levels of Ra226 in Bentley Lake over the period of time from May 1976 to the present are shown on Figure 8. In addition to monitoring the concentrations of dissolved Ra226 in the groundwater, samples of the clay grout were taken from upstream and downstream of the curtain and tested for total and dissolved Ra226. The results of these tests are shown on Table II.

With respect to the grout curtain, test results to date allow the following observations:

1. Although the grout curtain does not cut-off all the groundwater seepage from Tailings Pond No. 1 to Bentley Lake, it appears to be instrumental in reducing the level of dissolved Ra226 content to acceptable levels.

2. The reduction in dissolved Ra226 content can likely be attributed to the following:

   a) the grout curtain has increased the flow path thereby increasing the filtering action of the natural sands and gravels.

   b) the materials in the grout curtain react chemically or cause a chemical reaction with the Ra226. The test results on Table II appear to confirm this. However, a more detailed testing program is necessary.

   It is known that Ra226 is a cation and would be absorbed in an ion exchange process when in contact with clay. The natural clay used in the grout is likely a montmorillonite [1] characteristic of soils in less humid climates where soil solutions are slightly alkaline. Montmorillonites also have a high exchange capacity [2] which would encourage the absorption process.

   The inclusion of cement in the grout may have raised the pH around the clay crystals which prevents hydrogen ions from taking up positions on the ion exchange allowing more Radium 226 cations to be absorbed. Radium 226 has a relatively small hydrated size and is highly preferred on exchange sites on the clay.

   Since the process involves mass exchange and the volumes of grout are in tons versus grams for the dissolved Radium 226, it would appear that the absorption of the Radium 226 could continue for many thousands of years.

The current method of handling radioactive wastes by the Madawaska Mine has been extremely effective in reducing the amount of dissolved Ra226 to within current Federal guidelines. The levels of concentration of Ra226 in nearby water bodies have continued to decline despite the re-activation of the mine. The levels around this active mine site can be compared to those measured by the Ontario Ministry of Environment near inactive mines in the same area on June 12, 1976.

TABLE I

CONCENTRATION OF RADIUM 226 (DISSOLVED) UPSTREAM AND DOWNSTREAM OF GROUT CURTAIN (pCi/l)

DATE SAMPLED	Upstream			Downstream				REMARKS
	PZ3	PZ7	Sump	PZ4	PZ5	TW1	TW2	
May 26, 1977	3	4		2	1			MOE Report RPL 77-371 9/8/77
June 29, 1977	8	2		3	<1			MOE Report RPL 77-493 21/9/77
Aug. 8, 1977	3	3		3	2			Madawaska Mine
Sept. 1, 1977	-	-				1	1	Madawaska Mine
Sept. 19, 1977	5	14		1	2			Madawaska Mine
Oct. 24, 1977	29	38		5	1			Madawaska Mine
Nov. 29, 1977	40	38	2			1	7	Madawaska Mine
Dec. 13, 1977	6	7	13			<1	2	Madawaska Mine

TABLE II

TOTAL RADIUM 226 IN SOIL SAMPLES

B.H. #	SAMPLE #	DEPTH (ft)	DATE OF SAMPLING	DATE OF TEST	TOTAL Ra-226 (picog/g)	REMARKS
PZ2	4	25.0-26.5	14/7/77	28/10/77	7 ± 6	±6 - Standard Deviation
PZ2	5	30.1-31.5	14/7/77	28/10/77	8 ± 6	±6 - Standard Deviation
PZ9	14	40.0-41.5	27/7/77	28/10/77	5 ± 4	±4 - Standard Deviation
PZ9	23	64.5-66.0	27/7/77	28/10/77	6 ± 4	±4 - Standard Deviation

NOTES:

1. Tests were performed by Atomic Energy of Canada Limited

2. For each test, a sample weight of about 15 to 20 gm was sealed in a plastic capsule and stored for approximately 30 days to allow sample to reach an equilibrium state, then sample was subject to testing by a mass spectrometer.

3. According to A.E.C.L., the Radium 226 in these samples is about the lowest level the A.E.C.L. has tested. This accounts for the relatively large standard deviations in the measurements.

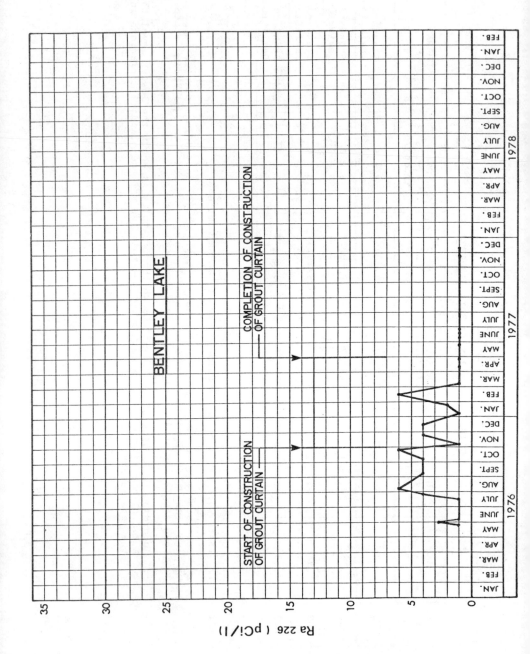

BENTLEY LAKE

Ra 226 (pCi/l)

Deere Creek, 16 miles west of Bancroft   7 pCi/l
Comrie Lake, 4 miles west of Cardif   14 pCi/l
Farrell Creek downstream of Dyno Tailings   15 pCi/l

The data available at this time would indicate the following factors contribute to this reduction in levels of dissolved Ra226 at the Madawaska Mine:

a)   treatment of the tailings pond and overflow with Barium Chloride before emission into the environment.

b)   filtering action of the natural sands and gravel in the area

c)   chemical reaction of the clay grout with dissolved Ra226

ACKNOWLEDGEMENTS

The authors are grateful to Madawaska Mines Limited for permission to publish this paper and for data and assistance provided in compiling the paper.

BIBLIOGRAPHY

1.   Krauskopf, K.B.:   "Introduction to Geochemistry", McGraw Hill, 1967.

2.   Lambe and Whitman: "Soil Mechanics", John Wiley & Sons Inc. 1969.   Pages 42 - 49.

DISCUSSION

J. MONTGOMERY, United States

Do you have any estimate of the costs to construct this ground curtain ; total cost ?

R.B. DODDS, Canada

According to the mine, too much. According to control agencies, maybe not enough. I do not have the figures, but I do know the amount of manpower that was put into it and the length of time and my own estimate would be about $ 750,000.00. The construction of an I-type of wall for a complete cut-off, our estimate at that time was about two million dollars for construction. It is a very difficult site.

W.E. KISIELESKI, United States

Have you made any radium measurements of the lake sediments in Bentley Lake to support the fact that the concentrations in the freewaters are that low, but you may have some low to high sedimentation concentration activity ?

R.B. DODDS, Canada

Yes, measurements have been taken downstream. Unfortunately, this mine was in operation for several years before any background data was starting to be picked up. There was not any treatment of the tailing water, but with any measurements now you have to temper with the fact that it could have been deposited a long time ago in a previous operation. As a point of interest, the government agencies have taken measurements in nearby streams of old tailings and also in old areas nowhere near mining operations and the levels have varied from 7-14 millicuries per liter.

W.E. KISIELESKI, United States

My concern was that this might be some physical or chemical action that would release this absorbed material rather than that which is actually being discharged into the free waters.

R.B. DODDS, Canada

That is possible. I think it would take a lot more monitoring, a lot more analysis, to determine if that is the case. I would not think so at this stage because the mine only started operation in 1976, and a grow curtain has only been in existence for almost a year, and the levels in the area are dopping despite the fact that the mine is back in operation. They have dropped from what they were in the previous operation.

A.B. GUREGHIAN, United States

Could you give us an idea about your talking about observation in your well ; at what level are you collecting your samples ?

R.B. DODDS, Canada

        The wells are located at different levels in the ground water. There should be more, we know that. Some of the picometers of the wells are located just into the ground water ; some are right down at the bedrock itself.

A.B. GUREGHIAN, United States

        Do you notice any change in concentration ?

R.B. DODDS, Canada

        No. In all the wells that we have been monitoring, we have been finding that the levels have been one or less picocurie per liter, so there has been no distinction showing up yet. That is around the gravel filter bed. On the grow curtain, upstream and downstream of the grow curtain, yes, there quite likely is. But, I have not enough data yet to show whether we are getting more down or up above.

THE LEACHING OF RADIUM FROM
BEAVERLODGE TAILINGS

D.R. Wiles
Chemistry Department, Carleton University
Ottawa, Canada, K1S 5B6

Abstract

A chemical study reveals that the leaching of Radium is greatly accelerated by the presence of dissolved salts of such ions as Na^+, NH_4^+, Ca^{++} and Ba^{++}. The reaction appears to involve an equilibrium adsorption of the cation on the tailings particles. This is followed by diffusion of the cation into the particle and a simultaneous diffusion of Radium out to the surface. Radium is finally desorbed into the solution. Reduction of the surface area by forming fused agglomerates has given a decrease of several orders of magnitude in the leaching rate.

Résumé

Une étude chimique a demontré que la vitesse de dissolution du radium dans des résidus miniers est accélérée par la présence de cations tels que Na^+, NH_4^+, Ca^{++} et Ba^{++} dans la solution. La réaction semble impliquer l'adsorption du cation sur la surface des particules de résidus miniers. Cette adsorption est suivie par une diffusion du cation dans le cristal et une diffusion du radium vers la surface. Enfin le radium se désorbe dans la solution. En réduissant la surface efficace, par fusion des agglomerats, il est possible de diminuer la dissolution du radium de plusieurs ordres de grandeur.

Introduction

It is well known that the radioactive residues left from extraction of Uranium from its ores are becoming a cause of environmental concern. This concern is not the same as that felt over nuclear reactor wastes, where the intense radiation gives an obvious hazard. With mine tailings, however, the hazard is mostly from the slow release of the radioactive components, through natural weathering to surrounding air and particularly to water. The danger then lies in the possible ingestion of the Radium released. The problem facing us is to learn how to prevent release to the environment of the radioactive elements - primarily Radium, but also Thorium, Lead and Radon. I shall be concerned here with Radium.

From the standpoint of an applied chemist, the problem of Radium in Uranium mine tailings devolves upon three questions:

1. What is the chemical form in which the Radium is bound in the tailings? The most obvious would be precipitated $RaSO_4$, but more likely would be Radium either adsorbed on some preformed silicate surface or incorporated into the structure of one or another compound formed in the mill circuit, such as an iron hydroxysulphate or the like.

2. What is the chemical reaction mechanism by which Radium is released from its position in the mine tailings? Is it by simple erosion? or by desorption from a surface? or by dissolution of the surface? or by some mechanism more clearly chemical in nature?

3. What can be done to reduce the release of Radium to the environment? To remove it from the tailings requires an answer to the first question, or a favourable answer to the second question. To prevent its initial incorporation into the tailings also involves knowing how it gets there and thus knowing the answer to the first question. To seal it in, or to "fix" it in situ, requires some knowledge of how it gets out - thus having an answer to the second question.

In our laboratory we are looking at all three questions, with little success to date on the first but very encouraging results on the second and third questions. I shall discuss primarily the work on the leaching mechanism.

Table 1. Analysis of a typical batch
of Beaverlodge tailings. (1)

size(μ)	Weight (%)	^{226}Ra (pCi g^{-1})
>210	0.01	170
150-210	3.75	
105-150	12.55	
75-105	15.17	265
40-75	12.70	
20-40	17.68	285
10-20	10.38	
<10	27.76	1655

Experimental Methods

The tailings we are working with are those from the Beaverlodge Mine of Eldorado Nuclear, Ltd. By the time we see the tailings, they have a size distribution as is given in Table I.

The distribution of Radium, shown also in Table I, is disproportionately in the finer particles, as is not uncommonly observed. The overall mineral composition involves mostly silica with some hematite, feldspar and chlorite.

Although we have done, and many others too have done (2), studies of the leaching of Radium by percolation of simulated rain water through a bed of the tailings, this method does not give very good information about the chemical mechanism involved. We have thus used the method of stirring the powder vigorously in an appropriate solution. Better yet would be to stir a solution past a fixed flat surface. However we felt that because of the inhomogeneity of the material and its low radioactivity, this would not be practicable.

Radium determination was done on a filtered aliquot of the reacting slurry by first removing Thorium and then precipitating $Ba(Ra)SO_4$. Measurement with a 2π proportional detector gave a gross α activity which was corrected for growth of Radon and its daughters. The accuracy of this method has been carefully calibrated by comparison with an α-particle spectrometer and commercial calibrated Americium sources.

Experimental Results

Typical results of a leaching experiment are given in Figure 1, which

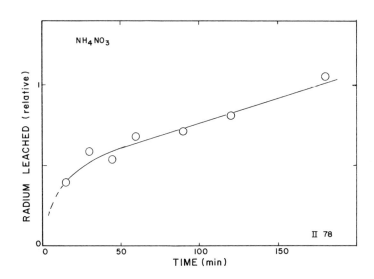

Figure 1. Typical results for leaching of Radium from Beaverlodge tailings by 0.010 M NH_4NO_3.

shows the amount of Radium leached as a function of time. For chemical purposes, the short times used are quite adequate although they may be of less immediate significance for natural leaching. The curves obtained follow no simple mathematical form although, since the particles are not homogeneous in size, a simple rate law is not necessarily expected. It is clear that there is a rapid initial leaching reaction followed by a slower leaching. The form of the slower reaction is shown more clearly in Figure 2, in which the concentration of Radium leached from fired pellets is plotted as a function of the leaching time.

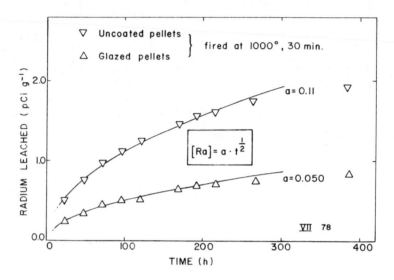

Figure 2. Leaching results from fired pellets of Beaverlodge tailings.

For these samples with small homogeneous surface, an initial fast reaction for which we have not many data is followed by a further reaction which is seen to obey a parabolic law:

$$[Ra] = C_i + C_t (time)^{\frac{1}{2}} \qquad (1)$$

where

C_i is the total concentration of Radium released by the initial fast reaction, and

C_t is a rate constant

Turning our attention to the fast initial reaction, we have made some simplifying assumptions in order to obtain a general picture of the process:

    i) The tailings are homogeneous
   ii) It is valid to make a single measurement (at t = 60 minutes) for comparing dissolution rates.

Typical results for these experiments are given in Figures 3,4 and 5. Figure 3 shows the leaching rate (measured at the end of the first 60 minutes) as a function of the concentration of added sodium,potassium and ammonium chlorides. It is apparent that the dissolution rate is directly proportional to the concentration, and that the three ions give different rates. This shows that

$$C_t = C_o [M] \qquad\qquad (2)$$

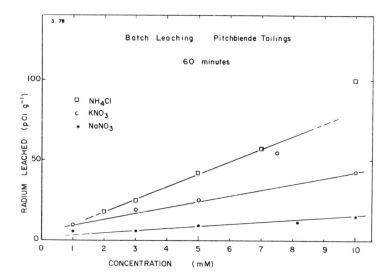

Figure 3. Leaching rates as a function of the concentration of Na^+, K^+ and NH_4^+.

where C_o is a constant and [M] is the concentration of the cationic species. We have found no significant difference between the chlorides and the nitrates.

The practical implication of this for environmental protection is immediately obvious in that ammonia liquors released into tailings ponds will clearly increase the rate of leaching of Radium into the pond effluent.

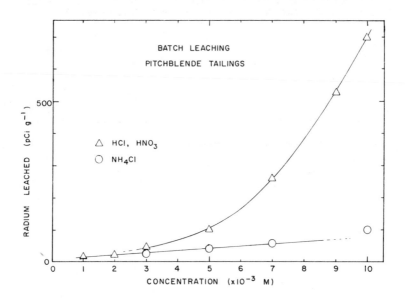

Figure 4. Leaching rates as a function of the concentration of H^+ (as HCl or HNO_3).

Figure 4 shows the same sort of result for pure acid leach solutions. One sees that while initially the reaction may be first order in H^+, at concentrations higher than 3×10^{-3} M (pH $\leq$ 2.5) there is a stronger dependence and a much more rapid reaction. It seems likely that this may be related to bulk dissolution of some critical component of the tailings, for example an Iron oxide.

Figure 5 gives some important results which really elucidate the gross features of the mechanism. The dependence of the leaching on Magnesium and Calcium is again first order. The significance of this is that whenever lime or dolomite is used to neutralize acidity in effluent waters, the point of addition of the lime must be carefully chosen to avoid increasing the liberation of Radium from sediments. The effect of Barium is interesting. Not only does

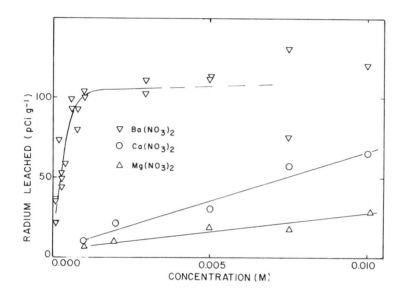

Figure 5. Leaching rates as a function of the concentration of Mg^{++}, Ca^{++} and Ba^{++}.

it show that addition of excess Barium should be avoided, but it shows also that the rate law resembles an adsorption isotherm. In this instance, we can deduce that the overall reaction must occur in three steps, starting with the rapid equilibrium adsorption of Barium (or other cation) on the particle surface. Equations 3,4 and 5 show this sequence, in which the second step is a solid state interchange between a Barium ion on the surface and a Radium ion in the crystal lattice:

$$Ba^{++} \text{ (sol'n)} + \text{surface site} \xrightleftharpoons{K_1} Ba^{++}\text{(surf.)} \qquad (3)$$

$$Ba^{++} \text{ (surf.)} + Ra^{++} \text{ (latt.)} \xrightarrow{k_2} Ra^{++} \text{ (surf.)} + Ba^{++} \text{ (latt.)} \qquad (4)$$

$$Ra^{++} \text{ (surf.)} \xrightarrow{k_3} Ra^{++} \text{ (sol'n)} + \text{surface site} \qquad (5)$$

Mathematically, this gives the equation:

$$\text{Rate} = k\,S.\ \frac{K_1[Ba^{++}]}{1+K_1[Ba^{++}]} \qquad (6)$$

where S is the total number of reactive surface sites and k is likely identical to k_2. At very small Barium concentrations, where $K_1[Ba^{++}] \ll 1$, this becomes

$$\text{Rate} = kS \cdot K_1[Ba^{++}] \qquad (7)$$

and the reaction is apparently first order in Barium. At higher Barium concentrations, $K_1[Ba^{++}]$ becomes larger than 1, so that

$$\text{Rate} = kS \qquad (8)$$

and the rate is no longer responsive to the Barium concentration because the surface is already saturated. This is in good agreement with our experimental results. Presumably, the condition given in Equation 7 describes the other metal ions studied.

The square root function of time (Equation 1, Figure 2) which describes the leaching process approximately suggests that the real rate-controlling step is a solid state diffusion so that k_2 in Equation 4 is related to the diffusion coefficient of Ba^{++} (or other ion) in the lattice containing Radium. This also tells us, incidentally, that the Radium is not simply adsorbed on a surface, but is rather incorporated or enclosed in some crystal lattice.

The results presented here must be regarded with some reservation, of course, inasmuch as the study was done over relatively short times and on rather poorly defined substances. However, the general aspects of the leaching mechanism seem clear and the observed effect of various dissolved cations may have practical importance.

Finally, a brief word about our pelletizing. The simple agglomeration of moist tailings gives pellets, roughly spherical, up to 1 cm in diameter. These can now be dried and fired to give a hard ceramic-like sphere. Alternatively, we can add various materials, such as powdered glass, which will form a protective glaze. The best conditions appear to be those which give a smooth surface and a well sintered mass. Typical results, showing the leaching of Radium from various agglomerates fired at different temperatures, are shown in Figure 6. It is seen that by this simple process the leaching of radium can be decreased by a factor of more than 2000.

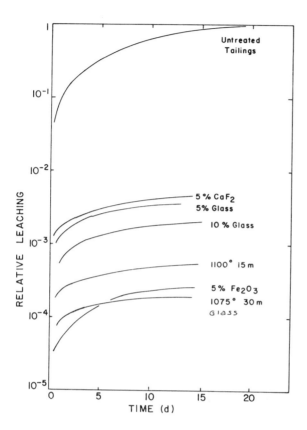

Figure 6.   Radium leaching from sintered pellets.

Acknowledgements

We are grateful for the support given by the National Research Council of Canada and by Eldorado Nuclear Ltd.

References

[1]   Khoja, Z.: M.Sc. Thesis, Carleton University, Ottawa, 1977.

[2]   Shearer, S.D. and Lee, G.F.: Health Physics, 10, 217 (1964).

DISCUSSION

J.D. SHREVE Jr., United States

    I see your curves went out to 20 days. What would your forecast be for 20 years ?

D.R. WILES, Canada

    We have not done the work yet.

J.D. SHREVE Jr., United States

    Do you have the intention to do some long-term studies ?

D.R. WILES, Canada

    Yes. These curves do. Some of these curves I showed you in my second slide which followed the parabolical law, but are sinking below, and if our mechanism is correct, they ought to follow that parabolic law so we could calculate pretty accurately 20 years. If they are falling below, we do not know why. The trouble is that we are finding the change as a function of time to be so small that we really cannot detect it, so we do not know. We have got some under way now which we intend to keep on studying until we get tired.

J.D. SHREVE Jr., United States

    I was picturing ultimate breakdown of the pellets themselves whereby you would revert to the original material. That seems quite possible to me, and I wanted your comment on that.

D.R. WILES, Canada

    It does. We have no idea. We have done compression testing of these and the big problem at the moment seems to be that once we take them out of our furnace, they cool too quickly and form cracks. We are now designing a small sized rotary furnace in which we will drive the pellets through the furnace and out the other side to cool them more quickly. We hope that this leaves a veneer and may give us better results. So, at the moment we cannot tell. The compression testing has been interesting, but not really informative. We are going to be doing abrasion testing and then we hope to pass it over to some people for power plant testing.

D.B. CURTIS, United States

    Would you care to give some opinions about the ramifications of your findings with regard to practices that have been talked about in the previous papers regarding the waste management ?

D.R. WILES, Canada

    It seems to me that with respect to backfilling into the mine, the easiest thing to backfill into the mine is the coarse particles. These are the particles that you are not worried about. The fine contain the radium. Now we can pelletize the fines and make them into nice spheres and they can then be put back into the mines, so even if the radium does continue to leach out, it makes them a whole lot easier to handle. Secondly, with respect to using these for building dams with, they are good structural material, and as long as they are buried adequately, they should be quite good. As to their being useful as building materials out in the wilderness, I do not know. Quite possibly, as long as nobody gets too close to them, I do not know. They will have one disadvantage and that is that being spheres, they will only pack as well as spheres will pack. They will have a higher volume than the stuff that comes out of the mine. I suppose that if you made spheres in a variety of sizes, they could be mixed a little more efficiently.

D.B. CURTIS, United States

What about the chemical work, barium and calcium and so forth in dissolution ?

D.R. WILES, Canada

We have not studied the barium and calcium stuff on these pellets. We have done that only on the bare powdered tailings. We have not got that far yet.

D.B. CURTIS, United States

What about these same studies of materials from different kinds of uranium ore bodies ? that might give you some insight.

D.R. WILES, Canada

Yes, we would like to start doing the same kind of study with some of the other types of tailings. But, at the present time, we have not done that.

J.R. BROWN, Canada

On Figure 5, what is the time that you leached these for ? With the barium, calcium and magnesium ?

D.R. WILES, Canada

Figure 5 was a 60-minute leach of the unpellitized powders.

J.R. BROWN, Canada

If you continued the concentrations for the calcium and magnesium, would you, do you think, reach the plateauing level of the barium ? And would that indicate that the radium is acting on the surface ?

D.R. WILES, Canada

I would like to think so. We have not done that as yet. We did try strontium and this is not very satisfactory experimentally, because strontium sulphate precipitates interferes with our experiment. The best result that we have to date, which I do not quite believe, is that the strontium curve is very, very close to that of calcium, but I do intend that we shall carry this on as far as we can go and, obviously, we want to test our mechanism.

J.R. BROWN, Canada

Just curious that when you say mineralized lattice radium, you are not sure. It could be surface lime ?

D.R. WILES, Canada

The fact that we get a pretty good parabolic behavior in the leaching suggests that it is not just a surface absorption. It does suggest very clearly that it is a matter of diffusion deeper and deeper into the matrix.

J.R. BROWN, Canada

A usual industry question. Cost per ton ?

D.R. WILES, Canada

I do not know. But, I do know that this pelletizing method is being used in iron pelletizing iron ore. There is a big plant in Poland that is doing 2,000 tons a day. It is being done with powdered coal and so the pelletizing itself is, I gather, not more than $ 3 or $ 4 a ton. The fire is up to 1,000 degrees. I gather, again, that that is not more than a couple of dollars a ton. So, compared with $ 750,000 to build a dam, or $ 15/ton to truck things from Port Hope back up to Chalk River, as the Canadian Government is spending, the cost of this method seems not to be exhorbitant.

V.I. LAKSHMANAN, Canada

My question is about exchange of barium and radium. When you did the studies, did you take the acidity out and follow the leaching rate ? Which means partially carried out the exchange between radium and barium, stopped the leach, take the leach out and start all over again. Did you find any agent reaction rate ?

D.R. WILES, Canada

Not any that we have seen. In fact, this is the method that we are using and we get this very nice parabolic behavior.

R.S. DANIELS, United States

I am wondering if you are familiar with the literature as far as high level waste disposals are concerned and the work that was done at Chalk River 20 years ago relative to fixing radioactive components.

D.R. WILES, Canada

I know the work of Bill Merrit and Mosan and people of that sort. I have not studied that literature in particular, however, because the problem is a little bit different. They have got to worry about a great deal of heat. We do not. They have got to worry about burial for maybe a hundred years. We have got to worry about burial for 10,000 years

R.S. DANIELS, United States

No. I was thinking about the problem of actually processing material to begin with where you are going up to 1,000 degrees anyway, and they were making some very tight lattices, very fine glasses and glazes with very, very low leaching rates and it just may be worthwhile exploring this. I know that you have attempted to reduce this to 3 to 4 degrees of magnitude very effectively just by firing the pellets and it may be worth exploring this literature which seems to have been forgotten, not by you, but by many others.

D.R. WILES, Canada

Well, the results of our finding it to be a diffusion mechanism is that we can now maybe make ourselves a very tight lattice so this diffusion cannot occur.

CONTROL OF RADIUM-226 RELEASES IN LIQUID EFFLUENTS

D. Moffett & A. J. Vivyurka
Rio Algom Limited
Elliot Lake
Ontario, Canada

ABSTRACT

Rio Algom Limited has used barium chloride treatment since 1965 at Elliot Lake to remove radium from effluents. The current system, however, is limited for the removal of suspended or particulate radium-226 and the recovery from the settling ponds of the barium/radium coprecipitate. A research program aimed at removing dissolved radium-226 and also at settling the resultant precipitates with flocculents has as its goal a final effluent of 3 pCi/L dissolved and 10 pCi/L total radium-226. Large scale laboratory tests and field investigations have shown the stringent nature of these targets.

CONTROLE DES REJETS DE RADIUM-226 DANS LES EFFLUENTS LIQUIDES

RESUME

La Société Rio Algom utilise un traitement au chlorure de baryum depuis 1965 dans son installation d'Elliot Lake afin d'éliminer le radium des effluents. Cependant, le système actuel est limité en ce qui concerne l'élimination du radium-226 en suspension ou sous forme de particules et la récupération à partir des bassins de sédimentation du coprécipité baryum-radium. Un programme de recherche visant à éliminer le radium-226 dissous, ainsi qu'à assurer la sédimentation des précipités qui en résultent à l'aide de produits de floculation, a pour objet d'obtenir un effluent final de 3 pCi/L de radium-226 dissous et de 10 pCi/L de teneur totale en radium-226. Des essais de laboratoire à grande échelle et des recherches sur le terrain ont montré le caractère impératif de ces objectifs.

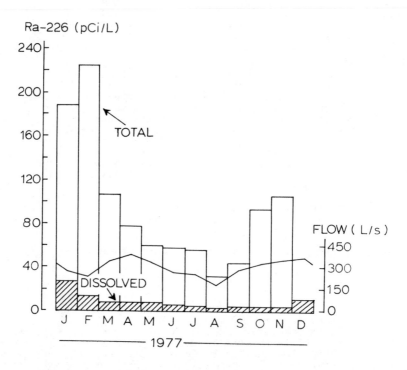

Figure 1:   Radium-226 Activities in Final Effluent
            from Operating Quirke Mine/mill.

1.   INTRODUCTION

The Elliot Lake uranium deposits are located just north of the mid point of the north shore of Lake Huron in Ontario, Canada. The climate is wet-temperate in nature. The yearly mean temperature is $4^{o}C$ and average total precipitation is 966 mm of which 210 mm is due to snowfall. This is Canadian Shield country with bedrock exposure at, or very near, the surface with a mantle of glacial deposits and alluvium being the unconsolidated cover. The ore body is a conglomerate of quartz pebbles set in a sericitic, pyritic matrix. Brannerite, uraninite and monazite are the economic minerals. An analysis of the present mill feed gives approximately 0.1% U_3O_8, 0.016% Re_2O_3, 0.03% ThO_2, 3.0% Fe, 3.5% S with the balance essentially silica.

The general mill process for extraction of uranium is by a sulphuric acid leach with concentration by ion exchange, using nitric acid elution, and precipitation of the final yellowcake concentrate with ammonia. Waste solids and liquids are treated with lime to a pH of 10-11 and pumped as a slurry of about 25% solids to a 100 hectare tailings basin. The clear effluent is essentially a saturated solution of calcium sulphate containing ammonia and nitrate. Radium-226 is the only radioactive constituent present which requires control; uranium, thorium, lead and polonium isotopes are all adequately controlled by the pH adjustment. The wet-temperate climate, however, means that large volumes of effluents must be discharged, and the only means for their disposal is to the surface watercourses.

2.   EFFLUENT TREATMENT

Rio Algom has one operating mine/mill, one currently being rehabilitated, and six idle properties. At the operating Quirke mine/mill, radium-226 control consists of addition of barium chloride to the clear effluent from the tailings basin. At the idle properties, a combined lime/barium chloride treatment is required because the pyrite in the inactive tailings is the source of acidic seepage and runoff.

Barium chloride has been used to remove radium from effluents in Elliot Lake since 1965. Unlined natural basins or lagoons are used to collect the radium-barium sulphate precipitates. Streams from both the operating mine and the idle properties contain an excess of sulphate which facilitates removal of the radium.

Current treatment practice at Quirke is designed only to remove dissolved radium-226. Radium removal is over 99% efficient; tailings effluent is reduced from an average dissolved (filterable) radium-226 activity of more than 800 pCi/L to about 7 pCi/L. It was recognized in 1974, however, that the removal of dissolved radium may result in effluents which contain suspended radium-226 activities which are higher than the dissolved values. The mean monthly final effluent quality for both total and dissolved radium-226 at Quirke through 1977 is given in Figure 1. This shows total radium-226 activities that are much higher than the dissolved values. Also, the control of suspended radium is particularly difficult during the winter months. There is a fair correlation between the total and dissolved radium-226 activities: totals are about 14 times the dissolved value.

Good effluent treatment practice must achieve virtually complete removal of dissolved radium and also ensure that the precipitated radium-barium sulphate settles sufficiently to give a final effluent which is low in suspended, and thus total, radium-226. The initial removal with barium chloride must be at least 99.6% efficient to meet the Provincial government's dissolved

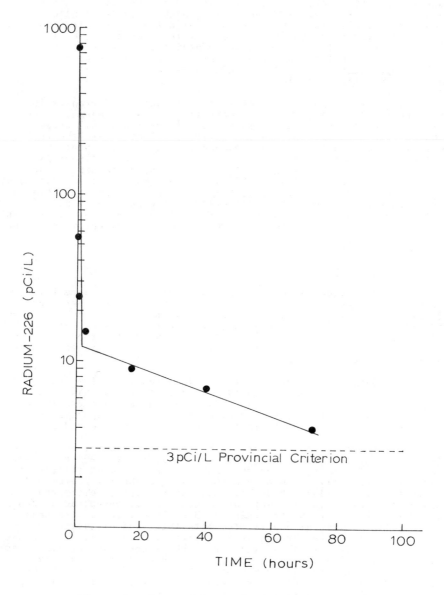

Figure 2:   Rate of Removal of Filtrate Radium-226
            from Solution

radium-226 activity criterion of 3 pCi/L. Furthermore, the Federal Atomic Energy Control Board's suggested target of 10 pCi/L total radium-226 requires a very high degree of suspended solids removal.

This paper describes laboratory and field investigations carried out at Rio Algom as part of a comprehensive research program aimed at keeping all radium-226 releases as low as possible.

3.   REMOVAL OF DISSOLVED RADIUM-226

Laboratory work has shown it is possible to remove radium-226 from solution and achieve a dissolved, or filtered, level below 3 pCi/L. The initial chemical separation step in an analysis for radium-226 shows that radium can be almost completely removed from solution by precipitation with barium chloride. All the laboratory work, however, was done on a small scale in 1 litre glass beakers.

The criterion for dissolved radium-226 in the Province of Ontario requires an effluent which contains no more than 3 pCi/L radium-226 when filtered through a 1.2 μm Millipore filter. Because it has not been possible to consistently produce effluents that low in dissolved radium-226 at the operating Quirke mine, and because of the limited use of data from small-scale laboratory exepriments, a number of large-scale tests were designed to demonstrate the filterable radium-226 activity which might be expected in the field.

80-gallon samples of clear, untreated solution from the Quirke tailings basin were placed in a 100-gallon baffled poly- ethylene tank equipped with a mixer. The desired dose of barium chloride was added and the solution given a rapid mix for 15 minutes. The mixer was then shut off and the solution allowed to settle. Grab samples were taken at predetermined times and immediately filtered through a 1.2 μm Millipore filter. This enabled the removal rate of radium-226 from solution to be followed. Typical results, for a dose of 15 ppm barium chloride[1], are given in Figure 2. There is a rapid initial removal of radium-226 from solution; the filterable radium-226 was reduced from 730 pCi/L to 15 pCi/L in only 90 minutes. Further removal, however, is a very slow process under the quiescent conditions of the test. A filterable radium-226 activity of 4 pCi/L required 72 hours, and Figure 2 indicates that the Province of Ontario's criterion of 3 pCi/L may require 90 hours, or more, detention time.

4.   REMOVAL OF SUSPENDED (TOTAL) RADIUM

Although the removal of dissolved or filterable radium-226 is difficult even under laboratory conditions, the Quirke effluent meets the Provincial criterion of 3 pCi/L for several months of the year (Figure 1). The reduction of suspended radium to the Atomic Energy Control Board's suggested target of 10 pCi/L total radium-226 activity appears to be an even more difficult task. Radium-barium sulphate is a very fine precipitate, and flocculating agents are required to aid in settling: Soaps [1], synthetic polymers and iron or aluminium salts [2,3] have all been suggested. Little comparative data is available, however, which would enable a choice of flocculent to be made. Prior to any large-scale field trial, it was decided to make a laboratory comparison of the usefulness of a number of flocculents in producing effluents low in total radium-226.

A variety of flocculents were tested: ferric chloride,

_____

[1]Barium chloride additions are expressed as mg/L $BaCl_2$

TABLE I: Settling Tests with Flocculents

TIME (hours) / FLOCCULATING AGENT	TOTAL RADIUM-226 pCi/L			
	NONE	FeCl$_3$[2] (10 ppm)	Pickle Liquor (40 ppm)	Alum (24 ppm)
6	294	138	134	238
18	43	42	65	155
30	28	23	27	110
48	18	14	18	-

[2]Ferric Chloride additions are expressed as mg/L FeCl$_3$

TABLE II: Quirke Treatment System - Solution Characteristics

	Point 1 TAILINGS BASIN EFFLUENT	Point 2 PRIOR TO 2nd BaCl$_2$ ADDITION	Point 3 FINAL EFFLUENT
Ra-226: TOTAL (pCi/L)	800	250	80
DISSOLVED	20	12	4
SUSPENDED SOLIDS (ppm)	2	12	5

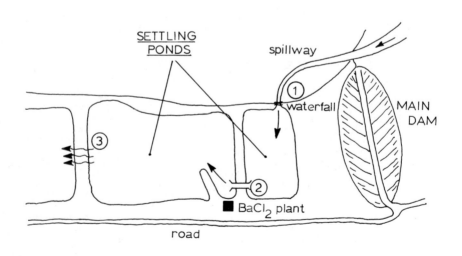

Figure 3: The Quirke Effluent Treatment System

ferrous sulphate, alum, a second dose of barium chloride, and waste
pickle liquor (mostly ferrous chloride) from a steel mill. The
settling characteristics of the flocs produced by the above reagents,
in combination with a primary treatment of 15 ppm barium chloride,
were investigated in long tubes, 3 m in length and 24 cm in diameter.
Samples were taken from the long tubes at suitable time intervals
from sampling ports provided 120 cm from the top. From this information a design settling curve, the S-curve, was developed which
allowed the total radium-226 activity and suspended solids concentration to be expressed as a function of settling time. The results
of a number of tests are given in Table I and these show that ferric
chloride, at a dose of about 10 ppm, is a good settling aid. The
data in Table I show the stringent nature of the suggested target
total radium-226 activity of 10 pCi/L.

5.   THE QUIRKE EFFLUENT TREATMENT SYSTEM

        A two stage barium chloride treatment is currently
employed at the operating Quirke mine for radium-226 control. The
system is shown schematically in Figure 3. The clear effluent from
the tailings basin discharges down a spillway alongside the main dam.
Barium chloride (9 ppm) is added continuously at point 1 just prior
to a waterfall into the smaller of two settling ponds. The overflow from this pond flows through a weir at point 2 where a second
smaller dose of barium chloride (2 ppm) is added. (The rate of
addition at both points is controlled from the plant located at
point 2.) The final effluent discharges by spilling over a rock
dam at point 3. Typical effluent characteristics at the three points
are given in Table II, and show how the settling ponds facilitate
the removal of dissolved radium by providing time for crystal growth,
and achieve a fair degree of settling of the suspended radium-226.

        All the laboratory work has shown the importance of
providing adequate time for settling processes to occur.

        Detention time measurements have been made on the larger
settling ponds under a variety of conditions: spring runoff, summer
and mid-winter ice cover. A slug of blue dye was dumped at point 2
and the effluent, at point 3, was continuously monitored for the
occurence of the dye. The pond has a volume of 56 800 m^3 and at
an average flow of 299 L/s should provide a detention time of 53
hours. The measured detention-time curve for July 1977 is shown in
Figure 4, and the actual mean detention-time (t_g, the centeroid of
the C-curve) is given at the time when half the tracer was detected
at the outflow. This occured after 30 hours and shows the settling
basin provides only this time for radium-barium sulphate precipitate
to settle. Figure 4 also shows a settling curve, the S-curve, for
an effluent which received a dose of 10 ppm ferric chloride in
addition to barium chloride. Since the C-curve falls over the flat
portion of the S-curve the total radium-226 activity in the effluent
can be read from the S-curve at time t_g. A total radium-226
activity of 25 pCi/L is thus predicted for the Quirke effluent. The
S-curve indicates that only a small improvement is possible if the
mean detention time of the pond could be extended to its theoretical
value of 53 hours.

6.   FIELD DEMONSTRATION OF TOTAL RADIUM-226 REMOVAL

        The reduction of total radium-226 in a final effluent to
25 pCi/L found in the settling tests represents a marked improvement
over the current values obtained at Quirke. Thus, it was decided
to carry out a full-scale field trial of ferric chloride at Quirke
to assess the ability of the laboratory settling curves to predict
final effluent quality in the field.

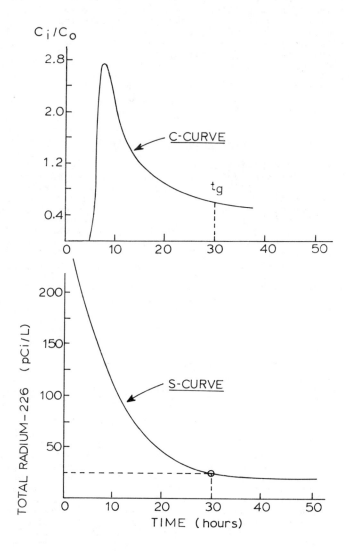

Figure 4: top: Detention Time Curve for Large Settling
Pond in July 1977.

bottom: Settling Curve for 10 ppm Ferric Chloride
Addition.

Baseline data for normal two-stage barium chloride treatment at Quirke (Figure 3) were collected for 12 days by monitoring the total radium-226 activity in the final effluent. The two-stage barium treatment was then altered so that additions were made only at the spillway. The radium-226 activity in the final effluent for this single-stage barium chloride treatment was monitored for 9 days. Once equilibrium had been reached in the system for the single-stage barium chloride treatment, ferric chloride (10 ppm) was added to the weir between the two settling ponds. The additions were continuous for a period of 12 days. During this time, the final effluent quality was regularly sampled and analyzed for radium-226 and suspended solids. A fourth and final, test-period entailed monitoring the final effluent immediately after the cessation of the ferric chloride additions. This was done for a period of 8 days.

The observed total radium-226 activities for the four test periods are given in Figure 5. There is a significant decrease in total radium-226 for a single-stage barium chloride treatment (56 pCi/L) compared to the normal two-stage system (86 pCi/L). Addition of ferric chloride further improved final effluent quality by reducing the total radium-226 acitivty of 20 pCi/L. The final test period showed the system returned to 70 pCi/L thereby demonstrating beneficial effects of ferric chloride. Suspended solids analyses of the final effluent quality showed corresponding changes for the four test periods.

In conclusion, there is little doubt as to the ability of even small doses of ferric chloride to substantially reduce the total radium-226 loadings to the surface watercourses. Implementation of ferric chloride treatment would, however, result in a considerable increase in the solids retained in the settling ponds and necessitates examination of the means for the collection and ultimate disposal of these radioactive sludges.

REFERENCES

[1]   Kremer, M., "Method of Separation of Radium", Patent granted to Commissariat at L'Energie Atomique, Paris, France. Canadian Patent No. 778.831, 1968.

[2]   Roake, A. F., and Schisler, J. M., "Removal of Ra-226 from Runoff at a Low-Level Radioactive Waste Management Area", Canadian Uranium Producers' Metallurgical Committee, Ottawa, 1977.

[3]   Wilkinson, P., and Cohen, D. B., "The Optimization of Filtered and Unfiltered Radium-226 from Uranium Mining Effluents", Canadian Uranium Producers' Metallurgical Committee, Ottawa, 1977.

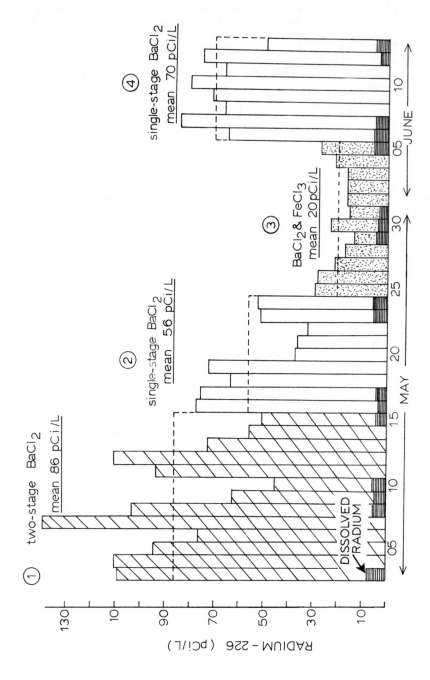

Figure 5: Total Radium-226 Activity in Final Effluent during Four Test Periods

DISCUSSION

I.J. ITZKOVITCH, Canada

Do you have any data on terminal densities with and without fair chloride addition ?

A.J. VIVYURKA, Canada

No. I do not believe anything has been done on that line at all.

I.J. ITZKOVITCH, Canada

Would not these be necessary to, in essence, determine your ultimate solids loading to your tailings ?

A.J. VIVYURKA, Canada

I do not see what relation that would have to it at all. We are dealing with treating clarified effluents, and I am not too sure whether densities would have anything to do with it at all.

I.J. ITZKOVITCH, Canada

Well, your build-up would increase if your terminal density were decreased. You start overflowing your substance base faster.

A.J. VIVYURKA, Canada

No, we have not done any work on that at all.

G.A. SEHMEL, United States

As I understand your Figure 3, you add your barium chloride at the top of the spillway, so you get intimate mixing in the spill-way and you then increase your reaction with better efficiency ?

A.J. VIVYURKA, Canada

Sorry, I did not want to give you the wrong impression. Yes, we do add it at the top of the waterfall. The slides that were taken about 5 years ago and we were not using that station at that particular time.

LEACHING OF RADIUM FROM URANIUM TAILINGS

D.M. Levins, R.K. Ryan and K.P. Strong
Australian Atomic Energy Commission, Research Establishment
Lucas Heights NSW 2232 Australia

ABSTRACT

Laboratory studies were undertaken to determine the factors affecting the leaching of radium-226 from tailings. Most of the experiments were carried out in 0.4 - 4 ℓ stirred vessels using tailings derived from an Australian ore containing 0.22% uranium. Fundamental studies identified the major factors as pH, liquid/solid ratio and the concentration of cations and sulphate ion in solution. Measurements of radium solubility in uranium milling, particularly during acid-leaching, waste neutralisation and storage in the tailings dam, were also carried out.

The feasibility of chemical extraction of radium from tailings by leaching in strong salt solutions was examined. Over 90 per cent of the radium was removed by four stage leaching in 5 M NaCl solutions at 25°C.

RESUME

Des études en laboratoire ont été faites pour déterminer les facteurs affectant le lessivage du radium-226 provenant des résidus. La plupart des expériences ont été faites dans des récipients de 0,4 à 4 l, utilisant des résidus dérivés d'un minerai australien, contenant 0,22 % d'uranium. Des études fondamentales ont identifié les principaux facteurs sensibles comme le pH, le rapport liquide/solide et la concentration de cations et d'ions sulfatés en solution. Les mesures de la solubilité de radium dans le bocardage d'uranium, particulièrement durant le lessivage acide, la neutralisation des déchets et le stockage au barrage des résidus ont aussi été réalisés.

Les possibilités d'extraction chimique du radium des résidus par lessivage dans des solutions fortement salées, ont été examinées. Plus de 90 % du radium était extrait par quatre lessivages dans des solutions à 5 M NaCl, à 25°C.

1. INTRODUCTION

The leachability of radium from tailings was studied to provide relevant data for assessing the environmental impact of proposed uranium mining and milling in Australia. Radium is radiologically important because, as a member of the alkaline earth family, it is readily incorporated into bone structure. The allowable activity of ^{226}Ra in drinking water is the lowest of all radioisotopes [1]. Radium can enter natural waters by a number of routes including leaching of ores by rainwater, discharge of mill waste streams containing dissolved or suspended radium, and seepage from the tailings dam.

This paper summarises the results of an ongoing research program on the leaching of radium which includes laboratory studies on the fundamental factors affecting the leaching of tailings and the fate of radium during milling, waste treatment and tailings storage. A process for chemically extracting radium from tailings is also being evaluated.

2. EXPERIMENTAL PROCEDURE

Fifty-kilogram samples of tailings were supplied by a number of Australian uranium mining companies. Experiments were largely confined to acid-leached tailings. Most of the fundamental data on radium leachability was obtained using tailings derived from an ore containing 0.22% uranium. The tailings had been prepared by leaching ore ground to 50 wt.% less than 75 μm in sulphuric acid at 55 wt.% solids and a temperature of 35°C. The duration of the leach was 24 hours and acid consumption was 40 kg t^{-1} at a pH of 1.8. After leaching, the tailings were washed three times by decantation with demineralised water, filtered and air dried. The sulphate content of the tailings was 8.7 mg g^{-1} of which only 0.4 mg g^{-1} was readily leachable in water.

Leach liquors and raffinates were required for some experiments. In such cases, the ore was leached under conditions recommended for commercial processing. The tailings were filtered and washed repeatedly with raffinate prepared by solvent extraction of uranium with Alamine-336. When prepared in this way, the sulphate concentration of tailings derived from ore containing 0.22% uranium was 14.3 mg g^{-1} of which about half was readily leachable in water.

Batch experiments were carried out in stirred vessels with capacities in the range 0.4 - 4 ℓ. Slurry samples were filtered through a coarse filter paper (Whatman No. 1) then through a 0.45 μm membrane filter to obtain samples for ^{226}Ra analysis. Blank experiments established that little dissolved radium was adsorbed on the paper filter but up to 15% was removed at neutral pH on the membrane filter. Presoaking these filters in 1 M MgSO$_4$ solution followed by washing with a little demineralised water reduced the amount of soluble radium retained on the membrane filter. Liquid samples were acidified and analysed for ^{226}Ra by an emanation technique that involved adsorption of ^{222}Rn on silica gel at -80°C followed by liquid scintillation counting [2]. Solid samples were totally dissolved by repeated fuming using nitric, hydrofluoric and perchloric acids. Leach liquors were analysed for ^{223}Ra and other radionuclides by γ-spectrometry of 250 ml samples using a Ge(Li) detector (efficiency - 15%; resolution - 2.1 keV for 1.332 MeV photons from ^{60}Co).

3. FUNDAMENTAL STUDIES

While there is a reasonable amount of data in the literature, only a few systematic studies of the natural leaching of radium from ore and tailings have been published [3-5]. Most of the experimental data was obtained on a very small scale. Shearer [3,4] reported that leaching was essentially complete after 15 min and was controlled primarily by the liquid-solid ratio. Of the salts tested, only barium chloride significantly promoted leaching of radium. Havlík et al. [5] and Seeley [6] reported that many salts are capable of releasing substantial fractions of the radium present in tailings.

3.1 Rate of Leaching

Figure 1 shows the change in concentration of ^{226}Ra and sulphate ion with time after contacting high sulphate tailings with demineralised water. The rate of leaching of radium was very rapid initially and difficult to measure because of the time required for filtration. (In these experiments, the time taken includes that required for filtration of half the sample.) The ^{226}Ra concentration reached its peak after about 15 min and declined thereafter. Other experiments established that there was little change in radium concentration beyond two hours. The maximum in radium concentration was reproducible and is also apparent in the data of Shearer [3] and Havlík et al. [5] but no explanation has been attempted. From Figure 1, it appears that the maximum is the result of two competitive processes, the dissolution of radium and a tendency towards its precipitation as sulphate concentration increases. For contact times longer than 15 min, the latter process apparently dominates.

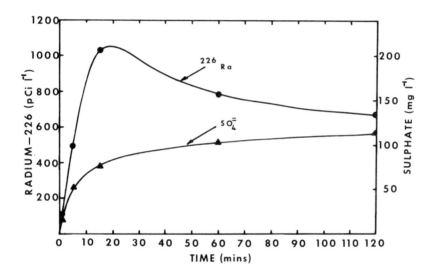

Figure 1. Rate of Leaching of Radium-226 from Tailings
(High Sulphate Tailings, Liquid/Solid = 30 ℓ kg^{-1}, Temperature = 25°C, pH = 3.8-4.1)

3.2 Effect of Liquid/Solid Ratio

The effect that dilution of tailings with demineralised water had on the extent of radium dissolution is shown in Figure 2. The percentage of radium leached increased almost linearly with liquid/solid ratio because there was only a slight reduction in the ^{226}Ra concentration in solution. This is analogous to the dissolution of a sparingly soluble salt, such as barium sulphate, in water. If a small quantity of solid BaSO$_4$ is added to water, the concentration of barium ions rises to its equilibrium value, i.e. its solubility in water. Dilution with more water dissolves additional BaSO$_4$ to restore the barium ion concentration to its original value. As long as there is sufficient solid BaSO$_4$ present, the percentage of the barium in solution will be proportional to the liquid/solid ratio. This analogy may well apply to radium except that the system is complicated by the presence of mixed sulphates of other alkaline earth elements. Consequently the concentration of radium in solution is much lower than that predicted from the published solubility product of radium sulphate.

The fact that significant quantities of radium can be leached by contact with large volumes of water is important in assessing the environmental impact of tailings. This was demonstrated, in early uranium milling operations, when tailings were sometimes discharged into rivers.

3.3 Effect of pH

The pH of the solution has a marked effect on the dissolution of radium. This effect is most apparent at high liquid/solid ratios and low sulphate

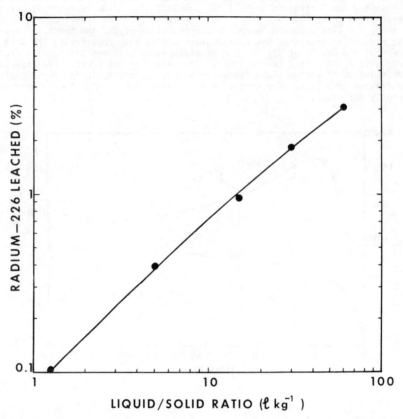

Figure 2.   Effect of Liquid/Solid Ratio on Leaching of Radium-226
(Low Sulphate Tailings, Contact time = 3 h, pH = 4.4-5.1, Temperature = 25°C)

concentrations. Figure 3 shows the percentage of [226]Ra leached for low sulphate
tailings at a liquid/solid ratio of 30 ℓ kg^{-1}. Hydrochloric, nitric and sulphuric acids were added to lower the pH from its initial value of 5.1 in demineralised water. No difference in leaching was detected between hydrochloric and
nitric acids but less radium was leached by sulphuric acid. The [226]Ra concentration reached its maximum in sulphuric acid at about pH 3 and thereafter declined
with further acid addition because of the increased sulphate concentration in
solution.

In the alkaline region, pH was adjusted by addition of NaOH. Radium dissolution was suppressed to pH 11 but at a higher pH a slight increase in leachability
was observed.

3.4 Effect of Salt Concentration

Figure 4 shows the effect of sodium salts on the percentage of [226]Ra
liberated at a liquid/solid ratio of 30 ℓ kg^{-1}. At the normal concentrations of
these salts in fresh waters, little radium is leached but over 80 per cent is
released by 1 M solutions of NaCl or NaNO$_3$. Tests with calcium and potassium
salts gave similar results. The anion in solution did not have a significant
effect except for sulphate which suppressed the dissolution of radium because of
the common ion effect. The enhancement of leaching by salt solutions could be due
to a decrease in adsorptive properties of tailings at high ionic strengths and/or
displacement of adsorbed radium by other cations. These effects apparently

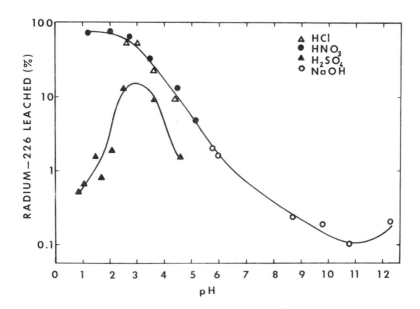

Figure 3. Effect of pH on Leachability of Radium-226
(Low Sulphate Tailings, Liquid/Solid = 30 ℓ kg^{-1}, Contact time = 1.5 h,
Temperature = 25°C)

dominate at high sulphate concentrations resulting in an increase in radium solubility for sulphate concentrations above 10 g ℓ$^{-1}$.

3.5 Other Variables

Low sulphate tailings were sieved and each size fraction leached in demineralised water at a liquid/solid ratio of 30 ℓ kg^{-1}. The equilibrium concentration of ^{226}Ra in solution was highest for the smallest diameter particles. However, the percentage of ^{226}Ra dissolved did not vary significantly with particle size because the finer particles contained more radium.

Temperature had only a small effect on radium solubility. Over the range 5-45°C, the extent of leaching increased slightly with temperature. Variations in stirrer speed within the range 200-800 rev. per min had no detectable effect on the rate of leaching of radium.

4. DISSOLUTION OF RADIUM IN URANIUM MILLING

The results reported in Section 3 can be used to interpret and understand the behaviour of radium in uranium milling circuits. The liquid/solid ratio and the sulphate concentration in mill process streams are the most significant factors affecting the extent of radium dissolution.

4.1 Leaching

Because radium has the most insoluble sulphate known, little is dissolved by conventional leaching with sulphuric acid. While this fact is well established, published figures for the percentage of radium dissolved vary considerably. For example, Winchester Laboratory [7] reported 3-5%, Tsivoglou and O'Connell [8] 0.4-0.7% and Clark [9] 0.2-0.4%. Most of these figures include not only the radium released during the leaching stage but also that solubilised in subsequent operations such as counter-current washing of tailings.

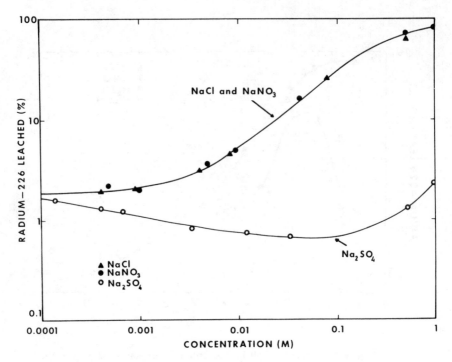

Figure 4. Effect of Concentration of Sodium Salts on Leachability
of Radium-226
(Low Sulphate Tailings, Liquid/Solid = 30 ℓ kg^{-1}, Temperature = 25°C
pH = 4.0-5.1)

While differences in ore type and leaching conditions might explain some of
the reported differences in radium leachability, the methods used for radium
analysis are the most likely cause of disagreement. Both radiochemical and radon
emanation methods have been used; the former measure all isotopes of radium
including ^{223}Ra (from the ^{235}U decay chain) while the latter generally measure
only ^{226}Ra. For ores in secular equilibrium, the activities of radionuclides in
the ^{238}U decay chain are 22 times those in the ^{235}U decay chain. Leaching destroys
this equilibrium owing to dissolution of most of the uranium and about half the
thorium [7]. This will have no effect on the concentration of ^{226}Ra because
equilibrium between ^{230}Th and ^{226}Ra will not be re-established for several thousand
years, but decay of ^{227}Th (half-life = 18.2 days) will result in growth of ^{223}Ra
activity. Since ^{223}Ra has a half-life of only 11.4 days, the long-term concentrations of ^{223}Ra in separated leach liquors and acid raffinates are determined by the
extent of leaching of ^{227}Ac (half-life = 22 years). The early work at Winchester
Laboratory [10] failed to recognise the importance of ^{223}Ra in acid leach circuits
but, in later reports [7,11], it was concluded that ^{223}Ra could contribute up to
15 per cent of the total radium activity in leach liquors and acidic waste streams.

In an attempt to resolve the conflicting data in the literature, a sample of
uranium leach liquor (from an ore containing 0.22% U) was analysed for total
radium by a radiochemical method [12] and for ^{226}Ra by a radon emanation method
[2]. In the radiochemical method, radium was co-precipitated as Ba(Ra)SO$_4$ and
counted over a period of one month. The α-activity in the sample decayed with a
half-life of 11.7 days (compared with the half-life of ^{223}Ra of 11.4 days). The
calculated ^{223}Ra activity at the time of radiochemical separation (fourteen days
after leaching) was 8600 pCi ℓ^{-1} (320 Bq ℓ^{-1}) while the ^{226}Ra activity, as determined by the radon emanation method, was only 170 pCi ℓ^{-1} (6.3 Bq ℓ^{-1}).

The ingrowth of ^{223}Ra into a uranium leach liquor was subsequently studied
using a Ge(Li) detector to measure the change in activity of ^{227}Th and ^{223}Ra with

time. This study showed that 58% of the thorium and 0.5% of the actinium were dissolved during acid leaching. Theoretical curves for the growth and decay of ^{223}Ra and ^{227}Th were calculated from these figures. Figure 5 shows excellent agreement of the theoretical curve with the experimental data. The ^{223}Ra activity reached its peak value of 12,000 pCi ℓ^{-1} (440 Bq ℓ^{-1}) after about three weeks. Even for short times, the ^{223}Ra activity exceeded that of ^{226}Ra.

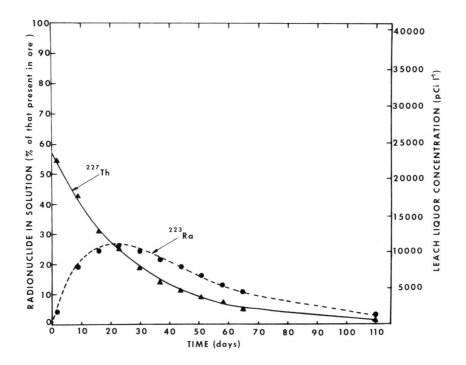

Figure 5. Concentration of ^{223}Ra and ^{227}Th in Typical Leach Liquor (0.22% U ore leached at 55 wt.% solids for 24 h at pH 1.8 and 35°C)

Several Australian ores with compositions ranging from 0.2-6.5% U (principally as uraninite) were leached in sulphuric acid at 55 wt.% solids under typical processing conditions. Acid consumption varied from 20-120 kg t^{-1}. In most cases, the extent of dissolution of ^{226}Ra, during leaching alone, was within the range 0.01-0.03%. By contrast, when a carnotite ore containing 0.32% U was leached under alkaline conditions, 0.9 per cent of the ^{226}Ra was dissolved.

4.2 Washing of Tailings

Because of the higher liquid/solid ratio, washing of tailings dissolves more radium than acid leaching. Table I summarises the effect of repeated washing of tailings with demineralised water at 17 per cent solids (approximately the solids concentration in the feed to a countercurrent decantation (CCD) circuit). After five contacts, roughly approximating five stages of a CCD circuit, about one per cent of the ^{226}Ra had dissolved. This probably represents the upper limit of radium solubility because build-up of sulphate in the CCD circuit, particularly in mills which recycle raffinate for washing, will tend to reduce the radium solubility owing to the common ion effect.

4.3 Waste Neutralisation

Neutralisation of the acidic tailings-raffinate slurry before impoundment is a simple, but effective, method of reducing the concentration of heavy metals and

Table I Radium Leached by Repeated Washing

Table I Radium Leached by Repeated Washing

(0.22% U ore, Contact time = 1 h, Temperature = 25°C)

Wash Number	^{226}Ra Concentration		% ^{226}Ra Leached
	pCi ℓ^{-1}	Bq ℓ^{-1}	
1	630	23	0.40
2	350	13	0.22
3	150	5.6	0.09
4	210	7.8	0.13
5	170	6.3	0.11
Total			0.95

radioactivity in seepage from the tailings dam [13]. The two largest proposed mills in Australia intend to neutralise tailings-raffinate slurries to about pH 8 using lime [14,15].

From previous published data [7], it might be expected that lime treatment would substantially reduce the concentration of ^{226}Ra. A recent assessment of the environmental impact of uranium milling stated that neutralisation of acid-leach effluents removes 90 per cent of the ^{226}Ra [16]. It is also commonly assumed, although no results have been published, that the presence of tailings has no effect on the residual radium concentration.

To determine the activity of ^{226}Ra in neutralised wastes, raffinates and tailings-raffinate slurries (40 wt.% solids) derived from two Australian ores were slowly neutralised by addition of hydrated lime over a period of six hours. Samples were periodically withdrawn and analysed for ^{226}Ra. Figure 6 shows that neutralisation of the raffinates to pH 8 removed 60-90 per cent of the ^{226}Ra in solution in agreement with earlier work [17,18]. But, surprisingly, when tailings were present, the concentration of ^{226}Ra actually rose, reaching a maximum at pH 6. This effect is partly explained by the fall in sulphate concentration as calcium sulphate is precipitated, leading to dissolution of radium from the tailings in accordance with the common ion effect. However, the rise in ^{226}Ra activity was greater than the fall in sulphate concentration on a percentage basis which suggests that other factors were contributing to the increased solubility. Calcium sulphate could be responsible because its incorporation into barium-radium sulphate precipitates has been shown to increase the solubility of radium [6].

Above pH 6, radium was removed from the tailings-raffinate slurries by adsorption, coprecipitation and hydrolysis processes. Even so, the concentration of ^{226}Ra immediately after neutralisation to pH 8 was still above that of the untreated acidic slurry.

5. LEACHING OF RADIUM IN THE TAILINGS DAM

5.1 Ageing of Neutralised Tailings Slurries

To simulate long-term containment in a tailings dam, 40 wt.% tailings slurries were neutralised to pH 7.5 and slowly agitated (0.2 rev per min) for over six months. Figure 7 shows the change in ^{226}Ra concentration with time for a tailings-raffinate slurry derived from an ore containing 0.22% uranium. An ageing effect is apparent with the ^{226}Ra concentration gradually decreasing to 120 pCi ℓ^{-1} (4.4 Bq ℓ^{-1}) over the first 60 days and then remaining constant.

In other experiments, tailings from a number of Australian ores were mixed with demineralised water at 40 wt.% solids and neutralised to pH 7.5. The liquid above the tailings was regularly sampled over a four month period. The initial and final radium concentrations are given in Table II and are clearly correlated with the uranium (and hence radium) content of the ore.

Table II   Concentration of ^{226}Ra in Solution above Tailings

% U in ore	After 2 days		After 4 months	
	pCi ℓ^{-1}	Bq ℓ^{-1}	pCi ℓ^{-1}	Bq ℓ^{-1}
0.10	40	1.5	15	0.6
0.22	100	3.7	60	2.2
0.33	350	13.0	140	5.2
1.76	580	21.5	140	5.2

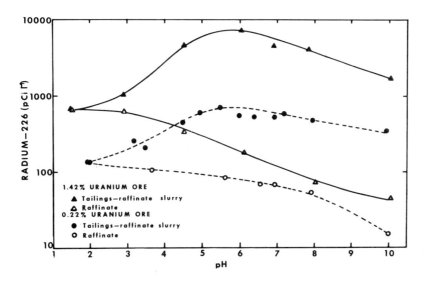

Figure 6   Concentration of Radium-226 in Lime-Treated Mill Wastes

5.2 Radium in Seepage

It is extremely difficult to simulate the seepage of rainwater and raffinates through a large tailings pile on a laboratory scale. Nevertheless, in order to obtain an estimate of the total radium that could be lost via seepage, a number of 25 mm diameter columns were packed to depths ranging from 0.1-0.5 m with tailings compacted to varying bulk densities. Liquid (either demineralised water or neutralised raffinate) was maintained at a constant head above the tailings. The volume and ^{226}Ra content of the seepage were regularly measured. The ^{226}Ra concentration in seepage decreased initially but after a few months approached a constant value of 25-300 pCi ℓ^{-1} (0.9-11 Bq ℓ^{-1}) depending on experimental conditions. This concentration was approximately the same as that measured in aged, neutralised tailings slurries. The volume of liquid that passed through the column was the most significant factor affecting the amount of radium leached.

6. CHEMICAL EXTRACTION OF RADIUM FROM TAILINGS

The ability of strong acids and salt solutions to dissolve radium (see Figures 3 and 4) suggests an alternative approach to problems resulting from the presence of radium in tailings. Radium could conceivably be removed from ore or tailings by chemical extraction during milling. This would substantially lessen the environmental impact of tailings because it would eliminate, or greatly reduce, the hazards arising from radon emission, leaching of radium, dust dispersal and gamma radiation.

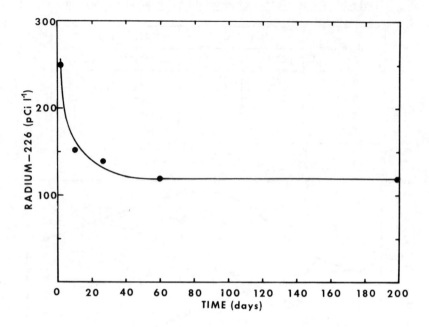

Figure 7   Ageing of Neutralised Tailings-Raffinate Slurry derived from
0.22% U Ore

Radium was commercially recovered from uranium ores before there was any significant demand for uranium itself. The methods used in these plants required large quantities   of chemicals and many processing stages, and would be exceedingly costly if used on the low grade ores mined today. In recent years, a number of processes have been considered involving the use of strong hydrochloric or nitric acids solutions [6,19,20]. Borrowman and Brooks [19] found that 1.5 M HCl was capable of removing most of the radium but they were unable to reach their target of 20 pCi g^{-1} (0.7 Bq g^{-1}). Ryon et al. [20] extracted up to 98% of the radium and thorium with 3 M HNO_3 at 70°C. Both HCl and HNO_3 are considerably more expensive than H_2SO_4 and acid consumption by gangue materials in strong solutions is likely to be excessive. Sears et al. [16] considered using nitric acid in a uranium mill but capital costs were estimated to be more than double those of a conventional mill.

Strong salt solutions have a number of advantages over acids particularly in regard to chemical consumption by gangue materials. Initial scoping studies indicated that operation at ambient temperature was feasible and release of radium was rapid. On the basis of cost and availability, sodium chloride was chosen for further study.

6.1 Rate of Leaching

As in the experiments with demineralised water (see Section 3.1), most of the radium was released within a few minutes in strong NaCl solutions and, thereafter, the radium concentration slowly decreased with time. Initial release of radium appears to be controlled by mass transfer from the surface of the particles or from readily accessible pores. Leaching times between 10 minutes and 2 hours appear suitable and, in practice, the solid-liquid separation step is likely to decide the contact time.

6.2 Effect of Salt Concentration

Figure 8 shows the effect of NaCl concentration on single stage extraction of

^{226}Ra at a liquid/solid ratio of 5. The extent of extraction increased with NaCl concentration but there did not appear to be a great incentive to operate above 3 M NaCl because the increase in radium solubility was relatively small. Solutions were analysed for other alkaline earth elements in addition to ^{226}Ra. The barium concentration closely paralleled that of radium but the calcium and strontium concentrations varied little with salt molarity.

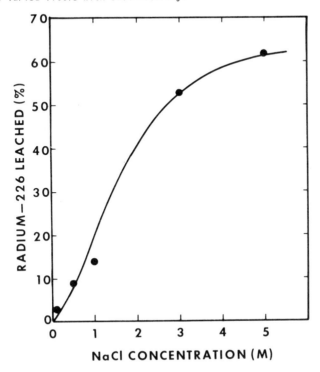

Figure 8   Effect of NaCl Concentration on Single Stage Extraction of Radium-226
(Low Sulphate Tailings, Liquid/Solid = 5 ℓ kg^{-1}, Contact time = 1 h,
Temperature = 25°C, pH = 5.7-6.1)

6.3 Effect of pH

Lowering the pH from 7 to 3 increased the extraction of radium slightly but the effect was much lower in NaCl than in demineralised water. Operation below pH 3 was considered undesirable because of the consumption of acid by gangue minerals.

6.4 Multistage Contacting at Varying Solid/Liquid Ratios

The extractions of radium obtainable in a single stage at a liquid/solid ratio of 5 were inadequate and below those measured in experiments with more dilute slurries. To determine the best leaching conditions, tailings were repeatedly leached at various solid/liquid ratios. Tailings were filtered after each leaching stage. Increase in liquid/solid ratio and multistage contacting favoured dissolution, as shown in Figure 9, because they effectively reduced the sulphate concentration in solution.

6.5 Effect of Sulphate Concentration

The role of sulphate in inhibiting radium dissolution was demonstrated by multistage leaching of low and high sulphate tailings from the same source of ore. The effect of changing the sulphate concentration was to shift the stage where maximum radium was extracted. Table III shows that, with low sulphate tailings,

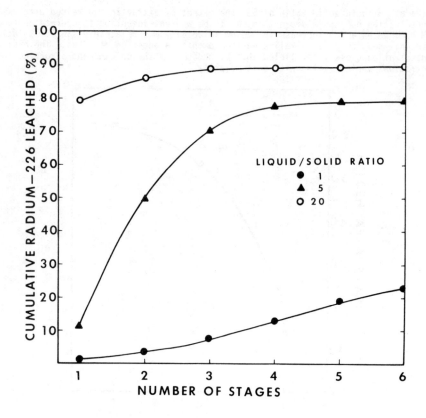

Figure 9   Multistage Extraction of Radium-226 with 1 M NaCl
(Low Sulphate Tailings, Contact time = 0.5 h, Temperature = 25°C, pH = 3.8-5.2)

maximum extraction was obtained in stage two whereas, with high sulphate tailings, this occurred in stage four. Clearly, the sulphate concentration in solution must be reduced before NaCl is fully effective in dissolving radium.

Table III   Percentage of Radium Leached by Multistage Contacting with 1 M NaCl
(Temperature = 25°C, Liquid/solid = 5 ℓ kg^{-1}, Contact time = 0.5 h, pH = 3.8-4.7)

Stage	Low Sulphate Tailings (8.7 mg SO$_4$ g^{-1})	High Sulphate Tailings (14.3 mg SO$_4$ g^{-1})
1	11.2	3.0
2	38.9	3.6
3	20.5	20.7
4	6.6	31.8
5	1.9	12.0
6	0.8	3.8
Total	79.9	74.9

6.6 Maximum Extraction of Radium

Tailings were repeatedly leached in 5 M NaCl at a liquid/solid ratio of 5 in order to determine the maximum percentage of radium that could be extracted. The total extraction was 92% and very little radium was dissolved beyond the fourth leaching stage. This compares with 89% extraction of uranium from the ore by conventional acid leaching. It appears likely that the radium that could not be extracted was associated with refractory uranium in the ore.

6.7 Feasibility of Process

Insufficient experimental data are available at present to fully evaluate the process and it is therefore not appropriate to prepare a detailed flowsheet or to estimate costs. The advantages of extraction by NaCl include low chemical costs, mild process conditions, rapid extraction and compatibility with conventional mill flowsheets. The disadvantages are the relatively large quantity of brine solutions, the introduction of chloride into waste streams which could lead to pollution problems, and the necessity of handling solutions containing high concentrations of radium.

Large quantities of brine are required for effective leaching of radium because of the need to reduce sulphate levels. Multistage leaching in CCD circuits is not a particularly efficient means of achieving this purpose and washing on moving belt filters is likely to require lower quantities of brine.

Unless there were a large source of salt water near the mill, the leach solution would have to be recycled after removal of radium and sulphate. Seeley [6] briefly investigated the leaching of tailings with 1 M NaCl - 0.1 M HCl and rejected this approach because of the gradual decrease in performance with repeated recycle. This was probably caused by the build-up of sulphate (and possibly radium) in the recycled leach solutions.

A few scoping tests were carried out to demonstrate the removal of radium and sulphate from NaCl solutions. Precipitation as barium (radium) sulphate or ion exchange was shown to be feasible but the effectiveness of radium recovery was reduced at high salt concentrations.

Radium recovered from the salt solution would have to be concentrated and stored as α-waste. Thorium-230 is not leached to any significant extent in salt solutions and radium would ultimately grow back into the tailings. However, about 60 per cent of the thorium is dissolved during acid leaching and could be removed from the raffinate.

7. CONCLUSIONS

- Significant quantities of radium can be leached by contact of uranium mill tailings with large volumes of water.

- Sulphates in solution tend to retard the release of radium.

- In the absence of high sulphate concentrations and other dissolved salts, pH has a marked effect on the leachability of radium from tailings.

- Little ^{226}Ra is dissolved during conventional acid-leaching and its activity in leach liquors and acidic raffinates is generally much lower than the activity of ^{223}Ra.

- Neutralisation with lime removes most of the ^{226}Ra from acidic raffinates but is not effective in the presence of tailings.

- The concentration of ^{226}Ra in neutralised tailings liquor and in seepage from tailings decreases initially with time but approaches a constant value after a few months in the range 15-300 pCi ℓ^{-1} (0.5-11 Bq ℓ^{-1}).

- Concentrated salt solutions promote rapid leaching of radium. This could be used to advantage if removal of radium from tailings was desirable for environmental reasons.

8.   ACKNOWLEDGEMENTS

The authors wish to thank Mary Kathleen Uranium Ltd, Noranda Australia Ltd, Pancontinental Mining Ltd, Queensland Mines Ltd, Ranger Uranium Mines Pty Ltd and Western Mining Corporation for providing representative samples of ore and tailings.

9.   REFERENCES

1.   ICRP Publication 2 : Report of Committee II on Permissible Dose for Internal Radiation. Recommendations of the International Commission on Radiological Protection. Pergamon Press, Oxford 1959.

2.   Darrall, K.G., Richardson, P.J. and Tyler, J.F.C. : "An Emanation Method for Determining Radium using Liquid Scintillation Counting", Analyst, Vol 98, p 640-645 (1973).

3.   Shearer, S.D. : The Leachability of Radium-226 from Uranium Mill Solids and River Sediments Ph.D. Thesis - University of Wisconsin, Madison (1972).

4.   Shearer, S.D. and Lee, G.F. : "Leachability of Radium-226 from Uranium Mill Solids and River Sediments", Health Physics, Vol 10, p 217-227 (1964).

5.   Havlík, B., Gráfová, J. and Nýcová, B. : "Radium-226 Liberation from Uranium Mill Solids and Uranium Rocks into Surface Streams", Health Physics, Vol 14, p 417-430 (1968).

6.   Seeley, F.G. : "Problems in the Separation of Radium from Uranium Ore Tailings", Hydrometallurgy Vol 2, p 249-263 (1977).

7.   Winchester Laboratory : Topical Report, January 1960, National Lead Company, WIN-112 (1960).

8.   Tsivoglou, E.C. and O'Connell, R.L. : Waste Guide for the Uranium Milling Industry, U.S. Dept. of Health, Education and Welfare, Technical Report W62-12 (1962).

9.   Clark, D. : State-of-the-Art : Uranium Mining, Milling and Refining Industry, EPA-660/2-74-038 (1974).

10. Winchester Laboratory : Interim Report on Investigations into the Problems of Radioactive Pollution of Uranium Mill Effluents, National Lead Company, WIN-101 (1958).

11. Winchester Laboratory : Summary Report 1959-61, National Lead Company WIN-125 (1961).

12. American Society for Testing and Materials : Standard Method for Radio-nuclides of Radium in Water. Annual Book of ASTM Standards. Part 31, p 677-681 (1976).

13. Levins, D.M., Ryan, R.K., Strong, K.P. and Alfredson, P.G. : "Improving Pollution Control in Uranium Mining and Milling" Chemeca 77, 5th Australian Conference on Chemical Engineering, p 309-314, Canberra, 14-16 Sept. (1977).

14. Ranger Uranium Mines Pty Ltd : Environmental Impact Statement Feb. 1974.

15. Pancontinental Mining Ltd : The Jabiluka Project - Draft Environmental Impact Statement Dec. 1977.

16. Sears, M.B. et al. : Correlation of Radioactive Waste Treatment Costs and the Environmental Impact of Waste Effluents in the Nuclear Fuel Cycle for Use in Establishing "as Low as Practicable" Guides - Milling of Uranium Ores, ORNL-TM-4903 (1975).

17. Ryan, R.K. and Alfredson, P.G. : "Laboratory Studies of the Treatment of Liquid Waste Streams from Uranium Milling Operations", Australasian Inst. Min. Met. Proc. No. 253, p 25-34 (1975).

18. Ryan, R.K. and Alfredson, P.G. : Liquid Wastes from Mining and Milling of Uranium Ores - A Laboratory Study of Treatment Methods Australian Atomic Energy Commission AAEC/E396 (1976).

19. Borrowman, S.R. and Brooks, P.T. : Radium Removal from Uranium Ores and Mill Tailings, U.S. Bureau of Mines BM-RI-8099 (1975).

20. Ryon, A.D., Hurst, F.J. and Seeley, F.G. : Nitric Acid Leaching of Radium and Other Significant Radionuclides from Uranium Ores and Tailings, ORNL/TM-5944 (1977).

DISCUSSION

I.J. ITZKOVITCH, Canada

What methods do you propose for recovery of radium from the sodium chloride leach liquor ?

D.M. LEVINS, Australia

We have actually looked at 2 approaches : one is ion exchange and the other is to look at barium radium sulphate precipitations. We have not got very far into this at the moment, but we have demonstrated that it can be done. But, it is a question of whether it can be done economically and we have not really worked that out.

I.J. ITZKOVITCH, Canada

What ion exchanges are you considering ?

D.M. LEVINS, Australia

You have to remove the sulphate as well. So we put both exchanges, just conventional strong acid.

V.I. LAKSHMANAN, Canada

I just would like to add one more possible disadvantage of a chloride system. The matter of corrosion.

D.R. WILES, Canada

I was curious about that radium dissolution of the pH of 12 that you mentioned. Did you filter those solutions before you analyzed the radium, or is it possible that this was really some sediment or fine powder or something that had been peptized with the high pH ?

D.M. LEVINS, Australia

We actually did filter the material, yes. I would not be completely convinced about that up in the curve. It was a small turn-up and the actual amount of leach was very low and, therefore, the analogy of the procedure was not as accurate as under the other conditions. So, I really would not like to say that it was a definite effect, but it perhaps ties in with some of the other leaching, alkaline leaching of ore which indicates a fairly large amount of leaching which is normal under those conditions. That is in a carbide system, of course.

J. MONTGOMERY, United States

That radium is probably real. We run a basic leach in a project up in Wyoming and we saw that when we took the pH above 12, we had a dip at around 11 ½ and the radium essentially follows the calcium, so if you go through to the higher pH, you are starting to dissolve COH twice. The radium generally follows it.

CONTROL AND PREVENTION OF SEEPAGE
FROM URANIUM MILL WASTE DISPOSAL FACILITIES

R. E. Williams
Professor of Hydrogeology
College of Mines
University of Idaho
Moscow, Idaho   83843
United States

INTRODUCTION. This paper constitutes an analysis of the technologies which are available for the prevention of movement of waste waters out of uranium mill waste disposal facilities via subsurface routes. This subject has been the object of considerable research and development throughout the mineral resource field. This paper brings together information from these research and development efforts and applies it to uranium mill wastes. A large number of techniques are available and have been investigated. These studies generally can be subdivided into efforts which attempt to take advantage of the hydrogeologic environment as a means of preventing or minimizing the entry of mill waste contaminants into ground water and subsequently into surface water, and efforts which attempt to provide man-made barriers to the entry of contaminants into ground water and surface water. Either of these two broad categories can be further subdivided into sub-categories which deal with very specific projects involving recharge areas, discharge areas, confining layers, permeabiliy studies, distribution coefficients, design techniques and the utilization of natural and synthetic liners for tailings ponds and evaporation ponds. This document presents an analysis of each of the techniques that have been proposed and makes recommendations designed to identify the mechanisms which can be expected to achieve maximum success with minimum risk of contaminant migration.

These recommendations are particularly pertinent to uranium mill waste disposal facilities because of the nature of the waste water contained in them. The waste water can be expected to have a pH ranging from 1.5 to 2 and it can be expected to contain a variety of dissolved solids depending in part on the mineralogy of the ore body, but also on the reagents used in the milling process. These dissolved constituents may include excessive concentrations of sulfate, nitrate chloride, arsenate, selenate, uranium, radium, and thorium. Excessive sulfate, chloride or nitrate will be produced by the acid leach process. Evidence available to date suggests that some cations can be expected to be removed from solution and retained for long periods of time (exact periods variable and unpredictable in hydrogeologically complex environments) in the subsurface earth materials beneath a tailings disposal site by exchange reactions if the pH of the waste water is near neutral or is neutralized soon after entering the porous medium. However, dissolved constituents in mill waste waters which are anions or are capable of forming complex anions are not readily removed from solution in the subsurface via exchange reactions. The apparent net result of these observations is that thorium (with possible exceptions), and radium can be retained at least temporarily within the subsurface materials near the waste disposal site but that the dissolved constituents which form anions can be expected to be transported almost conservatively away from the site along the flow paths of the underlying ground water flow system. Some uncertainty enters this picture in that thorium and uranium may form complex anions with chloride, fluoride, nitrate, sulfate and phosphate. Because of the nature of the leach process one or more of these anions will always be present in high concentrations in acid (pH 1.8) uranium mill waste waters. The behavior of such complex anions should be investigated more thoroughly. Very little laboratory data are available on their adsorption characteristics.

The aforementioned flow paths may be long or short. They may terminate near the waste disposal site at a pumping well, at a spring, or in ground water discharge areas such as lakes or springs. Anions moving along long ground water flow paths which do not terminate in the above discharge areas can be expected to remain in the ground water flow system for long periods of time. The length of these periods of time may be measured in years, decades, hundreds, or even thousands of years. The rate of movement of the anionic contaminants along ground water flow paths is controlled primarily by hydraulic gradient dispersion and saturated hydraulic conductivity. For this reason, detailed measurement of these parameters is initial to the analysis of the ultimate mill waste disposal facilities.

Utilization Of Natural Hydrogeologic Environments For The Minimization Of Leakage From Uranium Mill Tailings Ponds. The literature indicates that the solids in uranium, thorium, and radium by adsorption. Therefore, suitability of a site for disposal depends on the extent that the site can prevent the occurrence of release mechanisms. Criteria for the evaluation of the suitability of a site for land burial operations (shallow) have been presented by several authors. (1)

Hydrogeologic Criteria for Potential Uranium Mill Waste Disposal Sites and Mathematical Modeling of Contaminant Migration in Ground Water. The disposal of uranium mill wastes is sufficiently similar to the disposal of other types of low level radioactive wastes that similar site selection criteria can be utilized for both types of wastes. It is recognized, however, that these criteria will not exist near many uranium mill sites because economics dictates that mills be located near mines. In these cases, containment procedures will have to be designed into the disposal facility to prevent release to the environment.

Cherry et al (2) present the following hydrogeologic criteria for satisfactory waste disposal sites:

(1)   "Criteria for Intermediate-Term Burial Sites the land surface should be devoid of surface water, except during snowmelt runoff and exceptional periods of rainfall. In other words, the sites should not be located in swamps, bogs, or other types of very wet terrain.

(2)   the site should be separated from fractured bedrock by an interval of geologic deposits sufficient to prevent migration of radionuclides into the fractured zone. Except in unusual circumstances the direction and rate of ground water flow as well as the retardation effects are very difficult or impossible to predict in ground water regiment in fractured rocks. This lack of predictability necessitates that fractured rock be regarded as major hazard in terms of subsurface radioactive waste management. In fact, it is doubtful if contaminated ground water could be effectively detected and monitored in some types of fractured rock.

(3)   the predicted rate of radionuclide transport in the shallow ...deposits at the site should be slow enough to provide many years or decades of delay time before radionuclides would be able to reach public waterways or any other area which might be considered hazardous in the biosphere. In other words, considerable time would be available for detection of contamination and for application of remedial measures if necessary.

(4)   the site should have sufficient depth to water table to permit all operations to occur above the water table, or as an alternative the site should be suitable for producing an adequate water-table depth by flow sytem manipulation.

(5)   the site should be well suited for effective monitoring and for containment by flow-system manipulation schemes."

---

(1) Williams, R.E., 1979, A Guide to the Prevention of Ground Water and Surface Water Pollution by Uranium Mill Waste Disposal Facilities: Argonne National Laboratory (in preparation).
(2) Cherry, J.A., Grisak, G.E., and Jackson, R.E., 1973, Hydrogeological Factors in Shallow Subsurface Radioactive-Waste Management in Canada, in, Proc. Internatl. Conf. on Land for Waste Management, Ottawa, Canada (Oct. 1-3, 1973).

"Criteria for Long-Term Burial Sites
(1)   the land should be generally devoid of surface water and
      be relatively stable geomorphically. In other words
      erosion and weathering should not be proceeding at a rate
      which could significantly affect the position and
      character of the land surface during the next few hundred
      years.
(2)   the subsurface flow pattern in the area must be such that
      the flow lines from the site do not lead to areas con-
      sidered to be particularly undesirable, such as fractured
      bedrock, surface water bodies and aquifers used for water
      supply.
(3)   the predicted residence time of radionuclides within an
      acceptable part of the subsurface-flow system must be of
      the order of several hundred years. The hydrogeologic
      conditions must be simple enough for reliable residence-
      time predictions to be made.
(4)   the natural water table should be below the ground surface
      by at least several meters and the hydrogeologic setting
      should be such that large water-table fluctuations are
      very unlikely. This condition would provide assurance
      that leaching of radionuclides would not occur quickly
      when the site is abandoned and the ground water flow
      system regains equilibrium with the natural precipitation
      regime."

Popadopulos and Winograd (3) correctly point out that for a
complete evaluation of the suitability of a site for waste disposal
it is necessary to predict the ground water flow patterns and the
rate of transport of radionuclides and other dissolved constituents
in the regional hydrogeologic system.

Uranium mill wastes differ from ordinary low level radioactive
wastes in two major respects: 1) Uranium mill wastes are discarded
as a slurry. Therefore, the wastes themselves constitute a source
of artificial ground water recharge which can raise the underlying
water table if permeability distributions allow the waste water to
reach it. 2) The water in the slurry contains a variety of
dissolved contaminants in addition to radionuclides that also can
contaminate underlying ground water. (1) Consequently the site
selection criteria for uranium mill waste disposal sites are
slightly more restrictive (and less easily met) than the site
selection criteria for low level radioactive waste burial sites.
In order to avoid the highly expensive design mechanisms required
to contain uranium mill wastes artificially the waste disposal site
must be underlain by a stratum capable essentially of preventing
or greatly retarding the movement of the mill waste water. This
requirement necessitates a homogeneous stratum with a saturated
hydraulic conductivity of 10^{-7} to 10^{-9} cm/sec. This criterion is
necessary because ion exchange reactions cannot be expected to
retain some anions and some of the cations common to mill waste
waters. With this additional restriction the criteria listed above
can best be met by a site in a semi-arid to arid environment in a
ground water recharge area underlain by an unfractured shale or
clay layer with a permeability of 10^{-7} to 10^{-9} cm/sec. Mathematical
models have shown that even a thin sand dense of higher permeability
can divert the flow of infiltrating waste water toward a discharge
zone. (1) (4) (5).

(3) Popadopulos, S.S., and Winograd, I. J., 1974, Storage of Low-
    Level Radioactive Wastes in the Ground: Hydrogeologic and Geo-
    chemical Factors. EPA document EPA-520/3-74-009, 49p.
(1) See footnote 1, page 289.
(4) Freeze, R.A., 1972, Subsurface Hydrology at Waste Disposal Sites
    IBM Journal of Research and Development, vol. 16, no. 2, pp.
    117-129.
(5) Childs, K.E., Upchurch, S.B., and Ellis, B., 1974, Sampling of
    Variable Waste Migration Patterns in Ground Water: Groundwater,
    vol. 12, no. 6, pp. 369-376.

The work of Freeze (4) demonstrates conclusively the absolute necessity for obtaining extension accurate, detailed, hydrogeologic field data and for analysis of it if uranium mill waste waters are to be placed directly in contact with a sub-surface ground-water flow system. His work demonstrates conclusively that infiltrating waste waters can emerge at unexpected locations in unexpected time spans if this is not done.

Popadopulos and Winograd (3) present an excellent discussion of the data needed to use mathematical models for this purpose. They list the data needed in the following approximate order of increasing difficulty and (or) cost of acquisition.

(1) Depth to water table, including perched water tables, if present.
(2) Distance to nearest points of ground water, spring water, or surface water usage (includes well and spring inventory)
(3) Ratio of pan evaporation to precipitation minus runoff (by month for period of at least 2 years).
(4) Water table contour map.
(5) Magnitude of annual water table fluctuation.
(6) Detailed stratigraphy and structure to base of shallowest confined aquifer.
(7) Baseflow data on perennial streams traversing or adjacent to disposal site.
(8) Chemistry of water in aquifers and aquitards and of waste water emanating from the waste disposal facility.
(9) Laboratory measurements of hydraulic conductivity, effective porosity, and mineralogy of core and grab samples (from disposal site) of each lithology in unsaturated and saturated (to base of shallowest confined aquiver) zone--hydraulic conductivity to be measured at different water contents and suctions.
(10) Neutron moisture meter measurements of moisture content of unsaturated zone. Measurements to be made in especially constructed holes; at least 2 years' record needed.
(11) In situ measurements of soil moisture tension in upper 15-30 feet (4.5-9.0 m) of unsaturated zone; at least 2 years' record necessary.
(12) Three-dimensional distribution of head in all saturated hydrostratigraphic units to base of shallowest confined aquifer.
(13) Pumping, bailing, or slug tests to determine transmissivity and storage coefficients of saturated porous media.
(14) Definition of recharge and discharge areas for unconfined and shallowest confined aquifers.
(15) Field measurements of dispersivity coefficients.
(16) Laboratory and field determination of the distribution coefficient (K_d) for movement of critical dissolved ions through all hydrostratigraphic units.
(17) Rates of denudation and (or) slope retreat.

These data are necessary for a complete simulation of flow and nuclide transport through both the unsaturated and saturated zones.

Popadopulos and Winograd (3) correctly point out that in order to determine the extent to which sorption processes retard the movement of a given critical nuclide or other dissolved ion, the four principal types of information necessary are: a) the mineralogy (by size fraction including colloidal materials) of all hydrostratigraphic units traversed by the water, including units in the unsaturated zone; b) chemistry of water in aquifers and aquitards; c) chemistry of representative samples of leachate from the waste field; and d) laboratory experiments to determine the distribution coefficients (K_d) for the dissolved ion in representative samples of

---

(4) See footnote 4, page 290.
(3) See footnote 3, page 290.

each hydrostratigraphic unit.

Popadopulos and Winograd (3) conclude that for the disposal of low level wastes the ideal site would be overly simple hydrogeologic settings, i.e., unfaulted, relatively flat-lying strata of intermediate permeability in a region of low relief. However, these wastes are assumed to be discarded dry. In the case of uranium mill tailings, this is not possible because the wastes are discarded as a slurry and water is present in the disposal facility throughout the life of the mill. Thus, the ideal site for a mill must lie over an unfractured, relatively flat-lying stratum of very low permeability (for example an unfractured thick clay or shale stratum) in a region of low relief and arid climate.

But difficulties exist with respect to determination of coefficients required for mathematical models of waste water movement at potential uranium mill waste disposal sites. These difficulties are illustrated by D'Appolonia (6) in a report on Western Nuclear's Split Rock mill site near Jeffrey City, Wyoming. Page 4-12 contains the statement, "The difficulties associated with conducting and evaluating field dispersivity tests under an irregular ground water flow regime and heterogeneous, anisotropic conditions (which exist at the site) prevent performing such tests except for academic reasons. The accuracy of the results still depend on the test location, techniques of measurement, hydrogeology of the site, ground water quality and kind of tracer used."

LINERS. The risks and costs involved in evaluating and utilizing less than ideal sites in combination with the scarcity of ideal sites has lead to the use of pond liners to minimize the risk of ground water pollution by uranium mill waste disposal facilities.

Waste products from uranium milling and concentrating operations are produced and discharged in a slurry form. These waste slurries contain a solid and liquid fraction and are discarded in tailings ponds where the liquid fraction evaporates, seeps out of the pond or is discharged or recycled via a point source. Seepage may occur if the soil beneath a pond is sufficiently permeable to allow a portion of the liquid fraction to move into the subsurface. The pollution potential of seepage depends upon the ground water flow system and the constituents of the waste entering the system.

Under less than ideal hydrogeologic conditions a waste pond containing contaminants must be constructed with a reliable seepage control mechanism. Technological advances of the past decade have produced pond liners for this purpose. Piezometers are installed beneath and around the pond for purposes of monitoring and detecting leaks.

When a liner is selected for a project, there are several critical factors which must be considered; the liner material specified must be compatible with these factors. Three critical factors comprising the liner's environment are: 1) constituents of the waste retained by the liner, 2) pond design and subgrade conditions, 3) exposure of the liner to light, weather, or other potentially damaging conditions. In the case of uranium mill wastes the low pH of the waste water is the most critical factor with respect to liner selection. However, kerosene can be a factor also if it is allowed to enter the pond in large quantities. The subgrade must be constructed of compacted material and its surface must be free of irregularities.

Non-reinforced materials should be avoided where the slope of the subgrade is 2.5 to 1 or greater (horizontal to vertical). Liner materials containing a delayed cure time adhesive system should also be avoided on these slopes. The manufacturers of reinforced liners generally give specifications regarding slopes for their specific products. Soil moisture changes behind the liner and wave

(3) See footnote 3, page 290.
(6) D'Appolonia Engineers, Inc., 1977, Environmental Effects of Present and Proposed Tailings Disposal Practices at the Split Rock Mill, Jeffrey City, Wyoming, vol. 1, rept. 3, Project no. RM 77-419.

action on large ponds can produce sluffing which will form voids
behind the liner and result in failure. It is standard practice to
sterilize the area to be lined to prevent vegetative growth which
has the potential to penetrate the liner. The dead vegetation or
other organic matter in the soil produces gas which expands when
formed and is destructive to liners. The hydrostatic head produced
by the pond or a soil cover on the liner will not effectively hold
the liner in place. The released gas will form bubbles known as
"whalebacks" under the liner and cause failure. If gas from
organic decomposition or entrapped air resulting from a fluctuating
water table are anticipated, a vent system is necessary. (See 1
for details).

The various materials available for pond liners have different
aging and weathering characteristics when exposed to sunlight.
Some must be covered immediately. The manufacturers list of liner
characteristics and properties should be checked closely for these
details.

In addition to the points mentioned above, close attention
must be given to the jointing or seaming of the liner materials.
This is especially important for synthetic liners which are factory
seamed or field seamed by one of a variety of methods. The seaming
method should meet the requirements and recommendations specified
by the liner's manufacturer.

Admixed Liners. Liner systems which are formed in place from
a mixture of imported materials are called admixed liners. These
types of liners can be divided into hard surface liners and soil
sealants. The hard surface liners can be separated on the basis of
composition into two classes based on asphaltic composition and
portland cement composition. The soil sealants can be similarly
separated into natural materials and materials of synthetic composition. Williams (1) presents a detailed discussion of each
category.

CLAY SEALS. Certain naturally-occurring clays are used extensively as groundwater barriers to seal ponds, lagoons and landfills.
The sealing ability of clays is attributed to their adsorbed sodium
ions' ability to retain structural water and swell. The expanded
clay particles fill the soil's voids and yield permeabilities as
low as 10^{-7} cm/sec. These low permeabilities are adversely affected
by those industrial wastes whose high cation content or low pH
induces the process of exchange of the sodium ions. This ion
exchange process may greatly inhibit and even reverse the swelling
process by reducing the clay's ability to adsorb water, which
increases the permeability of the liner.

The presence of the clay mineral montmorillonite normally is
responsible for swelling in natural clays. Montmorillonite is the
main constituent in the commercially refined clay called bentonite.
The two principal types of bentonite are sodium bentonite and
calcium bentonite depending on the adsorbed ion. Sodium bentonite
characteristically possesses much greater swelling properties and
is most desirable for liners.

As stated above sodium bentonite is subject to a reduction in
swelling through the ion exchange process. For this reason, the
American Colloid Company has developed a number of proprietary
products that can be described as contaminant resistant bentonites.
These varieties of treated bentonite are manufactured under the
name of Volclay Saline Seal 100 and Volclay SLS. Salena Seal 100
is the most desirable for uranium mill tailings ponds. These
bentonite-based products are available in dry powder or granular
form and can be applied to surfaces using conventional lime or
fertilizer spreading techniques. After application, prehydration
with fresh water is necessary to bring about complete swelling.
Once hydrated, the molecular structure of the product prevents the
adsorbed water from being removed by the low pH cation-bearing
leachate or waste water. Another beneficial property which is a

---

(1) See footnote 1, page 289.

result of bentonite's plastic nature is its "self-healing" ability. When saturated, the consistency of bentonite is such that minor breaks or tears due to sharp objects, ground settling or other causes, will seal themselves. Permeabilities of 10^{-7} cm/sec. are achievable.

SYNTHETIC POLYMERIC MEMBRANES. Certain manufacturing steps greatly influence a synthetic liner's characteristics. The first step consisting of mixing the raw elastomer or resin is similar to the compounding procedure used to form the raw product. During mixing the liner manufacturer mixes the raw stock with various compounding ingredients, such as pigments, accelerators and antioxidants. Although the starting formula for each polymer is suggested by the raw stock supplier, the final selection is subject to the manufacturer's specifications for final properties. The product often varies depending upon the particular production equipment.

After the stock is mixed, it must be converted to rolled goods which are usually 1.2 to 1.8 m wide by 100 m long. This conversion is accomplished through a process called calendering. During calendering two important liner specifications are determined. The liner may be produced as reinforced or unreinforced, and the thickness is determined. During calendering, a "supported" liner is produced by sandwiching one or two layers of reinforcing fabric, called "scrim" between a top and a bottom elastomeric sheet. A laminate of this type consisting of one scrim layer between two sheets is called three-ply; two layers of scrim between three elastomeric sheets is called five-ply. Unsupported liners employ no scrim, but varying thickness is produced by laminating two layers of sheet rock. This procedure reduces the probability of pinholes in the finished product.

The final step in producing a finished membrane is fabrication. Wider panels are formed during this step by joining together the 1.5 m wide rolled stock, with or without scrim. The final width depends upon the manufacturer and job specifications. The type of seaming methods used depends primarily upon the material being seamed. Factory seams offer the advantage of greater quality control and reduce the amount of field seaming required. The finished product is available either vulcanized (cured) or uncured which directly affects the degree of success of field seaming methods.

Seaming of liners is critical because the liner's strength and sealing ability depends heavily upon the quality of the seams. Each of the four different methods is successful only on specific liners; each manufacturer specifies the method and technique best suited for his product. A brief description of the four methods follows.

The dielectric or radio frequency seaming method uses a high wattage radio frequency (30 to 40 MH_Z) to generate heat and cause the material to become thermoplastic and flow together producing a homogeneous seam. The sheets being seamed are passed between the seaming table with an inlaid transmitting antenna; a moveable bar contains the receiving antenna. Contact is maintained for 2 to 5 seconds to produce the seam. Due to the equipment required, this method is limited to factory, as opposed to field, fabrication.

The second method known as the thermal process is similar to the dielectric method. Heating elements are placed on either side of the liner; heat and pressure are applied to force the interface to flow together to produce the seam. Both the thermal and dielectric methods require close control of heat and dwell time to avoid critically thin areas. Limiting factors are liner thickness in the case of the thermal method (45 mils) and the high voltage arc produced from the dielectric method when excess carbon black is used in compounding.

A third method, solvent seaming, uses a solvent which attacks the inner surfaces of the overlapped liner. Heat and pressure are applied to evaporate the solvent and produce the seam. Inherent problems with the solvent method stem from solvent application.

Too little solvent produces "holidays" or unseamed areas while excess solvent degrades the sheet and/or scrim material. Degradation of the liner may also result from too powerful a solvent.

Various adhesives are used in the fourth method. There are two general types: 1) cold-setting, and 2) two-component systems. Often a special tie gum tape is incorporated into the seam by use of adhesives.

The selection of methods and adhesives depends upon the membrane. Adhesives generally are employed in field seaming.

Polyvinyl Chloride (PVC). The raw resin for PVC was first synthesized in Germany in the 1920s. It remained nearly useless until the early 1930s when a compound, generally called a "plasticizer," was developed by B. F. Goodrich Laboratories. Today many different plasticizers are used depending upon the application and service expected of the liner.

It is the loss of this plasticizer through mechanisms of water extraction or heat volatilization which gives PVC its rather poor aging characteristics. Loss of plasticizer causes stiffening, low resistance to tear and shock and some shrinkage. The liner becomes hard and an increase in tensile strength results. This aging and deterioration process is a major disadvantage inherent in the use of PVC.

Polyethylene (PE). In the compounding of polyethylene, no plasticizers are used. It cannot be plasticized to yield a soft, flexible material; therefore it is not calendered but is made into a film by an extruding process. PE is a much lower density material than PVC and because of the absence of plasticizers its thickness is limited to eight millimeters. Thicker sheets lose their flexibility.

Lower cost and a permeability of 1.5×10^{-9} cm/sec are PE's two major advantages. The disadvantages are that PE has a lower resistance to aging in the sun than its main competitor PVC and the thin PE is less resistant to abrasion, creasing and associated failures.

Chlorinated Polyethylene. (CPE) Dow Chemical Company introduced CPE as a brine pit liner in 1965. It was designed to fill the gap between the plastics discussed previously and butyl-type liners in cost. It also offered better selected properties. CPE is compounded without plasticizers but due to complex chemical reactions it still retains flexibility when manufactured in 20 mil, 30 mil and 45 mil sheets.

Since there are no plasticizers to degrade, CPE has good weathering characteristics. CPE is calendered so it can accept reinforcing material to support the liner. The calendered product has a three-ply construction with a reinforcing sheet of polyolefin. This design is resistant to delamination.

Permeability of CPE is on the order of 10^{-10} cm/sec. This permeability may be increased somewhat by the materials tendency to absorb certain types of leachate. Its manufacturer recommends a simple solvent sealing process for CPE. Despite its availability in a 45 mil thickness and reinforcement, the manufacturer does not recommend slopes steeper than 1.5 to 1. Haxo (7) (9) report that swelling occurs upon exposure to aromatic components; these compounds are detrimental to seams utilizing adhesives.

Hypalon. Hypalon was first developed in the 1940s by the DuPont Company, but was not used a liner material until 1969. Hypalon, whose proper name is chlorosulfonated polyethylene, is categorized with plastics because as initially manufactured it has

(7) Haxo, H.E., Jr., R.W. Haxo, and R.M. White. 1977. Liner materials exposed to hazardous and toxic sludges. First Interim Report. EPA No. 600/2-77-081. June. 63 pp.
(9) Haxo, H.E., Jr., and R.M. White, 1976. Evaluation of liner material exposed to leachate. Second Interim Report. EPA No. 600/2-76-255. September. 57 pp.

properties similar to the plastics. However, upon weathering and aging it slowly cures (vulcanizes) to an elastomeric material. The curing process begins at the outer surface and slowly proceeds inward.

Aging and weathering of hypalon are exceptional, due to the slow vulcanizing process. The final product is resistant to a wide variety of chemicals and corrosives. Some swelling has been reported to occur when exposed to aromatic wastes. This swelling is due to the increasing elastomeric properties as the liner cures. This swelling may place stress on the seams and increase the permeability which is reported to be 10^{-10} cm/sec.

Rolls are available in laminated sheets (to reduce pinholes) 30 mil and 45 mil in thickness as unsupported or supported liners. Unsupported hypalon is seldom used; it is not stable on slopes and lacks sufficient tear strength. Width is limited to 1.5 meters. Factory fabricated panels are available in various dimensions of over 24,000 square meters in size. Specifications are recommended to obtain the largest factory fabricated sheets; this minimizes field seaming. The detailed instructions for the solvent seaming procedures, provided by manufacturer, should be followed precisely. The vulcanizing of hypalon produces a film of material which is not compatible with the seaming solvents. For this reason, hypalon should be kept out of sunlight before seaming. The overall characteristics and lower cost, due to factory vulcanization, makes hypalon one of the most desirable liners. The major limitation is the detailed seaming technique which must be followed precisely.

Butyl Rubber. The first of the elastomeric compounds, butyl rubber, was developed during World War II. It was supplied principally by the Enjay Chemical Company and used predominantly in tire tubes. Since then major companies involved in polymeric chemicals have produced butyl and other similar synthetic rubbers.

Butyl rubber is a vulcanized copolymer containing isobutylene (97%) and minor amounts of ispene. The vulcanized product is calendered in either reinforced or unreinforced laminated sheets varying from 20 mil to 125 mil thickness. Aging characteristics are good, although a few compounds may ozone crack with age. Most new compounds are exceptionally ozone and weather resistant.

A permeability on the order of 10^{-10} cm/sec is reported and is primarily due to butyl's high density. This density is due to butyl's tightly packed molecular structure and not its weight per unit volume. This low permeability also impedes the action of solvents and other adhesives, resulting in a material that is difficult to seam. Seaming procedures are recommended by the manufacturer who also supplies the specially formulated unvulcanized butyl gum tape and adhesive. Because seams are made in 30.5 meter lengths (tape length) an especially smooth and flat pond surface is needed. Butyl has poor resistance to hydrocarbons and swells when exposed to aromatic chemicals.

Ethylene-Propylene Rubber (diene monomer). This elastomer is very similar to butyl rubber. EPDM is in fact a polymer "spin-off" of butyl rubber. Its development came about in an effort to improve the ozone resistant characteristics of butyl rubber. EPDM was developed in 1963 by U.S. Rubber at the same time a nearly identical polymer ethylene-propylene terpolymer (EPT-Nordel) was developed by DuPont. The primary difference between EPDM, EPT and butyl rubber compounds is the amount of loading materials each is capable of adsorbing when being compounded. The raw polymers of EPDM and EPT are more expensive to produce but because of their unique properties they adsorb a larger volume of compounding chemicals and thus produce a lower cost finished product. Chemically, the difference between straight butyl compounds and those containing EPDM and EPT is only slight. Because of this, processors often use the compounds interchangeably and this is resulting in a trend to eliminate distinctions based upon polymers in favor of defining the finished product by its properties.

The overall properties of aging and weathering of EPDM and EPT are slightly better than those of butyl rubber. Haxo (8) (9) reports a permeability on the order of 10^{-10} cm/sec. but notes a slightly higher rate of leachate and water absorption. Since EPDM and EPT are both hydrocarbon rubbers, they show a low resistance to attack from hydrocarbon compounds. The seaming difficulties of butyl exist for EPDM and EPT also. All three products are available as reinforced or unreinforced laminated sheets in thicknesses of 20 mil to 60 mil. All are protected by the manu-facturer's 20 year warranty.

Neoprene. Neoprene is a polymerization product of the monomer chlorprene. It was introduced by DuPont in 1932 as a synthetic having good resistance to oil, gasoline, heat, light and ozone. Its initial use was primarily in gaskets for oil pipelines. It came as a surprise that neoprene in sheet form was attacked by crude oil wastes. This, along with the considerably higher initial cost has seriously slowed neoprene's use as a liner material.

The aging of neoprene with respect to sun exposure is poor also. When the thin sheets in contact with ground are exposed to sunlight, the aging process increases substantially.

Elasticized Polyolefin. In the early 1970s, DuPont developed elasticized polyolefin; it is now marketed as pond lining system 3110. The sale of this product has been limited to qualified and experienced liner contractors directly from E.I. DuPont de Nemours Company.

DuPont's 3110 system has shown good sun aging characteristics and some resistance to oil wastes. It presently is in use holding such wastes as hydrofluoric acid, hydrochloric acid, dichlorl-methane and carbon tetrachloride. Tests have shown 3110 to be "mostly unaffected" with solutions of various chemicals ranging in pH from 2 to 13.5.

The principal sealing point of DuPont's 3110 is its field seaming method. The method was developed out of the need for a simple, reliable heat seaming technique for field applications. The 3110 system is sold with a hand-held heat welding gun which reportedly produces excellent homogeneous seams even when the seaming surfaces are smeared with wet clay. This seaming method has a top speed of 6 meters per second.

The 3110 system is available only in 6 meter by 61 meter sheets. The 20 mil unreinforced sheets are not recommended on slopes steeper than 2:1. DuPont's 3110 liner system is a relatively new product, its good aging and chemical resistant properties com-bined with the high quality, simple field seaming techniques make it preferable to many other liner materials.

Once the general characteristics and properties of all available liners are known, the selection of a liner for a specific project is enhanced greatly. The final choice of a liner should be first based upon the liner's reactions with the waste it will contain. This important property of liner compatibility should be weighed heaviest, for regardless of how impermeable or how economical in initial cost or how well the liner ages in outdoor exposure, if it deteriorates when in contact with the waste effluent, it is useless as a liner.

The various characteristics of uranium mill wastes suggests that liner selection be limited to the following: two plastic compounds, CPE and hypalon, four elastomeric compounds, butyl rubber EPDM, Neoprene, DuPont's 3110 system, and American Colloid's Saline Seal 100. Williams (1) presents additional information on these recommendations.

(8) Haxo, H.E., Jr. 1976, Assessing synthetic and admixed materials
     for lining landfills. Pages 130-156 in Gas and leachate from
     landfills formation, collection and treatment. Proceedings of a
     research symposium held at Rutgers University, New Brunswick,
     New Jersey, March 24-25, 1975. EPA No. 600/9-74-004.
(9) See footnote (9), page 295.
(1) See footnote (1), page 289.

REMOVAL OF ^{226}Ra FROM TAILINGS POND
EFFLUENTS AND STABILIZATION OF URANIUM MINE
TAILINGS - BENCH AND PILOT SCALE STUDIES

N.W. Schmidtke, D. Averill, D.N. Bryant,
P. Wilkinson and J.W. Schmidt

Wastewater Technology Centre,
Environmental Protection Service,
Environment Canada,
Burlington, Ontario.

ABSTRACT

Increased world demand for uranium has resulted in recent expansion of Canadian uranium mining operations. Problems have been identified with the discharge of radionuclides such as ^{226}Ra from tailings pond effluents and with the stabilization of mine tailings. At Environment Canada's Wastewater Technology Centre (WTC) two projects were undertaken in cooperation with the Canadian Uranium Mining Industry and other federal government agencies to address these problems. The first project reports on the progress of bench and pilot scale process simulations for the development of a data base for the design of a full scale mechanical physical/chemical ^{226}Ra removal waste treatment system with an effluent target level of 10 pCi ^{226}Ra total per litre. The second project addresses problems of the leachability of radionuclides and the stabilization of both uranium mine tailings and BaRaSO$_4$ sediments from the treatment of acid seepages.

RESUME

L'accroissement de la demande mondiale d'uranium a amené dernièrement une augmentation de son exploitation minière au Canada. Ces problèmes ont été reliés au rejet de radionucléides tels que le ^{226}Ra des effluents des bassins à stériles et à la stabilisation des stériles. Au Centre technique des eaux usées d'Environnement Canada, nous avons entrepris, en collaboration avec l'industrie de l'exploitation minière de l'uranium au Canada et avec d'autres organismes du Gouvernement fédéral, deux projets pour résoudre ces problèmes. Le premier projet consiste à faire le compte rendu des progrès réalisés avec la simulation en laboratoire et en unités pilotes en vue de réunir un ensemble de données permettant la conception et le calcul d'un dispositif à plus grande échelle pour l'élimination du ^{226}Ra par des moyens mécaniques et physico-chimiques, avec comme objectif une activité de 10 pCi de ^{226}Ra total par litre. Le deuxième projet traitera des problèmes de la percolation des radionucléides et de la stabilisation des stériles d'uranium et des sédiments de BaRaSO$_4$ provenant du traitement des percolats acides.

INTRODUCTION

Increased world demand for uranium has resulted in the expansion of existing uranium mining operations, the reopening of old mines and the development of new mining areas in Canada.

Insofar as water pollution potential from these operations is concerned, there are two major areas which must be addressed. One is the discharge of radio-nuclides such as ^{226}Ra from tailings pond effluents, the second area of concern deals with the stabilization of uranium mine tailings.

This paper will highlight the research and development activities of the Wastewater Technology Centre (WTC) in these areas.

REMOVAL OF ^{226}Ra FROM TAILINGS POND EFFLUENTS

BACKGROUND

In Canada, the Provinces of Ontario and Saskatchewan have established effluent objectives of 3 pCi/L for filtered ^{226}Ra for discharges from uranium mining tailing ponds. However, levels exceeding this objective have been detected in waterways adjacent to some mining operations (1). As well, it has been recognized that significant quantities of suspended ^{226}Ra may be released to the aquatic environment undetected by the present surveillance method of analyzing ^{226}Ra. This has led to a proposal by the Atomic Energy Control Board (AECB) to tighten the effluent guideline "to a target of 10 pCi/L total ^{226}Ra" (2).

In Canada, the current method of treating uranium mine effluents involves the sedimentation of "limed" solids in a tailings settling pond. The decant from the tailings pond is then treated with $BaCl_2$ to remove ^{226}Ra by co-precipitation with $BaSO_4$. The resultant $BaRaSO_4$ is a fine, slow settling precip-itate and normally large settling ponds with long residence times are utilized to effect solids/liquid separation. Recent research (3) has indicated that due to changes in pH and sulphate levels in the overlying waters, $BaRaSO_4$ sediments may re-release ^{226}Ra to the environment. Technology, as currently practiced is unable to meet an effluent total ^{226}Ra objective of 10 pCi/L.

In view of this gap in ^{226}Ra removal technology, the WTC initiated a bench scale project to assess the feasibility of developing a mechanical waste-water treatment system using the physical/chemical unit operations of precipita-tion, coagulation, flocculation and sedimentation for the removal of ^{226}Ra to target levels as proposed by AECB.

In March, 1976, preliminary bench scale experiments were initiated to better understand and improve the ^{226}Ra removal process. During 1977, AECB assisted in accelerating the program by providing additional funds.

Based on these preliminary studies a jointly managed and funded govern-ment/industry pilot scale field project was developed. The project is supported with resources from Rio Algom Mines Limited, Denison Mines Limited, Eldorado Nuclear Limited, the AECB and WTC. In addition the Federal Department of Energy Mines and Resources, Madawaska Mines and Gulf Minerals are also participants. Two other companies, Amok Limited and Uranerz Exploration and Mining Limited have indicated an interest in participating. The project goals and program are set out in Tables I and II, respectively.

PRELIMINARY BENCH SCALE EXPERIMENTS

Objectives

The objective of these experiments was to examine precipitant dosage and mixing requirements as well as the effects of inorganic flocculants and organic polymers on the removal of ^{226}Ra and the settleability of the $BaRaSO_4$ precipitate from effluents from the tailings ponds of several uranium mines.

TABLE I.   JOINT GOVERNMENT/INDUSTRY PROGRAM GOALS
FOR ^{226}Ra REMOVAL FROM URANIUM MINING EFFLUENTS

1. To develop at pilot scale, a physical/chemical process to reduce the ^{226}Ra content of uranium mining and milling effluents.

2. To demonstrate at pilot scale, a reasonably achievable level of ^{226}Ra in the effluent, with target levels of 10 pCi/L total ^{226}Ra and 3 pCi/L dissolved (filterable) ^{226}Ra.

3. To establish a data base for the design of a full scale treatment system.

4. To evaluate process alternatives for dewatering the sludge produced in the physical/chemical treatment process.

5. To establish a data base for the design of a full scale sludge dewatering process.

6. To estimate costs for physical/chemical treatment and sludge dewatering for a full scale system.

TABLE II.   PHASED JOINT GOVERNMENT/INDUSTRY PROGRAM
FOR ^{226}Ra REMOVAL FROM URANIUM MINING EFFLUENTS

PHASE I   -   Bench Scale Physical/Chemical Tests

1. Replication of selected previous tests for confirmation.

2. Reproduction of jar test results using a mixer permitting direct calculation of power input.

3. Optimization of BaRaSO$_4$ precipitation.

4. Optimization of coagulation/flocculation.

5. Investigation of possible advantages of sludge recycle.

6. Dynamic (flow-through) testing.

7. Determination of floc settling rates.

8. Investigation of low temperature performance.

PHASE II   -   Pilot Scale Development Tests

1. Pilot scale investigation and optimization of physical/chemical process alternatives.

2. Bench scale sludge dewatering studies.

PHASE III -   Pilot Scale Demonstration Tests

1. Pilot scale demonstration of optimum physical/chemical process conditions.

2. Production of sludge for pilot plant dewatering studies.

PHASE IV   -   Pilot Scale Sludge Dewatering Studies

1. Development and demonstration of sludge dewatering techniques.

Decant Characteristics

Samples of various tailings pond effluents were collected, characterized as shown in Table III, and subjected to various precipitation, coagulation and flocculation tests in the laboratory.

TABLE III.   TAILINGS POND DECANT CHARACTERISTICS (4)

Decant Source	pH	SS (mg/L)	SO$_4^=$ (mg/L)	Fe (Total) (mg/L)	**Filtered ^{226}Ra (pCi/L)	^{226}Ra (Unfiltered) (pCi/L)
A-1	8.6	2	1 400	<0.1	971 ± 24	1 085 ± 22
A-2	8.5	<1	1 450	<0.1	1 097 ± 23	1 096 ± 33
B-1	8.5	4	1 500	<0.1	572 ± 17	764 ± 22
B-2	8.7	21	1 700	*NA	386 ± 6	474 ± 7
C-1	8.4	17	239	<0.1	20 ± 1	45 ± 2

* Not analyzed.
** 0.45 μ.

Initially ^{226}Ra levels were determined by a chemical separation method. In order to reduce sample turnaround time this method was changed in January, 1977 to a modification of the radon de-emanation technique (5).

Precipitation Experiments

Precipitation tests were conducted in the laboratory to simulate field conditions. Based on the results of these tests and information received from the uranium mining companies, a standard precipitation procedure was adopted. The precipitating agent, barium chloride, was added to the tailings pond effluent samples at a Ba^{2+} dose of 10 mg/L with mixing at 100 rpm for 30 minutes. The resulting $BaRaSO_4$ suspension was coagulated and flocculated using a standard jar test apparatus.

Flocculation Screening Tests

A series of flocculation screening tests were conducted with a variety of coagulants and coagulant aids. Anionic and non-ionic polymers (0.1 to 10 mg/L), ferric sulphate (5 to 50 mg/L), a commercial soap solution (10 to 100 mg/L) as well as bentonite and kaolin clays when added to precipitated effluents with or without flash mixing, all proved to be *unsuccessful* in achieving the target total ^{226}Ra objective of 10 pCi/L. The addition of 5 mg/L of ferric chloride or 20 mg/L of alum or 0.5 ml/L of a high iron content acid seepage resulted in the formation of a bulky floc for effluents A and B which originated from acid-leach processes (Table III). Floc formation for effluent C, from an alkaline-leach process, was successful at alum dosages exceeding 20 mg/L.

Precipitate Settleability

The settleability of $BaRaSO_4$ precipitate, with and without coagulant addition was examined by measuring the unfiltered ^{226}Ra content of the top half of one-litre batch samples after various periods of quiescent settling. Each test included a rapid mix period of three minutes at 100 rpm and a flocculation period of one-hour at 40 rpm. Some typical $BaRaSO_4$ precipitate settling data illustrating the effect of ferric chloride addition are shown in Figure 1. *The use of ferric chloride markedly enhanced the $BaRaSO_4$ precipitate settleability.* However, the settling rates remained relatively low in terms of practical clarification considerations.

Both filtered and unfiltered radium analyses were conducted during the settleability experiments. Tailings pond effluents A and B were observed to meet the criterion of 3 pCi/L filtered ^{226}Ra without the addition of a coagulant but required a coagulant to meet the proposed level of 10 pCi/L total ^{226}Ra. For sample C, alum addition was required to meet both the present and proposed effluent guidelines.

Another series of experiments was conducted to investigate the factors affecting floc settleability. *In general, settleability improved with increasing coagulant dosage, flocculation time and to a lesser extent, flocculation speed. A typical example of the flocculation time and speed effects is shown in Table IV.*

TABLE IV.   EFFECTS OF FLOCCULATION SPEED AND TIME
ON UNFILTERED ^{226}Ra IN DECANT B.1 (4)

Flocculation Time (h)	Flocculator Speed (rpm)		
	20	40	60
0.25	53 ± 1.8	–	11.0 ± 0.5
0.5	17 ± 0.6	5.1 ± 0.4	6.7 ± 0.3
1.0	4.8 ± 0.3	5.7 ± 0.2	3.1 ± 0.2
3.5	3.7 ± 0.2	–	–
1.0*	178 ± 2.1	146 ± 3.9	122 ± 3.2

* No flocculant addition.
Flocculant: $FeCl_3 \cdot 6H_2O$ (10 mg/L).
Settling time: one hour.
After Wilkinson and Cohen (1977).

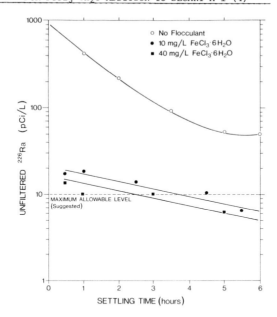

FIGURE 1.   UNFILTERED ^{226}Ra VALUES VS SETTLING TIME
FOR $FeCl_3 \cdot 6H_2O$ ADDITION TO DECANT A-1 (4)

○ No Flocculant
● 10 mg/L $FeCl_3 \cdot 6H_2O$
■ 40 mg/L $FeCl_3 \cdot 6H_2O$

UNFILTERED ^{226}Ra (pCi/L)

MAXIMUM ALLOWABLE LEVEL
(Suggested)

SETTLING TIME (hours)

JOINT GOVERNMENT/INDUSTRY PROGRAM - Bench Scale Physical/Chemical Tests

As indicated previously this program was divided into four phases (Table II).

Phase I involved a detailed laboratory program which was designed to assign some degree of confidence to the bench scale data obtained in the preliminary study and to develop design criteria for the construction of a physical/chemical pilot plant.   To achieve these objectives, both static (batch) and dynamic (continuous-flow) bench scale tests were conducted.

Decant Characteristics

As the site for the pilot scale program, the joint government/industry committee selected Rio Algom's Quirke Mine at Elliot Lake, Ontario. Consequently, all bench scale studies at the WTC were conducted using tailings pond effluent from this mine.   The characteristics of the tailings pond effluent are summarized in Table V.

TABLE V.   TAILINGS POND DECANT CHARACTERISTICS

Decant Designation	Filt. ^{226}Ra (pCi/L)	Unfilt. ^{226}Ra (pCi/L)	pH	SS (mg/L)	Alkalinity as $CaCO_3$ (mg/L)	SO_4 (mg/L)	NO_2^- (mg/L)	NO_3^- (mg/L)	NH_3 (mg/L)	TKN (mg/L)
RA1	670	670	9.6	<2	120	1 100	16.2	59.0	9.3	12.2
RA2	680	716	10.2	<2	170	1 300	17.0	33.0	21.0	23.8
RA3	791	912	10.4	<2	180	1 300	63.0	102	105	-
RA4	652	698	9.5	-	-	-	-	-	-	-
RA5	721	773	10.4	36	210	1 200	16.0	68.0	28.0	31.8
					110	1 300	16.0	143	21.0	28.9
RA6	795	845	10.2	6.9	190	1 300	16.0	154	26.0	30.2
RA7	339	412	9.9	<2	130	710	7.0	40.0	19.0	24.2
RA8	542	544	9.9	<2	-	1 200	130	60.0	29.0	31.8

Bench Scale Tests

        The batch experiments were performed with a versatile mixing apparatus[1] and a baffled two-litre beaker which could be used to simulate the proposed pilot plant precipitation reactors and flocculators. The geometric similitude thus offered, greatly facilitated scale up. Power input could be measured thereby enabling the calculation of velocity gradients which are extremely important in precipitation/flocculation processes.

        The continuous-flow experiments utilized a 20 ml/min, bench scale treatment plant (Figure 2). This apparatus was designed on the basis of the batch experimental results, but with sufficient flexibility to permit the optimization procedure to be continued in dynamic tests.

FIGURE 2.   SCHEMATIC OF BENCH SCALE CONTINUOUS FLOW APPARATUS

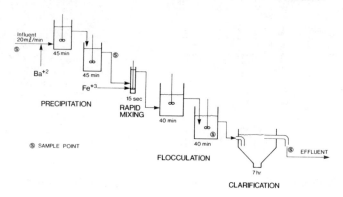

The precipitation and flocculation optimization experiments made extensive use of factorial designs (6) to screen variables efficiently and to give estimates of the independent effect of each variable.

Results

        The results of the bench scale experiments, both batch and continuous-flow, can be summarized as follows:

Confirmation Tests (1 & 2)

· *The confirmation experiments with the jar test apparatus and the ELB mixer gave results comparable to those obtained in previous WTC jar tests with a decant sampled at a different time of year.*

Precipitation Optimization (3)

· *A barium chloride dose equivalent to 16 mg/L as Ba^{2+} was selected.*

· *A batch reaction time of 15 minutes appeared to be satisfactory and measurable post-precipitation did not occur up to an hour after the end of precipitation.*

· *The initial rate of radium precipitation was rapid and appeared to be a first order reaction, as expected from theoretical considerations (Figure 3). Radium precipitation proceeded slowly after residual filtered ^{226}Ra levels of 10 pCi/L had been achieved.*

· *The theoretical continuous flow contact time, using two stirred reactors in series, was determined to be 90 minutes, based on first-order reaction kinetics and a 15 minute batch contact time. However, bench scale continuous flow tests indicated that a contact time of at least 135 minutes was required in two or three series-connected reactors.*

[1]Model - ELB Experimental Agitator, Bench Scale
    Equipment, Dayton, Ohio, U.S.A.

FIGURE 3.   C/Co VS TIME FOR FILTERED ^{226}Ra AT Ba^{+2} DOSE OF 24 mg/L

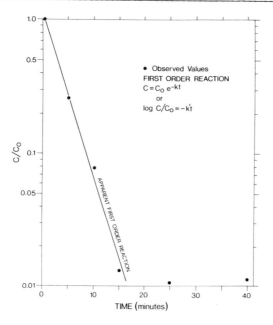

- *The mixing speed during precipitation was not critical over the range investigated. Similarly, the impeller type was not significant.*

- *Baffle plates in the reactors may have been beneficial and were recommended for the pilot scale apparatus.*

- *The use of seed crystals had no effect, and two-stage barium addition appeared to be detrimental.*

Flocculation Optimization (4)

- *Flocculation optimization experiments demonstrated that ferric chloride was a better coagulant than alum. A dosage of 4 mg/L as Fe^{+3} appeared to be sufficient and the strength of the coagulant feed solution had no effect over the range investigated.*

- *The optimum rapid mix energy input, expressed as the dimensionless product Gt, was found to be within the range 5 000 to 15 000, with rapid mix times between three seconds and two minutes giving equally good results at this energy input.*

- *The optimum batch flocculation time was 20 minutes and kinetic calculations showed that two flocculators in series would be required for continuous flow operation.*

- *A Gt product in the range of 20 000 to 30 000 was found to be optimum for flocculation.*

- *No optimum pH for flocculation was demonstrated over the range tested.*

Sludge Recycle (5)

- *Simulated batch recycle experiments indicated that recycle in the flocculation stage provided some improvement in unfiltered radium levels. Further experimentation is required to confirm these preliminary results.*

Floc Settling Rates (7)

- Sedimentation results to date are of a preliminary nature because the optimization of the precipitation and coagulation/flocculation operations is continuing.

- *Even though the settling of large flocs is rapid, low clarifier loading rates of less than 5 $m^3/m^2 \cdot d$ (100 Igpd/ft^2) are required for the small flocs to settle in order to achieve a target ^{226}Ra total level of 10 pCi/L.*

- *It was also observed that flocs were sensitive to shear, therefore headloss between the flocculators and the clarifier in the pilot plant must be minimized.*

Summary

Table VI shows some typical ^{226}Ra removal process results. These results suggest that pilot scale operation could produce an effluent meeting the total ^{226}Ra target level of 10 pCi/L, although at relatively low clarifier overflow rates.

TABLE VI. UNFILTERED ^{226}Ra RESULTS FROM CONTINUOUS FLOW SYSTEM

Date (1978)	Effluent		Condition
	2nd Flocculator*	Clarifier	
May 13	18.6		Baseline operation.
14		4.6	
16	25.4	5.9	
17	24.5	11.0	
		6.8	
18	19.0	5.9	
	23.6	8.0	
19	30.1	3.9	
		3.4	
23	40.7	6.6	Factorial precipitation experiment begun.
		17.0	
24	22.2	5.3	
25	24.0 (7 min)	11.9	
26	22.9 (9 min)	3.9	
27	38.6	14.5	
28	22.3	7.7	
29	17.7	8.4	
30	54.2	4.5	
31		13.2	
June 1		12.7	

* Settling time five minutes except as noted.

Future Experiments

The batch and continuous flow bench scale experiments are continuing. Future experiments will investigate process sensitivity to temperature and changes in influent characteristics, the feasibility of sludge recycle in the precipitation operation, and the use of polymers.

JOINT GOVERNMENT/INDUSTRY PROGRAM - Pilot Scale Study

On the basis of the bench scale experiments a mobile 33 m^3/d physical/chemical pilot plant has been designed and constructed to meet the following objectives:

1. demonstration of scale up of the process(es) developed at bench scale

2. optimization of the process with respect to various operational parameters

3. demonstration of the effluent filtered and total ^{226}Ra activity levels which can be achieved.

Effluent polishing operations and alternative treatment processes will be studied as an adjunct to the principal process development study.

Pilot Plant Design

The pilot plant, housed in an enclosed highway van and accompanying flatbed trailer, contains equipment arranged in the process configuration shown in Figure 4. In addition, there are sufficient process units to run two precipitation operations or two coagulation/flocculation and sedimentation operations in parallel. The mobile plant also contains ancillary equipment such as process control systems, chemical slurry tanks and metering pumps, columns for filtration and adsorption tests and a long-tube settling column.

FIGURE 4. SCHEMATIC OF PILOT SCALE PROCESS CONFIGURATION

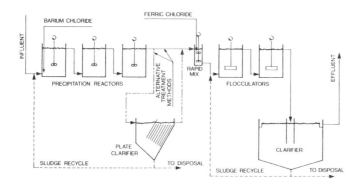

Pilot Plant Program

The pilot plant program has been designed to consist of four operational runs. Each run will last between four and six weeks and is separated by two-week intervals during which maintenance work and supplementary testing will be under-taken. These intervals will provide time for data analyses to be used in the detailed design of subsequent experiments.

The experimental program will make use of the parallel operations flex-ibility of the pilot plant for the investigation of several operational parameters. The process modes to be investigated will include once-through treatment, sludge recycle in the coagulation/flocculation step, sludge recycle in the precipitation step with a variation of the number of process units in either step.

Current Status

The first pilot scale experiment, initiated on July 17, 1978, will investigate the performance of the pilot plant under the operating conditions identified in the bench scale program, using the once-through treatment process (Figure 4). This experiment will provide information with respect to scale up from the bench and process variability. With sample replication, the experiment will also provide an analytical data base which will facilitate discrimination between various operational conditions in subsequent experiments.

STABILIZATION OF URANIUM MINE TAILINGS

The second program area under investigation at the WTC addresses the problem of uranium mine tailings stabilization. A program consisting of two components was initiated as early as 1974 with the following objectives:

1. a long term study concerning the leachability of such radionuclides as ^{226}Ra, ^{210}Pb, ^{228}Th, ^{230}Th and ^{232}Th from uranium mine tailings and co-precipitated $BaRaSO_4$ sediments (from the treatment of acid seepages),

2. examination of ultimate disposal techniques for these tailings and BaRaSO₄ sediments (from the treatment of acid seepage).

Based on the program objectives six experiments were designed as illustrated in Figure 5 addressing the following:

1. leachability of radionuclides from abandoned tailings with and without BaRaSO₄ sediments,

2. stability of BaRaSO₄ sediments in simulated settling ponds (same sediment as in 1.),

3. fresh tailings - chemical/biological stabilization tests,

4. effect of microbial degradation on the leaching of radionuclides from chemically fixed and non-fixed tailings and BaRaSO₄ sediments,

5. leachability of radionuclides - abandoned tailings, various fixation depths,

6. abandoned acidic tailings - chemical/biological stabilization tests.

Effluent and leachate toxicity to fish following chemical fixation of uranium mine water residues will also be assessed.

FIGURE 5. EXPERIMENTAL PROGRAM FOR STABILIZATION
AND LEACHABILITY ASSESSMENT OF URANIUM MINE TAILINGS

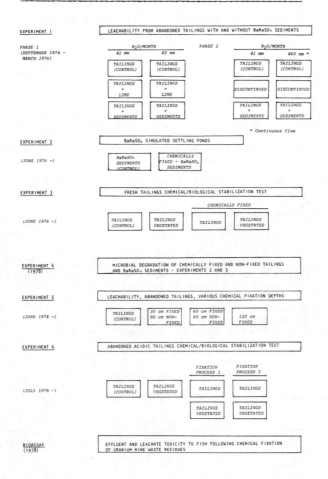

EXPERIMENT 1. - LEACHABILITY OF RADIONUCLIDES FROM ABANDONED TAILINGS - WITH AND
WITHOUT BaRaSO₄ SEDIMENTS

This experiment was initiated in 1974 using tailings and sediments in
lysimeters as illustrated in Figure 6.

FIGURE 6.   TYPICAL ACRYLIC LYSIMETERS USED IN EXPERIMENT 1. (7)

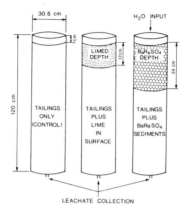

All lysimeters received two water application rates at 10-day intervals
equivalent to rainfalls of 41 and 82 mm/mth.

Early data analyses showed that the lysimeter receiving lime at rates
recommended for revegetation (≈50 tonnes/ha) resulted in leachate characteristics
which were similar to the control lysimeter. Consequently this treatment was
discontinued.

Because of the long response times involved it was deemed expedient to
accelerate these times by increasing the 82 mm/mth water rate applied every 10
days to a constant flow equivalent to 460 mm/mth. (The low rate remained
unchanged). A summary of various radionuclide levels in the leachate from these
experiments is shown in Table VII.

TABLE VII.   RADIONUCLIDE LEVELS IN LYSIMETER LEACHATES
(EXPERIMENT 1:   PHASE 1 and 2)

Radionuclide	Date	Tailings (Control)		Tailings + BaRaSO₄ Sediments	
		Water Application Rate		Water Application Rate	
		Low*	High**	Low*	High**
^{226}Ra	Aug. 1974	81	62	79	60
	June 1976	10	28	11	67
	Dec. 1977	5	20	7	56
^{210}Pb	Aug. 1974	860	1 100	880	833
	June 1976	1 980	1 305	2 100	1 425
	Sept. 1977	3 040	970	3 265	920
^{230}Th	Aug. 1974	1 565	245	310	1 440
	June 1976	1 234	224	58 520	23 805
	Dec. 1977	693	68	35 130	16

NOTE: Initial Radionuclide Activity Levels (pCi/g)

	^{226}Ra	^{210}Pb	^{228}Th	^{230}Th	^{232}Th
Tailings	122	102	6.5	3.2	1.6
BaRaSO₄	1 360	2 935	162	3 790	332

* Low = 41 mm/mth)
** High = 82 mm/mth) 10-day intervals - August 1974 to June 1976.

= 460 mm/mth   continuous - June 1976 to December 1977.

<u>Radionuclides</u> - For ^{226}Ra, the data show higher leachate concentrations for higher water application rates. This did not hold for ^{210}Pb as the levels in the leach- ate decreased with increasing water application rate thereby showing the effect of dilution. Figure 7 shows the effect of water rate for ^{230}Th in the leachate peaking at 2 x 10^5 pCi/L after five months of leaching and then declining as a function of water application rate. The higher rate resulted in lower leachate levels. This is again symptomatic of a dilution or flushing action. Radionuclide mass balances around the lysimeters have not yet been completed.

FIGURE 7. LEVELS OF ^{230}Th IN LEACHATE FROM ABANDONED TAILINGS

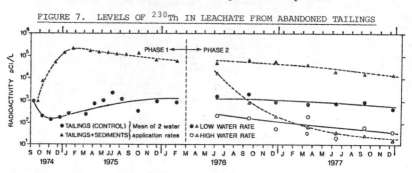

EXPERIMENT 2. - BaRaSO$_4$ SIMULATED SETTLING PONDS

The objective of this experiment, initiated in 1976, was to establish the degree and rate of radionuclide release to surrounding waters from co-precipitated BaRaSO$_4$ sediments and to reduce this rate by chemical fixation.

The simulated settling ponds are illustrated in Figure 8. The experiment uses one pond as a control (sediments), the other contains chemically fixed sediments. Both are under a constant 41 cm head of water with an effluent rate of one litre/day.

FIGURE 8. TYPICAL LYSIMETERS FOR SIMULATED SETTLING POND
AND STABILIZATION TEST EXPERIMENTS

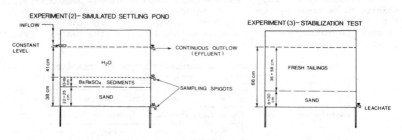

Radiochemical analyses for ^{226}Ra, ^{210}Pb, ^{228}Th, ^{230}Th and ^{232}Th were performed on effluent samples representing two-month leaching periods.

^{226}Ra - Over an 18-month period the level of filtered ^{226}Ra in the effluent from the chemically fixed sediments settling pond was <3 pCi/L as opposed to an average of 7 pCi/L from the non-fixed sediments.

^{210}Pb - The ^{210}Pb detectable limit (15 pCi/L) was not exceeded until the leaching period July to August, 1977 when levels of 26 pCi/L were measured in the effluent from non-fixed sediments.

<u>Thorium</u> - Levels of ^{228,232}Th were <5 pCi/L, while levels of ^{230}Th were <50 pCi/L for all treatments during the July to August, 1977 leaching period. The low levels

of thorium in the effluent were attributed to the pH being neutral to alkaline.

pH and Conductivity - The pH of leachate from the BaRaSO₄ sediments remained
relatively stable at around 10 for the chemically fixed sediments and 8 for the
control sediments (Figure 9).

FIGURE 9. LEACHATE pH FROM ABANDONED TAILINGS, BaRaSO₄
SEDIMENT AND FRESH TAILINGS

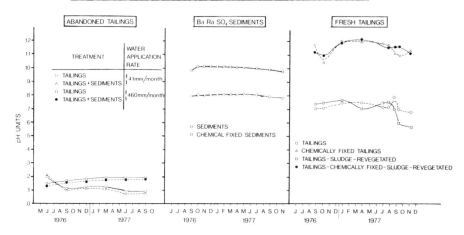

Figure 10 illustrates the considerable release of ions to the effluent
following chemical fixation.

FIGURE 10.   CONDUCTIVITY OF BaRaSO₄ SEDIMENT EFFLUENT

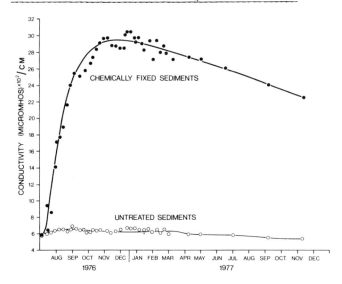

Conductivity in effluent from chemically fixed sediments peaked at 3 000
micromhos/cm after five months, gradually declining to 2 300 micromhos/cm by
November, 1977. Conductivity of effluent from the non-fixed BaRaSO₄ sediments
remained stable at about 600 micromhos/cm.

Bioassay Tests - Bioassay tests conducted on effluent during April, 1977 from
chemically fixed sediments showed it to be 100% lethal to rainbow trout. Effluent
produced from the same treatment during November, 1977 was 12.5% lethal; however,

prolonging the standard 96 h test to 155 h again resulted in 100% mortality. Bioassay tests are being continued in an effort to determine the cause of this toxicity.

EXPERIMENT 3. - FRESH TAILINGS - CHEMICAL/BIOLOGICAL STABILIZATION TESTS

This particular experiment was designed to establish radionuclide leaching rates from freshly milled tailings and attempt to reduce the leaching rate by:

1. chemical fixation,

2. sewage sludge application to obtain 3% organic matter in the surface 15 cm followed by revegetation,

3. a combination of 1. and 2. above.

The experiment consisted of placing four treatments (Figure 5) in lysimeters as shown in Figure 8. A seed mixture of red top and creeping fescue was planted. The frequency of leachate collection and analyses depended on normal rainfall and the amount of irrigation required to maintain adequate soil moisture for optimal plant growth. Approximately seven leachate collection periods per year were anticipated. The grass crop was harvested three times during the 1977 growing season.

^{226}Ra - The levels of ^{226}Ra in the leachate and the quantity leached during the 1977 growing season are shown in Table VIII.

TABLE VIII. APRIL AND NOVEMBER ^{226}Ra LEVELS IN LEACHATE
AND LEACHATE VOLUMES AS WELL AS TOTAL ^{226}Ra LEACHED
DURING THE GROWING SEASON - 1977

Parameter \ Treatments		Tailings (Control) (A)	Chemically Fixed (CF) Tailings (B)	Tailings Sludged, Vegetated (C)	CF Tailings Sludged Vegetated (D)
^{226}Ra April 1977	pCi/L	67	12	84	22
^{226}Ra November 1977	pCi/L	180	45	155	55
Volume leached per lysimeter	L	364	372	213	213
^{226}Ra leached per lysimeter	pCi	66 804	17 029	29 133	14 582

The data illustrate the effect of evapotranspiration from revegetation (C & D). Consequently only about 60% of the leachate volume was collected from this treatment in comparison to that from the non-vegetated tailings (A & B). Chemical fixation also resulted in 75 and 50% less ^{226}Ra leached than non-fixed tailings (A) and non-fixed vegetated tailings (C), respectively.

^{210}Pb - The levels of ^{210}Pb in the leachate have remained at <15 pCi/L for all treatments.

Thorium - Thorium levels in leachate were low at <5 pCi/L for ^{232} and ^{228}Th, and <50 pCi/L for ^{230}Th in November, 1977. It is anticipated that thorium levels in the leachate will remain negligible until the tailings pH drops to at least 5.0.

pH - Leachate pH values as shown in Figure 9 were between 10 and 12 for chemically fixed tailings, whereas non-fixed control and revegetated tailings remained at a pH greater than 7 for 15 months after which the effects of bacterial oxidation resulted in a drop in pH to less than 7 and as low as 5.5.

Grass Yields and Radionuclide Content - Chemical fixation did not have any adverse effect on grass yield provided sewage sludge was applied. Yields were 13 to 15 tonnes per hectare which are comparable to those of medium productive soils.

Levels of ^{226}Ra in the plant tissue (third cut) were relatively high, 55 pCi/g from sludged tailings, compared to 10 pCi/g from chemically fixed sludged tailings. Levels of thorium and ^{210}Pb in plant tissue remain to be determined.

<u>Bioassay Tests</u> - Static bioassay tests using rainbow trout were conducted in 1976. These tests showed that chemical fixation resulted in effluents toxic to fish. However, leachate collected in November, 1977 was 100% lethal from non-fixed treatments compared to 14% from chemically fixed treatments. Tests are continuing in an effort to elucidate this matter and determine the causes of toxicity.

PROGRAM EXPANSION

A series of new experiments have been initiated in 1978 (Figure 5; Expts. 4-6). At this time the objective of these experiments can be stated:

· Microbial degradation of tailings and $BaRaSO_4$ sediments following chemical fixation treatments (4)

<u>objective</u>: *to determine the impact of microbial action on chemically fixed and non-fixed tailings and $BaRaSO_4$ sediments (follow-on to Expts. 1 & 2).*

· Leachability from abandoned acid tailings following various depths of chemical fixation (5)

<u>objective</u>: *to ascertain the depth of fixation required to reduce leaching of radionuclides to acceptable levels.*

· Abandoned acidic tailings - chemical/biological stabilization tests (6)

<u>objective</u>: *to assess the effect of chemical/biological stabilization on abandoned acidic tailings as contrasted to fresh tailings - Expt. 3.*

In conjunction with the tailings experiments a series of bench scale tests are in progress which are designed to elucidate some of the observed phenomena which are not fully understood.

For example:

Levels of ^{226}Ra leaching from tailings have increased in various experiments following high water application rates. This has occurred in both abandoned and freshly disposed tailings as well as chemically fixed tailings. Monitoring of receiving streams in the disposal areas have indicated higher levels of ^{226}Ra when increased rates of percolation through the tailings were caused by spring thawing. Results to date show similar results regardless of the tailings pH which ranged from 2 to 11.

It is suspected that ^{226}Ra is relatively insoluble in the tailings complex provided that sufficient sulphate is present to form $BaSO_4$. Increased water percolation through the tailings may solubilize and flush the sulphate at rates greater than it can be generated, hence solubilization of the ^{226}Ra. Wetting and drying cycles may also affect the sulphate solubility.

This bench test will attempt to determine the effects of various rates of water percolated through tailings at both wetting and drying periods compared to equivalent amounts of water applied during the same periods as a constant flow.

A further example consists of the following:

The long-term stability of $BaRaSO_4$ in settling ponds is questionable. One factor which may cause solubility of ^{226}Ra and its migration to adjacent water courses may be agitation. This agitation may be caused by the type of in-flow to the settling area or drainage from adjacent slopes. Of particular importance is the post abandonment period when various causes of turbulence, not yet determined, may have to be controlled. This bench test will attempt to determine the effects of agitation of $BaRaSO_4$ sediments from a settling pond of an active uranium mine at Elliot Lake, Ontario.

As illustrated in this paper, the water pollution control problems in the uranium mining industry are extremely complex. This concern for the environment has led to accelerated programs to develop and demonstrate water pollution control technology which will minimize the environmental impact of uranium mining operations. This concern is equally shared by the Canadian Uranium Mining Industry as reflected by the substantial environmental assessment and control programs which they are undertaking themselves and in their participation in the cooperative technology development and demonstration program with Environment Canada and other federal government agencies.

ACKNOWLEDGEMENTS

The authors wish to express their appreciation to Mr. D. Ide, WTC and Mr. J. Fitzgerald, Rush Engineering Services Limited for their assistance in these projects; particularly the analytical work. To assist WTC (in the ^{226}Ra removal project), IEC International Environmental Consultants Limited were contracted to provide advice and assistance in the accelerated bench scale program and in the design of the pilot scale process flow sheet. Their contribution and in particular that of Dr. P.M. Huck is gratefully acknowledged. The analytical and technical support of the Laboratory Services and Facilities Services Sections, WTC are also acknowledged.

REFERENCES

1. Roy, A.C., and W. Keller, "Water Pollution in the Serpent River Basin, Status Report", Ontario Ministry of the Environment, Northeastern Region, 1976.

2. Coady, J.R., (AECB, Ottawa) Personal Communication, June 2, 1977.

3. Bland, C.J., and D.E. Gonzalez, "A Report on a Study to Determine the Chemical and Physical Stability of Barium/Radium Sulphate Sludges Under Varying Conditions of Temperature, pH and Dissolved Sulphate Concentrations". Performed for Energy Mines and Resources, Canada, Contract #0SU76-00147, March, 1977.

4. Wilkinson, P., and D.B. Cohen, "The Optimization of Filtered and Unfiltered ^{226}Ra Removal from Uranium Mining Effluents, Status Report (1976-1977)". Presented at Canadian Uranium Producers' Metallurgical Committee Workshop on ^{226}Ra Control, Ottawa, Canada, October 17, 1977.

5. Rushing, D.R., W.J. Garcia and D.A. Clark, "The Analysis of Effluents and Environmental Samples from Uranium Mills and of Biological Samples for Radium, Polonium and Uranium". Proceedings, Radiological Health and Safety in Mining and Milling Nuclear Materials, International Atomic Energy Agency, Vienna, 1964.

6. Davies, O.L., "The Design and Analysis of Industrial Experiments", Oliver and Boyd, Edinburgh (1956).

7. Bryant, D.N., D.B. Cohen and R.W. Durham, "Leachability of Radioactive Constituents from Uranium Mine Tailings - Status Report (1974 to 1977)". Presented at Canadian Uranium Producers' Metallurgical Committee Meeting, Ottawa, Canada, May 19-20, 1977.

DISCUSSION

D.R. WILES, Canada

I have for quite some time been interested in knowing what is the nature of your chemical fixation ?

J.W. SCHMIDT, Canada

The chemical fixation that has been used in these experiments is called "chem-fix" which is a proprietary process. There is another process available by a Canadian company which uses presumably naturally occuring materials for fixation. In these additional experiments we anticipate looking at that and perhaps making a comparison between the two processes. The suggestion is that it is much less expensive than the "chem-fix" process. That remains to be seen.

D.C. McLEAN, United States

What do you propose to do with the radium sludge that you produce in your physical chemical process ?

J.W. SCHMIDT, Canada

We do not know. Basically, we are hoping that we can dewater in to an extent, and put it in a form in which it can be handled. How it will be ultimately disposed of has not yet been decided. The idea here is to contain that sludge, rather than dispersing it on the lakes, as is the generally occuring practice, for example.

REMARQUES AU SUJET DE LA STABILISATION DES STERILES DE
L'INDUSTRIE DE L'URANIUM PAR IMPLANTATION DE VEGETATION

R. Bittel, N. Fourcade, P. Zettwoog
Commissariat à l'Energie Atomique
Fontenay-aux-Roses, France

RESUME

        L'industrie de l'uranium (extraction et traitement des
minerais) laisse un volume important de déchets ou stériles, inex-
ploitables pour la récupération de l'uranium, mais qui constituent
des risques pour l'environnement et qui l'enlaidissent. Une solution
à ce problème est l'implantation de végétation sur ces amoncellements. Les conditions à respecter pour cette opération sont décrites.
Il apparait qu'il n'existe pas de technique universelle, mais que,
parmi ce qu'autorisent différentes disciplines scientifiques, on doit
s'inspirer de la connaissance des paramètres locaux de manière à
obtenir une intégration de l'ancien site industriel dans le paysage
général.

ABSTRACT

        The uranium industry (extraction and treatment of the
mineral ores) produces a large volume of wasten unusable for recupe-
rating uranium but which nevertheless constitutes an environmental
risk and which disfigures the latter. One solution to this problem
is the planting of vegetation on the waste rips. The conditions to
be fulfilled in this operation are described. It appears that no
universal technique exists, but that from among those who are responsible for different scientific disciplines, we ought to gain knowledge of local parameters in such a way as to acheive the integration
of an old industrial site in the general landscape.

1. INTRODUCTION

L'industrie de l'uranium laisse des volumes très importants de déchets "stériles" (roches à teneur en U négligeable ou non économiquement exploitables et déchets de traitement) qui constituent un risque environnemental [15].

Les amoncellements de ces déchets solides, abandonnés sur les carreaux des mines et des usines n'ont évidemment aucune esthétique. Il est donc souhaitable de trouver des solutions, d'une part pour pallier les dégradations des sites et réintégrer ceux-ci dans le cadre de leur environnement, d'autre part pour remédier à tout risque de pollution. La solution, qui est analysée et qui est utilisée dans beaucoup de cas est la stabilisation des amoncellements par leur mise en végétation. D'autres solutions sont évidemment possibles et employées en fait actuellement, par exemple : le remblayage d'anciennes mines et l'utilisation pour la voirie sur le carreau minier.

2. CONDITIONS EXISTANT DANS LES AMONCELLEMENTS DE DECHETS SOLIDES

Les tas de déchets provenant de l'industrie minière de l'uranium et qu'on peut désigner sous le vocable général de "stériles" présentent des caractéristiques très défavorables à toute implantation de végétaux supérieurs.

1) conditions radiologiques : fortes teneurs en radium et en ses descendants, donc risque de pollution par le radium et le radon ;

2) conditions physiques : très grande hétérogénéité, parfois absence de composants fins (limons, argiles), absence de matières organiques ;

3) conditions agrochimiques et toxicologiques : absence de complexe adsorbant, déséquilibres minéraux, en particulier déficience en azote, concentrations trop élevées en métaux lourds, en particulier en cuivre, plomb et zinc, comme c'est le cas pour d'autres exploitations minières [15].

3. REMEDES AUX CONDITIONS DEFAVORABLES

3.1. Isolation des radioéléments

On peut utiliser les méthodes employées pour l'isolement des sources de radioactivité naturelle dans le but d'inhiber l'émanation de radon, en particulier couvrir les stériles par une couche compacte suffisamment épaisse de différents matériaux, allant du ciment à la terre argileuse tassée, surmontée d'un sol agricole. Le radium est bloqué par syncristallisation avec le sulfate de baryum [20].

3.2. Amélioration des caractéristiques physiques du sol

Il importe surtout d'apporter au milieu les phases fines, sable, limon, argile, humus, qui, dans les sols normaux, sont les intermédiaires entre les phases grossières (gros sable, cailloux, débris de roches), les eaux circulant dans les sols et les racines des végétaux.

La solution qui paraît la plus naturelle est d'apporter sur les entassements la terre superficielle des carreaux qui, bien souvent, a été enlevée mais gardée à l'écart, ou bien la terre prélevée dans les régions voisines. On effectue ensuite des semis de végétaux convenablement choisis (paragraphe 3.3.), des implantations d'arbustes prélevés dans des pépinières, ou bien on utilise la technique employée pour gazonner les talus bordant certaines autoroutes en revêtant la terre rapportée par des plaques de gazon accompagné de l'horizon superficiel du sol d'origine (où se trouvent les racines). Ces solutions, qui souvent se sont révélées coûteuses (transport de volumes importants de sol), lentes (développement très lent de la végétation sur des terres rapportées) et même aléatoires [1-11], ont donné des résultats encourageants en France [7].

On peut songer à employer les boues résultant du traitement des effluents agricoles, industriels, urbains : on apporte ainsi une quantité appréciable de matière organique, mais son humidification reste très aléatoire en l'absence d'un support jouant le rôle des horizons profonds des sols en place et en raison de la forte concentration des boues en métaux lourds toxiques. Il est d'ailleurs à craindre que cette pratique n'enrichisse en métaux lourds indésirables un milieu qui en est déjà trop pourvu [16-17]. A titre d'exemple, les boues résiduaires de la ville de Scranton (Pennsylvanie, USA), utilisées pour stabiliser les terrils de l'industrie charbonnière, contiennent en moyenne par kg 260 mg de zing, 80 mg de cuivre et de plomb, 3 mg de cadmiun.

On peut évidemment attendre que les phénomènes naturels pallient les conditions physiques défavorables, lixiviation par les eaux météoriques, apports de sable, de limon et d'argile par le vent, effets successifs d'humidification, de dessiccation ou de gelées pour faire éclater les débris rocheux, installation, à la longue, d'une végétation naturelle, donc adaptée aux conditions défectueuses de milieu. Alors se développera, petit à petit une vie végétale qui fournira la première source d'humus, ce qui, par la suite, permettra d'installer d'autres plantes plus exigeantes. Il n'est pas exclu, pour accélérer les processus naturels, d'ensemencer les stériles par des végétaux inférieurs très résistants aux conditions physiques défavorables : micro-organismes intervenant dans les métabolismes du fer, du soufre et de l'uranium, sorédies des lichens qui, in situ, revêtent parfois les roches uranifères et dont l'action pédogénétique [2-3-14] facilite l'installation ultérieure de végétaux plus exigeants.

3.3. Adaptation aux conditions agrochimiques et toxicologiques

Il est nécessaire d'apporter les éléments indispensables aux végétaux sous une forme assimilable. Des essais en séries, effectués sur des graminées cultivées sur des déchets d'industries minières (métaux lourds), ont permis de prévoir les avantages d'apports d'engrais classiques. Le test auquel on a eu recours est le développement racinaire. En général, le facteur agrochimique limitant est la déficience en azote. On y remédie bien entendu par des apports azotés ; mais en raison du manque d'éléments fins dans la structure des entassements, les formes solubles d'azote sont entraînées avec les eaux météoriques avant d'être utilisées par la végétation. On doit donc préférer les formes organiques de l'azote lentement minéralisables. L'emploi des boues résiduaires peut être préconisé, car elles apportent à la fois de l'azote organique et un substrat qui limite les pertes de nitrates par drainage.

Malheureusement, le supplément de métaux lourds qui en résulte est susceptible de diminuer l'intensité de la nitrification [21]. La réduction des nitrates après absorption par les végétaux peut être gênée par de trop fortes concentrations en métaux lourds. L'arrosage par les eaux résiduaires contenant à la fois de l'azote et divers complexants des métaux lourds peut améliorer l'assimilation des nitrates, à condition que ces eaux ne soient pas elles-mêmes trop pourvues en éléments toxiques (inhibition de nitrate et nitrite-réductases [6]. L'utilisation dans la couverture végétale de mélanges de graminées et de légumineuses est favorable [11] puisque les légumineuses fixent l'azote atmosphérique. On a préconisé l'emploi du mélange suivant :

graminées
Lolium perenne (ray-grass) : 25 % du poids du semis
Dactylis glomerata (Dactyle) : 20 % du poids du semis
Phleum pratense (fléole) : 20 % du poids du semis
Festuca rubra (fetuque) : 20 % du poids du semis

légumineuses
                    Trifolium repens "Kent" (un trèfle) : 10 % du poids
                                                              du semis
                    Lotus corniculatus (Lotier) : 5 % du poids du semis.

        L'existence de plantes résistantes aux radioéléments natu-
rels, en particulier à l'uranium et au radium n'a pas été démontrée.
Cependant Marchantic polyphurma (une Hépathique) voisine des mousses,
présente des anomalies sur terrains radioactifs et sa domination vis-
à-vis des autres Bryophytes y devient très clairement évidente [8].
En général, ce sont les éléments associés aux radioéléments et qui
sont parfois présents en fortes concentrations dans les stériles,
qui, dans le cas des mines d'uranium comme dans celui des mines de
métaux lourds, conditionnent, en raison de toxicités chimiques, la
remise en végétation.

        Il est bien connu que les différents végétaux sont plus ou
moins sensibles à la présence de métaux lourds et les études phyto-
sociologiques dans des régions fortement minéralisées fournissent
des renseignements précieux pour orienter le choix des espèces à
implanter sur les entassements de "stériles" [9]. On peut distinguer
parmi les végétaux herbacés résistants [1] :

        - les métallophytes qu'on ne trouve que sur les terrains
très minéralisés, par exemple : de nombreux lichens, mousses et
algues, Viola calaminaria (une espèce appartenant au genre Violette)-
Thlaspi alpestre calaminare (Tabouret, Crucifère) - Armeria maritima
(voisin de la Primevère commune) ;

        - des pseudométallophytes parmi lesquels on trouve des
plantes sauvages très fréquentes, en particulier : Agrostis tenuis
(Agrostide), Polygala vulgaris (ordre des Géraniales), Campanula
rotundifolia (une Campanule) Rumex acetosa (l'oseille), Avena
pubisens (Avoine pubescente) Holcus lanatus (La Hougue), Plantago
lanceolata (le Plantain), Genista tinctoria ("Genêt des teinturiers"),
Linum catharicum (un Lin). Sous un climat tempéré subhumide, comme
celui de l'Europe Occidentale, d'autres végétaux peuvent s'implanter
sur des sols presque squelettiques et riches en métaux lourds. C'est
le cas du Ray-grass (Lolium pratense), du Dactyle (Dactylis
glomerata), de la Fetuque (Festuca rubra), de certains trèfles
(trifolium repens) et du Lotier (Lotus corniculatus) [11]. Les végétaux inférieurs enfin, sont en général, résistants aux métaux lourds
[18].

        Les arbres sont en général moins résistants que les plantes
herbacées. Cependant certains genêts, qui sont des arbustes, sont
des pseudometallophytes. Les conifères ont une faible résistance,
sauf le Pin Noir d'Autriche, Pinus Laricio.

        Cependant beaucoup d'espèces d'arbres peuvent être implan-
tées si ce n'est sur les stériles, du moins dans les espaces pertur-
bés par les opérations minières. C'est, en particulier, ce qui est
réalisé en Irlande dans les zones uranifères au nord ouest de Dublin
(T.A.R.A.) [10]. Il est alors souhaitable de transplanter de jeunes
arbres préalablement élevés en pépinières.

        La résistance des végétaux aux métaux lourds est liée à
l'incorporation de plasmides aux noyaux cellulaires [12-13]. L'obtention d'espèces et de variétés résistantes peut se faire par sélec-
tion à partir de graines de végétaux supérieurs des sols fortement
minéralisés.

        De toute manière, le choix d'une espèce ou d'une variété
dépendra de divers facteurs externes, en particulier du pH et de la
teneur en calcium d'un milieu. Ainsi, sur un milieu acide décalcifié,
on préfère Agrostis tenuis, plutot qu'Agrostis stolonifera, autre
espèce d'Agrostide, qui s'adapte très bien aux fortes teneurs en
calcium [15]. Il est évidemment impératif de tenir compte des
conditions climatiques.

4. EFFETS DE LA MISE EN VEGETATION DES STERILES

Il s'agit tout d'abord d'une amélioration de l'aspect
général du site, mais il faut souligner un certain nombre d'autres
effets : la présence de racines implique un "effet rhizosphérique"
dû au grand développement des micro-organismes, à la production de
composés à action acidifiante et souvent à action complexante [4].
Ces processus conduisent en général à accroître la mobilité des
éléments du milieu, notamment des métaux lourds. L'apport d'engrais
phosphatés, parfois apportés pour que les végétaux puissent profiter
au mieux de l'enrichissement du milieu en azote, peut au contraire
provoquer une immobilisation momentanée des polluants métalliques.
Enfin, la vie végétale qui implique l'incorporation des substances
minérales dans des composés biochimiques organiques (incorporation
dans les tissus végétaux, sorption ou complexation par les composés
humiques ou fulviques résultant de la décomposition des litières
végétales) conduit évidemment à des changements d'états physico-
chimiques pour tous les éléments présents, et, par suite, à une
nouvelle répartition des radioéléments et des métaux lourds dans le
milieu physique et le milieu organisé. Il semble d'ailleurs que les
végétaux très résistants prélèvent ces éléments avec la même intensité que d'autres végétaux moins bien adaptés. Les végétaux qui sup-
porteront bien les conditions toxicologiques qui leur sont imposées
sont donc des vecteurs importants de la contamination, notamment
lorsque, dans le milieu, sont en trop faible proportion les colloïdes
argileux et humiques qui, dans les sols normaux en place, s'opposent
en partie aux prélèvements racinaires. Il faut rappeler ici que
l'acidité du milieu qui prévaut souvent en raison de l'oxydation
microbiologique des sulfures, augmente l'assimilabilité par les
plantes des divers métaux lourds du milieu, à l'exception du molybdène et peut être du radium si les concentrations en sulfate sont
suffisantes. Dans le cas de sols podzoliques probablement comparables
aux sols qu'on obtiendra sur les stériles après une longue période
de végétation, les facteurs de transferts du radium du sol aux
cendres de végétaux sont de l'ordre de 10 pour les lichens et les
mousses, de l'ordre de 100 pour le sorbier et de 50 pour la myrtille.
A titre de comparaison, les facteurs de transferts de l'uranium
varient de 0,1 à 1 (lichens et mousses) et sont donc d'un ou deux
ordres de grandeur inférieurs à ceux du radium [19]. Si les sols sont
relativement acides, l'absorption racinaire de tous les métaux
lourds, sauf le molybdène, est favorisée. Par contre, si les sols
sont basiques, à la suite d'apports massifs d'amendements calcaires,
seul le molybdène, s'il est présent dans le milieu, passerait en
trop grande quantité chez les végétaux et conduirait à des phénomènes de toxicité chez le bétail [5]. De toute manière, les conséquences des opérations de mise en végétation à moyenne ou lointaine
échéance, ne pourront être prévues que grâce à un contrôle agro-
pédochimique de longue haleine sur des sites dont les paramètres
ont été évalués avec soin.

5. CONCLUSION

Les études réalisées dans le but de stabiliser les stériles
sont essentiellement multidisciplinaires ; elles nécessitent le
concours de minéralogistes possédant des connaissances approfondies
dans le domaine de la radioprotection, des agro-chimistes et des
biochimistes bien informés des problèmes récents posés par l'éco-
toxicologie. Outre son intérêt pratique et l'urgence de traiter
certains aspects sur le plan local, l'intérêt fondamental des
problèmes évoqués nécessite donc d'être souligné.

Différentes solutions sont possibles, elles viennent d'être
brièvement analysées. On en saurait trop insister sur le fait qu'il
n'existe pas de panacée universelle susceptible de résoudre toutes
les difficultés. Il est indispensable de traiter, "coup par coup"
le cas des différents sites, en tenant évidemment compte des consi-
dérations générales qui ont été développées précédemment, mais en se

reportant sans cesse aux conditions particulières du site. C'est
ainsi, et il existe déjà des exemples concrets de réussite, qu'on
parviendra à "intégrer" un site minier abandonné dans le cadre de
l'environnement général. Mais il ne faudra pas oublier l'histoire
des sols ainsi reconvertis. Les conséquences à plus ou moins lointaine échéance du repeuplement végétal nécessiteront des contrôles
prolongés avant et pendant la nouvelle utilisation du milieu. En
effet, la végétation peut faire apparaître une nouvelle voie de
transfert de la contamination car les radioéléments ont alors accès
aux espèces biologiques supérieures.

REFERENCES BIBLIOGRAPHIQUES

[1] ANTONOVICS J., BRADSHAW A.D., TURNER R.G.
     Heavy metals tolerance in plants
     Adv. in Ecological Res., 7, 1-85, 1971.

[2] ASCASO C., GALVAN J.
     Studies on pedogenic, action of lichen acids
     Pedobiologia, 16, (5), 321-332, 1975.

[3] ASCASO C., GALVAN J., ORTEGA C.
     The pedogenic action of Parmelia Conspersa
     Lichenologist, 8, 151-171, 1976.

[4] BITTEL R., MAGNAVAL R.
     A propos de quelques aspects de l'étude des transferts des
     polluants radioactifs jusqu'aux sites critiques des organismes
     Radioprotection, 12, (4), 329-343, 1977

[5] BLACK C.A.
     Soil acidity
     dans : BLACK C.A., Soil Plant Relationships.
     New York, J. Wiley and sons, 273-355, 1968.

[6] BROWN C.M., JOHNSON B.
     Inorganic Nitrogen Assimilation
     dans : Advances in acquatic microbiology (DROOP M.P. and
     JANNASH H.W. éd.) Londres, Academic Press, 1, 49-114, 1977

[7] DELAGE R., ROUSSEL V.
     Environnement des petites découvertes du Morvan
     Industrie Minérale - Mines 1-5, 1977.

[8] DELPOUX M.
     Etude expérimentale des effets de la radioactivité naturelle
     tellurique sur les végétaux. Hypothèse sur l'influence de
     l'environnement fortement énergétique sur les êtres vivants
     Thèse, Doctorat ès Sciences Naturelles, Université Paul
     Sabatier, Toulouse, n° d'ordre 631, 25 septembre 1974.

[9] ERNST W.
     Okologisch - soziologische Untersuchumgen auf Schwermetall -
     pflanzengesellschaften Südfrankreiches und des Östlichen
     Harzvorlandes.
     Flora (Iena) 156, 301-318, 1966.

[10] IRLANDE - Tara Mines Limited
     The Landscape plan.

[11] JOHNSON M.S., Mc NEILLY T., PUTWAIN P.D.
     Revegetation of metalliferous mines spoil contaminated by lead
     and zinc
     Environ. Pollut., 12, 261-277, 1977.

[12] JOWETT D.
     Populations of Agrostis spp. tolerant of heavy metals
     Nature, 182, 816-817, 1958.

[13] LEFEBVRE C.
Evolution in heavy metals tolerant *Armenia maritima*
dans : International Conference on Heavy Metals in the
Environment. Toronto (Ontario, Canada), (27-31 octobre 1975),
C 148-C 150.

[14] OZENDA P., CLAUZADE P.
Les lichens. Etude biologique et Flore illustrée
Paris, Masson (1970).

[15] SMITH R.A.H., BRADSHAW A.D.
Stabilization of toxic mine wastes by the use of tolerant plant
populations.
Trans. Inst. Min. Metall. 81, A 230-A 238, 1972.

[16] SOMMERS L.E., NELSON D.W., YOST K.J.
Variable nature of chemical composition of sewage sludges.
J. Environ. Qual., 5, (3), 303-306, 1976.

[17] SOPPER W.E., KARDOS L.T., EDGERTON B.R.
Reclamation of a burned anthracite refuse bank with municipal
sludge.
Compost. Sci., 17, (2), 12-19, 1976.

[18] STOCKES P.M.
Adaptation of green algae to high levels of copper and nickel
in aquatic environments.
dans : International Conference on Heavy Metals in the
Environment. Toronto (Ontario, Canada), (27-31 octobre 1975),
C 146-C 147.

[19] VERKOVSKAYA I.N.
The content and translocation of natural radioactive elements
in the system soil - plants - animals under natural and
experimental conditions.
dans : Centre d'Etudes Nucléaires de Cadarache, Symposium International de Radioécologie (8-12 septembre 1969), 781-832.

[20] VIVYURKA A.J.
Rehabilitation of uranium mines tailing areas
Can. Mining J., 1975.

[21] WILSON D.O.
Nitrification in soil treated with domestic and industrial
Sewage sludge.
Environ. Poll., 12, 73-81, 1977.

DISCUSSION

J. BAUER, France

You talked about the problem of contamination of higher levels of plant life. This is for the wild life you are talking about ; the problem of preventing contamination is difficult ? Europe is already peopled enough with animal species that are just about everywhere and it is through these animal species that we could get contamination. Are you planning in this case to enclose the reconstituted or contaminated zones ?

N. FOURCADE, France

Well, we will sample some species and analyse them for radium. If there is too much radium we should stop consumption of such species.

J. BAUER, France

I was talking about mercury, for example, in the tuna fish.

N. FOURCADE, France

Everything will depend on the standards we apply, then. But I think the problem is one that ought to be dealt with by radiobiologists.

i

CANMET'S ENVIRONMENTAL AND PROCESS RESEARCH ON URANIUM[1]

by

D. Moffett*, G. Zahary**, M. C. Campbell+ and J. C. Ingles++

ABSTRACT

Environmental research related to uranium tailings is being carried out within the Mining Research Laboratories and Mineral Sciences Laboratories of CANMET, EMR. Field-related research on uranium tailings has been conducted at Elliot Lake for over five years. Much of the work has been focused on a program to rehabilitate pyritic tailings. This has resulted in developing a practicable technology for growing vegetation on such wastes. Limitations have been small size of the test-plots and the relatively short period during which experiments have been carried out.

A research program aimed at identifying and reducing acidic and radioactive effluents is also underway at Elliot Lake. These liquid effluents have been identified as the most serious threat to the environment.

Research at the Mineral Sciences Laboratories and its predecessor divisions relating to the processing of uranium and thorium ores is outlined. A process has been developed for recovering thorium which could reduce the overall radioactive load in the tailings. Much of the current work is related to developing new technology for recovering uranium from lower-grade ores which, however, is unlikely to be implemented within the next ten years. A significant effort is also being made in removing pyrite, preconcentrating radioactive minerals, and identifying and removing chemical compounds that carry radium in solid tailings. All of these investigations could have impact on near-term solutions to some environmental problems. Mineral Sciences Laboratories has the analytical capacity to identify and characterize both the mineralogical and most radiochemical constituents of the Elliot Lake ores.

* Research Scientist, ** Manager, Elliot Lake Laboratory, Mining Research Laboratories.
+ Manager, Ore Processing Laboratory, ++ Assistant Chief, Mineral Sciences Laboratories, Canada Centre for Mineral and Energy Technology, Department of Energy, Mines and Resources.

Key words: Elliot Lake, Uranium Ores, Tailings Disposal, Vegetation, Radioactive effluents.

[1] Originally prepared as a brief to the Ontario Environmental Assessment Board on the uranium mine expansion in the Elliot Lake area, July 1977 (Report MRP/ERP/MRL/MSL 77-84 (R)).

RECHERCHES RELATIVES A L'ENVIRONNEMENT ET AU TRAITEMENT DE L'URANIUM (CANMET) [1]

par

D. Moffett*, G. Zahary**, M.C. Campbell+ et J.C. Ingles++

RESUME

Les recherches environnementales relatives aux résidus d'uranium se poursuivent aux Laboratoires de recherche minière et à ceux des sciences minérales de CANMET (EMR). Des recherches pertinentes sur le terrain se poursuivent à Elliot Lake depuis plus de cinq ans. La majorité des travaux a porté sur un programme visant à reconditionner les résidus pyritiques, et ont abouti à la mise au point de techniques pratiques de culture de la végétation sur ces résidus. La portée des expériences est toutefois limitée, puisque ces cultures ont été effectuées sur des lots expérimentaux de petites dimensions et au cours de périodes de durée relativement courte.

Un programme de recherches destiné à identifier et à réduire les effluents acides et radioactifs se poursuit présentement à Elliot Lake. On a en effet reconnu que les effluents présentent une menace des plus sérieuses pour l'environnement.

On trouvera ici les recherches, effectuées aux Laboratoires des sciences minérales, et dans les divisions qui l'ont précédé, et qui portent sur le traitement des minerais d'uranium et de thorium. Un procédé de récupération du thorium vient d'être mis au point: il permettrait de réduire la charge globale de radioactivité dans les résidus. La majorité des travaux actuels portent sur l'élaboration de nouvelles techniques pour la récupération de l'uranium à partir des minerais de basse teneur qui, toutefois, ne pourront vraisemblablement être appliquées avant les dix prochaines années. D'importants efforts sont également faits pour prélever la pyrite, préconcentrer les minéraux radioactifs, et identifier et retirer les composés chimiques qui véhiculent le radium dans les résidus solides. Toutes ces recherches pourraient avoir des répercussions sur les solutions, à court terme, de certains problèmes environnementaux. Les Laboratoires des sciences minérales peuvent procéder à des analyses pour identifier et caractériser les composés minéralogiques et pour la plupart radioactifs des minerais d'Elliot Lake.

---

*Chercheur scientifique, **Directeur, Laboratoire d'Elliot Lake, Laboratoires de recherche minière.
+Directeur, Laboratoire de traitement des minerais, ++Chef adjoint, Laboratoires des sciences minérales, Centre canadien de la technologie de l'énergie et des minéraux, ministère de l'Energie, des Mines et des Ressources.

Mots-clés: Elliot Lake, Minerais d'uranium, Elimination des résidus, Végétation, Effluents radioactifs

[1]Original préparé, sous forme de mémoire, à l'intention de la Commissio d'évaluation environnementale de l'Ontario, relativement à l'expansion des mines d'uranium de la région d'Elliot Lake, juillet 1977 (Rapport MRP/ERP/MSL 77-84 (R)).

INTRODUCTION

The Canada Centre for Mineral and Energy Technology (CANMET) of the Department of Energy, Mines and Resources (EMR) conducts research in the extraction, processing and use of minerals in Canada. An essential aspect of this work is investigating new technology to improve working conditions and productivity, and to reduce the impact of mineral production on the natural environment.

CANMET, which until recently was known as the Mines Branch, has been associated with the processing of uranium ores since 1930. Its initial involvement was the development, by the late R.J. Traill, of a process for recovering radium from pitchblende. When the Eldorado uranium mine at Great Bear Lake was reopened after World War II, CANMET was asked to develop a process for recovering uranium from the gravity plant tailings, and the Radioactivity Division was established in 1946-47 to handle the assignment. A successful process was developed, and as the industry expanded in the early 50's, modifications as well as new processes were developed for treating the newly-discovered low-grade ore bodies.

Operating staff for the new mills built during this period were trained by the Radioactivity Division in the new technology, participating in process development in the pilot plant at Ottawa and treating ore from their respective properties. As a result, there has always been close rapport between CANMET and industry personnel. This rapport which has been maintained and reinforced by semi-annual meetings of a CANMET-sponsored forum known as the Canadian Uranium Producers' Metallurgical Committee, initiated in 1960.

With the commissioning of the new mills, research on uranium processing was de-emphasized, and the Radioactivity Division was absorbed into the Mineral Sciences and Extraction Metallurgy Divisions. Some development work on the recovery of thorium from plant process solutions continued through this period, parallelling efforts of the mining companies.

Research on environmental impact in the mineral industry is currently organized under the Mineral Sciences Laboratories and Mining Research Laboratories of CANMET. The Elliot Lake Laboratory was established in 1965 as part of the division now known as Mining Research Laboratories to conduct mining research under field conditions, principally in underground mines. Environmental research has been under way at this laboratory since its establishment (1). The focus originally was on developing methods of monitoring health hazards in the underground environment. The scope was expanded in 1971 to include the reclamation of mining wastes on surface. Both research areas have been expanded as concern has grown for the effects of mineral production on worker health and the environment. Research has recently been increasingly focused on issues directly related to the uranium industry.

Tailings at Elliot Lake over the past twenty years have been discharged into surface impoundment basins. Although alternative disposal schemes such as the use of tailings as backfill, separate impoundment of pyritic material, and deep-lake disposal have been proposed, CANMET research at Elliot Lake has been directed towards rehabilitating surface impoundment basins by vegetation. Work has concentrated on this aspect of the problem because there are already over 80 million tonnes* of tailings in these surface basins, and because this mode of disposal is unlikely to be altered in the near future. The other major investigation is to identify and control the acidic and radioactive seepage from abondoned tailings areas.

Research in mineral processing by the Mineral Sciences Laboratories has developed in response to two main factors:

(1) Newly discovered ores are proving to be more complex than those from previous producing areas.

(2) Ore treatment procedures continue to provide environmental problems.

---

* metric units are used throughout this report. 1 tonne = 1.10 short tons.

REHABILITATION BY VEGETATION

A survey in 1971 (4) revealed that tailings containing sulphides were a significant mining-waste problem and that establishing a vegetative cover was the most acceptable reclamation procedure, but was not always technically or economically practical. Environmental research was thus initiated. The Nordic tailings basin of Rio Algom Ltd. was chosen as a site for field experiments because of its proximity to the Elliot Lake Laboratory.

Research began with laboratory studies of the physical and chemical properties of the tailings (5), followed by vegetation growth tests in environmental chambers (6) and finally by field trials (7). This led to the development of an effective and practical vegetation scheme (12),estimated to cost from $4,000 to $6,000 per hectare* or $1,600 to $2,500 per acre.

CANMET Vegetation Scheme

The procedure for establishing a vegetative cover on the Elliot Lake tailings is summarized in Table 1. It involves an initial treatment to neutralize acidity in the intended root zone and to provide a basic load of nutrients. This is followed by cultivation and seeding with a mixture of two grasses. Fertilizer is subsequently applied monthly during the growing season for a five-year period after which the vegetative cover is considered self-sustaining.

Two basic criteria for tailings rehabilitation are met by establishing a healthy grass cover:

(1) Erosion by wind or water is resisted by a physically stable surface.

(2) The appearance of the tailings basin is improved.

---

* 1 hectare = 2.48 acres.

Research Review

Research leading to the development of the vegetation scheme is
described in laboratory reports and publications listed in the bibliography
(5-13). The final and critical experiment began in 1973 on a 0.40-hectare
(1-acre) area. This area was prepared as outlined in Table 1 except that
26 species of plants, 14 of which were perennials, were sown.

Table 1 - Scheme for Vegetating Uranium Tailings

Year 1	Year 2	Years 3 to 5
Clear area of debris	Fertilize with 33-0-0 at 0.011 kg/m² (100 lb/acre)	Fertilize with 33-0-0 at 0.011 kg/m² (100 lb/acre)
Apply limestone at 2.24 to 6.72 kg/m² (10 to 30 tons/acre)*	Reseed bare patches	Fertilize with 5-20-20 at 0.022 kg/m² (200 lb/acre)+
Fertilize initially with 5-20-20 at 0.022 kg/m² (200 lb/acre)	Fertilize with 5-20-20 at 0.022 kg/m² (200 lb/acre)+	
Rototill top 15 to 20 cm		
Seed** and fertilize with 0-46-0 at 0.011 kg/m² (100 lb/acre)		
Fertilize with 5-20-20 at 0.022 kg/m² (200 lb/acre)		

* Rate of limestone application is specific to each tailings area.

** Seed recommended is a 40 : 60 mixture of Redtop and
Creeping Red Fescue at 0.0055 kg/m² (50 lb/acre).

+ Apply at monthly intervals during growing season.

It was concluded at the end of the first year that a wide variety of
plants will survive on tailings although many do not grow well. At the
beginning of the fourth growing season, in 1976, nine of the fourteen
perennials were judged to have persisted with acceptable health and vigour.
The measured plant yield and coverage of ground for each of the species is
shown in Table 2. Ground coverage is a primary objective and on that basis
Reed Canarygrass, Redtop, Kentucky Bluegrass, and Creeping Red Fescue
have been recognized as the best species. Legumes performed poorly and

Table 2 - Yield and Ground Coverage for Plants
Grown on Uranium Tailings

Species	Plant Yield* (kg/m^2)**	Coverage %
Alfalfa+	0.128	45
Reed Canarygrass	0.107	100
Redtop	0.066	100
Kentucky Bluegrass	0.059	92
Climax Timothy	0.056	81
Tall Fescue	0.046	88
Creeping Red Fescue	0.038	95
Birdsfoot Trefoil+	0.018	-
Ottawa Red Clover+	0.016	10

* Dry matter.
** To convert kg/m^2 to tons/acre multiply by 4.46.
+ Legume.

are not considered essential to a basic seed mixture. Because of its superior

sod-forming properties and aggressive nature, a 40 : 60 mix of Redtop and

Creeping Red Fescue is used in the scheme although there is some loss in

overall plant yield.

Soil Development

For a vegetative cover to be self-sustaining, material in the root zone

must perform the basic functions of a soil. One of these functions is to

degrade dead plant material and recycle nutrients. This is performed by a

variety of living organisms, e.g., bacteria, fungi, insects, etc. The

number and diversity of micro-organisms that exist in the vegetated tailings

are a measure of the occurence of this process. Table 3 gives the observed

mean values for vegetated and bare tailings and shows that sufficient numbers

of micro-organisms exist beneath the grass to support the recycling

process (15). A local forest soil is included for comparison.

Table 3 - Number of Micro-organisms in Vegetated
and Bare Tailings

Soil Type	Treatment	Mean Number of Micro-organisms per Gram of Soil (dry mass)
Tailings	None	0.01×10^6
Tailings	Vegetated	49.0×10^6
Forest (control)	-	2.2×10^6

Role of Trees

It has been observed that deciduous trees such as birch, poplar and willow have rooted naturally or volunteered on parts of the vegetated areas of the tailings. This encroachment of the natural forest is very desirable in the restoration and rehabilitation. In an attempt to accelerate the reforestation process, some 500 two-year-old cedar, spruce and pine seedlings were planted in 1973. A formal assessment of the experiment is underway (16), but the initial impression is that coniferous trees are not easily established even though, on the basis of physical properties, the tailings would appear to support their growth.

Radioisotope Uptake

Although radioactivity in the tailings presents no obstacle to establishing a good grass-cover, the creation of pathways for the transport of this radioactivity to the biosphere is important. In the summer of 1976, a study was undertaken which sought to determine the uptake of radioisotopes by grasses growing on uranium tailings (14). The concentration of uranium, thorium, radium-226, lead-210 and polonium-210 in the four most productive grass species growing on the tailings at the Nordic tailings basin were measured. The observed mean values for grass tissue from the tailings and from a control potting soil are shown in Table 4. Uranium, radium-226 and lead-210 contents were found to be significantly higher in the grasses grown on the tailings than in the control soil. No estimate of the potential uptake of radioactivity by animals grazing on the vegetated tailings has been made.

Table 4 - Radioisotope Concentration in Grasses Growing on
Uranium Tailings

| | μg/g | | pCi/g | | |
	U	Th	^{226}Ra	^{210}Pb	^{210}Po
On tailings	.030*	.014	5.48*	2.32*	.090
Control	.010	.017	.24	.12	.018

* significant increase over control at 95% confidence
level (t-test).

"Black Grass"

The black discolouration observed in some of the test plots has
received publicity and a measure of notoriety. Largely for that reason it
is discussed here. Limited field and laboratory studies were begun in 1975 (11)
in an attempt to identify the cause of the problem. It was hoped that the
black discolouration could be duplicated in the laboratory, but this has not
been possible. Fungi, bacteria or metal toxicity have been ruled out as
causes. The discolouration does not kill the grass, although it does inhibit
growth. A similar discolouration has been observed in grasses grown on
other tailings containing sulphides; the best solution appears to be to avoid
Timothy since it is the species most susceptible to discolouration.

Some Limitations

In spite of general optimism about the program it would be
misleading not to identify some of the limitations of our work. Soil development
is a process that takes place on a geologic time scale. Biological life is
subject to the principle of limiting factors, i.e., its existence and performance
is determined by the most limiting of the essential environmental factors.
Results would be more conclusive if the field plots were older and larger,
and plans have been made to vegetate the 2.5-hectare Spanish-American
tailings area in 1977.

All the field work done so far has been on the areas covered by coarse
tailings representing about 70% of the total surface area. Very little work
has been done in areas covered by slimes, largely because this material will
not support mechanical equipment.

It would be inaccurate to describe the tailings as anything but a hostile environment for plant life. The success achieved has been possible only because of the large investment in improving the soil environment and in accepting a low level of productivity from the biological system, i.e., accepting a low rate of survival and reseeding. Thus, critical constraints are more likely to be of an economic rather than of a technical nature.

TAILINGS EFFLUENTS

A survey of waste disposal practices at Canadian uranium mines in 1975 (2) indicated that water pollution represents the major environmental impact of uranium tailings basins. A research program was defined with the following objectives:

(1) To identify and characterize the acid-producing mechanisms which operate within the tailings.

(2) To investigate the paths by which deleterious substances - acid, metals and radioisotopes - are leached from the tailings and reach the watercourses.

(3) To propose remedial action to reduce pollutant discharge to the watercourses.

Acid Production

The work so far has consisted of field investigations of 17-year-old tailings at the west arm section of the Nordic tailings basin (20) which have provided an opportunity to monitor the acid-producing reactions in typical sulphide material. The coarse tailings or sands show markedly greater evidence of pyrite removal by oxidation at shallow depth than does the fine material, or slimes. Figure 1 shows this for two typical samples. Once acid is produced near the surface, it permeates through the tailings, leaching metals and radioisotopes into the groundwater. The existence of bacteria capable of accelerating the breakdown of pyrite has been established; in most cases the oxidation takes place very close to surface.

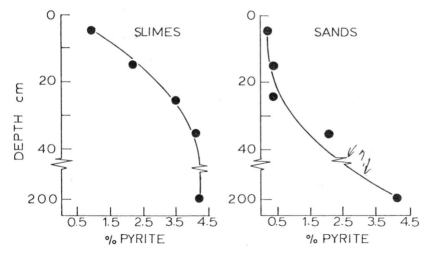

Figure 1: Pyrite content of tailings with depth.

Almost 20 years after the basin was abandoned, about 60% of the pyrite originally present in the top 20 cm of the tailings has been oxidized to sulphuric acid. This corresponds to a sulphuric acid production of 2000 tonnes for this 14-hectare tailings basin. There is evidence of less oxidation with increasing depth - the next 20 cm oxidizing only to the extent of producing a further 1000 tonnes of acid.

The parameter most often used to characterize acidity is pH. However, the pyrite oxidizing reactions that take place within the tailings often produce effluents that have an acceptable pH and yet may be extremely acidic. A better gauge of the inherent acidic nature of a liquid sample is a titration of total acidity with sodium hydroxide (21).

Pollutant Path

The depression of pH in a number of lakes in the Serpent River system is attributed to acidic seepage from tailings basins. A network of wells and weirs has been established on the west arm of the Nordic tailings and flow patterns within the tailings have been identified. Water from below the water table has been sampled each month from these wells for almost two years. Monitoring at the weirs has provided information on the quality and volume of seepage. Table 5 gives an example of the level of contaminants

Table 5 - Water Quality in the Tailings Basin

Sample location	mg/L			pCi/L					µg/L
	Acidity	TDS**	Fe	^{226}Ra	^{210}Pb	^{232}Th	^{230}Th	^{228}Th	U
Beneath Sands	12,120	24,200	1,744	27	4,067	1,400	16,000	500	5,200
Beneath Slimes	< 50	4,840	68	137	42	-	-	< 15	198
Seepage	1,790	4,570	604	13	32	80	800	150	370
MPC$_W$ *	0	1,500	0.3	10	100	2,000	2,000	7,000	40,000

* Maximum permissible concentration in water.
  The given MPC$_W$'s are compiled from a variety of regulatory agencies.

** Total Dissolved Solids.

in seepage. From this data collection system, an understanding of the paths by which rain or snow falling on the tailings is translated into acidic seepage is being developed.

Because radioisotopes present in the tailings can be detected at very low concentrations, they provide a good means of following the leaching process and subsequent seepage into the surrounding watercourses. A detailed radiological investigation of the west arm tailings basin has been completed (24) and has shown the relationship between the zones of oxidation in the tailings and water pollution. Table 5 gives a summary of the water quality observed at various points in the tailings and also for seepage water. In particular, there are two distinctly different zones of oxidation within the tailings, which appear to be determined by the size classification of the tailings. The sands overlie highly acidic and radioactive water, whereas the water under slimes is much less contaminated. The quality of seepage through the dam is intermediate between these extremes. It would appear that careful disposal practice during the active life of the tailings basin to minimize the area covered by sands might reduce effluent pollution.

An important anomaly in Table 5 is that the observed ^{226}Ra concentration in water beneath the slimes (137 pCi/L) is higher than in the acidic water under the sands (27 pCi/L). Since radium leaching from tailings is limited by the high sulphate concentration in the acidic water the observed water quality beneath the slimes may represent the effect of neutralization treatment undertaken to establish vegetation.

Radium Redissolution

The tailings contain a large reservoir of radium, and dissolution and leaching of this radium to the surrounding watercourses is potentially the most hazardous aspect of uranium tailings. One concern is the dissolution of the sludges and precipitates produced in the barium chloride treatment of effluents. A preliminary study to determine the stability of radium-barium sulphate precipitates was begun in the fall of 1976. This laboratory investigation examined the release of radium from a synthetic sludge into overlying solutions of differing pH and sulphate concentration. The results, summarized in Table 6, are erratic and raise more questions rather than provide answers. Only the sulphate concentration of the solution was found to influence radium release significantly, with an overall minimal release at 800 ppm.

Effect of Vegetation

The reclamation and vegetation of strip coal mines results in substantial reductions in acid loadings to the watercourses. Whether a vegetated tailings area would have a similar beneficial effect on water pollution from uranium tailings is unknown. A cooperative experiment

Table 6 - ^{226}Ra Activity (pCi/L) in Solutions Overlying Sludges

pH	Sulphate Concentration (ppm)			
	0	250	800	1500
2		3.0	1.2	8.2
4		5.2	1.2	6.2
6	56.8*	39.8	38.2	4.4
7		61.0	2.2	2.8
8		41.6	1.4	26.0
10		66.0	1.2	36.6

* mean of 3 values in distilled water (pH 6.5).

with Rio Algom Ltd. was begun in 1975 in an effort to assess the effects on water quality of various forms of surface treatment, including vegetation. Four model tailings pits were constructed on the Quirke mine site, each 9.14 m x 9.14 m by 1.52 m deep (30 ft x 30 ft by 5 ft deep). The loading of pollutants from each of these pits is monitored bi-weekly. One pit has been vegetated according to the CANMET scheme and a second remains untreated as a control. The others are covered with alluvial till, sawdust and 1.52 m (5 ft) of water. The experiment will run for three more years to allow sufficient time for the pyrite to oxidize.

Future Direction of Research

The necessary fragmentation in the work described in the foregoing sections of this report is currently being resolved as results are applied to the restoration of an entire tailings basin. Vegetation of the Spanish-American tailings basin represents the first attempt at rehabilitation of an entire uranium tailings basin by vegetation without top-soil cover. The consequences of a continuous vegetative cover on the acidic and radioactive seepage will be monitored.

A number of important milestones in the research program will take place in 1977-78: our test plots will cease to be fertilized as the 5-year management program closes; additional data will become available on the effects of vegetation on seepage from the cooperative experiment with Rio Algom Ltd.; and monitoring at the Nordic tailings basin over the past two years has provided a wealth of information which will be applied to the development of further control systems and procedures.

PROCESS RESEARCH

Thorium Recovery

The primary purpose of the work on thorium recovery was to improve the profitability of the Elliot Lake operations, but it also has an obvious application in eliminating an undesirable contaminant from effluents. In initial studies, only thorium was removed (35), but the work was later expanded to include the removal and recovery of rare earths which are in some demand for use in TV picture tubes and fluorescent lamps (36-39). While rare earths are not presently considered environmentally undesirable, their recovery and sale could help offset the additional cost of tailings treatment.

These treatment processes, even if they resulted in complete recovery of the thorium in solution, would not eliminate all thorium from the tailings.

Some is contained in monazite and in uranothorite grains locked in the monazite. These minerals are not attacked even by the most drastic treatment likely to be used on the ore, and hence the thorium retained is unlikely to present an environmental problem. Thus, answers to technical problems of thorium control in effluents are to be found both from the work at CANMET and that done by the companies themselves.

Water Treatment

An investigation was made in the early seventies of a proposed process for removing sulphate, nitrate, and ammonia from Elliot Lake effluents (33). This process employed ion-exchange to remove the sulphate and nitrate, followed by pH adjustment and air-stripping to remove ammonia. It was not found to be economically viable.

Recommendations for alternative treatment processes having a greater potential for success were also made. One was the separate treatment of mine water by liming and aeration to precipitate metals and strip the ammonia. Another was the elimination of nitrate and ammonia usage by substituting chloride elution and either magnesia or hydrogen peroxide for precipitation-processes that have been investigated at CANMET in the past.

Ore Treatment

The principal cause of continuing acid seepage and subsequent leaching of heavy and radioactive metals from Elliot Lake tailings is the presence of pyrite. Bulk removal of pyrite from tailings could alleviate the seepage problem and also make the tailings more amenable to revegetation. Preliminary test work on old tailings from Elliot Lake indicated that up to 96% of the pyrite could be removed by flotation (34). Pyrite-free tailings would be more easily vegetated and preliminary results have indicated that a substantial concentration of radium-226 is achieved by flotation.

Work on tailings treatment will continue in fiscal year 1977-78 with more emphasis being placed on fresh tailings. In addition, work is planned on pyrite removal from raw ore, and on the effects of preconcentration on distribution of the various radioactive species in the separate fractions of beneficiated ore. The results from this work, which will require substantial analytical and mineralogical support, could provide a pyrite-free tailing.

Extraction: Radium distribution in the mill

Certain characteristics of radium-226 in tailings at Elliot Lake and some supportive evidence in laboratory leaching tests, indicate it is potentially soluble during sulphuric acid leaching, but is retained or precipitated out on carriers during the process and is discharged with the mill tailings. Such a carrier, a member of the jarosite family, with a radium content of 80,000 pCi/g, has been formed in the tailings and is believed to be amenable to removal by flotation. Mineral Sciences Laboratories has completed a sampling campaign of the Quirke mill of Rio Algom Ltd. to trace the distribution and disposition of radium-226. The samples are currently being analyzed and an assessment report prepared. Knowing that radium-226 is present, development work can be undertaken to better control the disposition of this and perhaps other radioisotopes and to minimize their discharge with the tailings. This concept must be considered somewhat speculative, of course; a more definitive assessment will depend on further work expected to be completed by the end of the calendar year 1978.

Extraction: Chloride leaching

The long term resolution of environmental problems related to uranium extraction may well lie in the development of an entirely new process. The conventional sulphuric acid process has pyrite, sulphate, radium and heavy metal problems. Nitric acid leaching has been investigated, but poses even greater anion disposal problems. Carbonate leaching is limited to alkaline ores. It appears that chloride chemistry provides the most promising avenue for developing new technology. In 1976, a comprehensive literature search was conducted (41-46), and laboratory work on chloride chemistry was begun along two routes: hydrochloric acid leaching (47) and dry chlorination.

Hydrochloric acid leaching would use conventional unit operations for uranium ore processing in that it would involve leaching, solution purification by solvent extraction or ion-exchange, and precipitation. Some initial test work has been done, but an effective process system has yet to be developed. The second route being investigated involves the high temperature chlorination of radioactive ores with the potential of recycling chlorine and minimizing the problem of anion disposal. To date, one report has been completed on dry chlorination with results of only a very few tests (50). Both processes have advantages, but major process design problems remain to be overcome. A high recovery of uranium, thorium, radium and rare earths would be effective in meeting environmental requirements. Objectives have been set to recover at least 95% of the uranium, 80% of the thorium, 80% of the rare

earths and to produce tailings with less than 20 pCi/g of radium-226. The schematic flowsheet of the dry chlorination process (40), shown in Figure 2, gives some concept of the complexity of development work required prior to even an initial economic assessment of such new technology.

It should be noted that a new generation of processes, even with highly successful test work, must be considered to be a decade away, and could well meet with difficulties such as unavailability of suitable construction materials in that time frame. Nevertheless, initial test work has been encouraging.

ANALYTICAL RESEARCH

The requirements for analytical methods to determine radioelements for process research differ from those employed in environmental surveillance. For the former, speed in turn-around time is essential, whereas this requirement is usually much less stringent in environmental work. Environmentally-oriented radioelement analyses can thus rely on existing methods based on chemical and physical separations followed by measurement of more easily counted daughter elements, while for process research, direct methods are preferred. Such direct methods generally require more sophisticated and expensive equipment.

Mineral Sciences Laboratories already has much of the needed equipment as well as staff trained in its use, and has developed a rapid method for radium-226 analysis (58) which has greatly advanced current process research on this radioelement. Progress towards development of similar methods for other radioelements, adaptable to existing resources within MSL, is handicapped by manpower limitations.

COORDINATING COMMITTEES

CANMET has played a leading role in establishing technical committees or forums for the exchange of research information. The Canadian Uranium Producers' Metallurgical Committee has already been mentioned and reference need only be made to a sub-committee recently established by this group to coordinate research on radiological analysis.

A second organization is the Joint Panel for Occupational and Environmental Research for Uranium Production. It was established in late 1976, and seeks to compile an inventory of research and to encourage the development of a comprehensive program of research. Presently some ten organizations from industry, government and labour are contributing information to this panel.

CONCLUSION

This report has reviewed the contributions made by CANMET scientists in developing technology to further reduce the environmental impact of uranium mining and milling at Elliot Lake. An attempt has been made to place this work in perspective. Much of the basic information is drawn from research reports of other scientists, and where fine technical distinctions are to be made, the original reports should be consulted. Furthermore, much of the work is in the development stage and has to be tested in actual tailings basins and in pilot plants before results can be transferred to production operations. Results will be publicized through scientific publications and by field demonstrations to aid in the transfer of technology to industry and regulatory bodies.

In the short term CANMET research will continue more or less as described. The present distribution of manpower within the two relevant CANMET laboratories is shown in the appendix. In the longer term, conclusions reached by the Environmental Assessment Board will undoubtedly have an impact on future research.

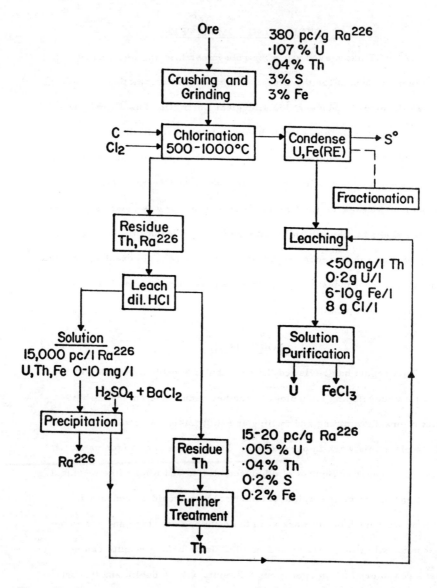

Figure 2: Schematic flowsheet for dry chlorination process.

REFERENCES

General

1. * CANMET "Submission to the Royal Commission on the Health and Safety of Workers in Mines of Ontario"; CANMET (formerly Mines Branch) Director's Services Report DS 75-3 (ADM); 1975.

2. * Moffett, D. "The disposal of solid wastes and liquid effluents from the milling of uranium ores"; CANMET REPORT 76-19; 1976.

3. * Sage, R. editor "Pit slope manual, Chapter 1 - Summary"; CANMET REPORT 76-22; 1976.

Reclamation by Vegetation

4. * Murray, D.R. "Vegetation of mine waste embankments in Canada"; CANMET IC 301; 1973.

5. Murray, D.R. "Factors affecting revegetation of uranium mine tailings in the Elliot Lake area; Part 1 - Physical conditions"; Mines Branch Divisional report IR 71-104; 1971. "Part 2 - Chemical conditions"; Mines Branch Divisional report IR 72-55; 1972.

6. Murray, D.R. "Growth chamber experiments, Progress Report 1972"; Mines Branch Divisional report IR 73-16; 1973.

7. Murray, D.R. "Field trials, Progress Report 1972"; Mines Branch Divisional report IR 73-19; 1973.

8. Murray, D.R. "Mineralogy of waste materials as influencing plant growth and soil development"; Mines Branch Divisional report IR 73-33; 1973.

9. * Zahary, G., Murray, D.R. and Hoare, B. "Reclamation - A challenge to open-pit mining operations in Canada"; World Mining Congress, Peru; 1974.

10. Purych, N. "Soil microbiology in relation to the revegetation of mine tailings"; Mines Branch Divisional report DR 74-116; 1974.

11. Murray, D.R.,Moffett, D. and Shearer, D. "Progress report on black discolouration of vegetation grown on sulphide-containing uranium tailings"; CANMET Laboratory report MRP/MRL 76-78 (TR); 1976.

12. * Murray, D.R. and Moffett, D. "Vegetating the uranium mine tailings at Elliot Lake, Ontario"; Journal of Soil and Water Conservation; v. 32, no. 3, pp 171-174; 1977.

---

* Reports presented or published.

13. * Murray, D.R. "Reclamation by vegetation"; CANMET Laboratory report MRP/MRL 76-166 (OP); 1976. Oral presentation at the Coal Industry Reclamation Symposium, Banff, Alberta; February 1977.

14. * Moffett, D. and Tellier, M. "Radioisotope uptake by vegetation growing on uranium tailings"; Canadian Journal of Soil Science; In Press.

15. McCready, R.G.L. "Microbial assessment of the vegetated test plots on the pyritic uranium tailings in the Elliot Lake area"; CANMET Laboratory report MRP/MRL 76-161 (TR); 1976.

16. * Murray, D.R. "The influence of uranium tailings on tree growth"; CANMET Laboratory report MRP/MRL 77-80 (OP); Oral presentation, Canadian Land Reclamation Association; 1977.

17. Moffett, D. "Land reclamation at Canada's uranium capital"; CANMET Laboratory report MRP/MRL 77-19 (J); 1977.

Tailings Effluents

18. Murray, D.R. and Melis, L.A. "Core sampling of Elliot Lake tailings"; Mines Branch Divisional report IR 73-170; 1973.

19. Moffett, D. and Murray, D.R. "Tailings disposal, hazards and treatment in the Elliot Lake area"; CANMET Laboratory report MRP/MRL 76-9 (TR); 1976.

20. Moffett, D. "Preliminary investigations of an inactive uranium tailings basin"; CANMET Laboratory report MRP/MRL 76-25 (TR); 1976.

21. * Moffett, D. "Acidity and pH : useful indicators of water pollution?" CANMET Laboratory report MRP/MRL 76-34 (OP); Oral presentation at CIM Algoma Branch; 1976.

22. McCready, R.L.G. "A microbiological investigation of uranium tailings water samples"; CANMET Laboratory report MRP/MRL 77-5 (TR); 1977.

23. * Moffett, D. "Rapid analysis of uranium mill effluents using specific ion electrodes"; CANMET Laboratory report MRP/MRL 76-60 (OP); Oral presentation at Canadian Uranium Producer's Metallurgical Committee; 1976.

24. Moffett, D. and Tellier, M. "Radiological investigations of an inactive uranium tailings basin"; CANMET Laboratory report MRP/MRL 77-54 (J); 1977.

25. * Moffett, D. "Research on tailings disposal at Elliot Lake"; CANMET Laboratory report MRP/MRL 76-169 (OP); Oral presentation at the Canadian Mineral Processors' Conference; 1977.

26. Moffett, D. "Radiochemical analytical capacity available to uranium producers"; CANMET Laboratory report MRP/MRL 77-74 (TR); 1977.

27. * Moffett, D. "Environmental aspects of thorium"; CANMET Laboratory report MRP/MRL 77-43 (OP); Oral presentation at the Canadian Uranium Producers' Metallurgical Committee; 1977.

28. Jongejan, A. and Wilkins, A.L. "An evaluation of the effect of penta-chlorophenol on acid drainage in Elliot Lake"; in preparation.

29. Jongejan, A. "Physical and chemical aspects of weathering of pyrite/carbonate mixtures in areas of water pollution from tailings areas"; CANMET Laboratory report MRP/MSL 75-185 (R); 1975.

30. Jongejan, A. "Summary of microbiological methods used in the study of the weathering of tailings to 1976"; CANMET Laboratory report MRP/MSL 76-75 (TR); 1976.

31. Skeaff, J.M. "Distribution of radium-226 in uranium tailings"; CANMET Laboratory report MRP/MSL 76-137 (J); 1976.

32. * Moffett, D. "Uranium waste researchers consider alternate means of disposal"; Canadian Mining Journal; v. 98, No. 1, pp 48-50; 1977.

33. * Gilmore, A.J. "The removal of sulphate, nitrate and ammonia from mine and mill effluents"; MRP/MSL 75-115 (OP); Oral presentation at the Canadian Uranium Producers' Metallurgical Committee; 1975.

34. Raecevic, D. "A progress report on the removal of pyrite from Elliot Lake uranium tailings"; CANMET Laboratory report MRP/MSL 76-357 (TR); 1976.

Process Research

35. Simard, R. "The recovery of metal-grade thorium concentrate from uranium plant ion exchange effluents by amine solvent extraction"; Mines Branch report TB 13; 1960.

36. Ritcey, G. and Lucas, B. "Stagewise separation of uranium, thorium and rare earths by liquid-liquid extraction"; Mines Branch report TB 113; 1969.

37. * Ritcey, G. and Lucas, B. "A proposed route for the co-extraction of uranium and thorium from sulphuric acid solutions and recovery by selective stripping and denitrification"; Oral presentation at AIME general meeting, New York, 1971.

38. Ritcey, G. and Lucas, B. Canadian patent 976363; 1975.

39. * Lucas, B. and Ritcey, G. "Examination of a rare earth solvent extraction circuit for the possible upgrading of the yttrium product"; CIM Bulletin, v. 68, no. 753, pp 124-140; 1975.

40. * Ritcey, G.M. "Treatment of radioactive ores at CANMET"; CANMET Laboratory report ERP/MSL 77-139 (TR); Oral presentation at the Canadian Uranium Producers' Metallurgical Committee; 1977.

41. Parsons, H.W. "Chlorination of uranium ores literature review"; CANMET Laboratory report ERP/MSL 76-86 (LS); 1976.

42. Lucas, B.H. "Literature survey for the uranium program (chloride metallurgy of uranium) period 1950 - 1954"; CANMET Laboratory report ERP/MSL 76-102 (LS); 1976.

43. Parsons, H.W. "Chloride metallurgy of uranium literature review 1955 - 1959"; CANMET Laboratory report ERP/MSL 76-112 (LS); 1976.

44. McNamara, V.M. "The chlorination of uranium and thorium ores and of uranium-containing minerals - a search of the available literature for the period January 1, 1960 - December 30, 1964"; CANMET Laboratory report ERP/MSL 77-134 (LS); 1977.

45. Gilmore, A. J. and Skeaff, J. M. "The chloride metallurgy of uranium and thorium - a review, 1965 - 1969"; CANMET Laboratory report ERP/MSL 76-113 (LS); 1976.

46. Saint-Martin, N. "Literature survey on chloride metallurgy of uranium 1970 - 1975"; CANMET Laboratory report ERP/MSL 76-121 (LS); 1976.

47. Saint-Martin, N. "Hydrochloric acid leaching of an Elliot Lake uranium ore - a preliminary study"; CANMET Laboratory report ERP/MSL 77-39 (TR); 1977.

48. Saint-Martin, N. "Preliminary laboratory HCl-acetone leach tests on an Elliot Lake uranium ore"; CANMET Laboratory report ERP/MSL 77-40 (TR); 1977.

49. Saint-Martin, N. "Leaching of various radioactive ores - a literature survey"; CANMET Laboratory report ERP/MSL 77-89 (LS); 1977.

50. Skeaff, J. M. "High temperature chlorination of an Elliot Lake uranium ore - preliminary results"; CANMET Laboratory report ERP/MSL 76-283 (TR); 1976.

51. Skeaff, J. M. "A proposed process for the extraction of uranium and thorium from refractory uranium ores by nitric acid and the removal of radium-226 from the residue"; in preparation.

52. Skeaff, J. M. "Solvent extraction of uranium from uranium leach liquors"; CANMET Laboratory report MRP/MSL 75-276 (TR); 1975.

53. Skeaff, J. M. "Nitric acid leaching of uranium ore"; CANMET Laboratory report MRP/MSL 75-165 (TR); 1975.

54. Lucas, B. H. and Prudhomme, P. "Progress report on the separation of uranium from leach liquors by single-stage deep fluidized bed ion exchange"; CANMET Laboratory report MRP/MSL 77-53 (TR); 1977.

55. McCreedy, H. H. "Co-ordination meeting of IAEA on bacterial leaching research"; CANMET Laboratory report ERP/MSL 77-77 (FT); 1977.

Analytical Research

56. Kaiman, S. "Mineralogical investigations of radioactive ores"; CANMET Laboratory report ERP/MSL 76-251 (J); 1976.

57. Reynolds, V. G. and Zimmerman, J. B. "Modifications of optical fluorescence meter used for the determination of low concentrations of uranium"; CANMET Laboratory report MRP/MSL 76-206 (TR); 1976.

58. * Zimmerman, J. B. and Armstrong, V. C. "The determination of radium -226 in uranium ores and mill products by alpha energy spectrometry"; CANMET REPORT 76-11; 1976.

59. Kaiman, S. "Mineralogical examination of old tailings from the Nordic mine, Elliot Lake, Ontario"; CANMET Laboratory report ERP/MSL 77-190; 1977.

APPENDIX

Distribution of Manpower Within CANMET

	Man-Year	
	Mining Research Laboratories	Mineral Sciences Laboratories
Rehabilitation by vegetation	1.20	-
Tailings effluents	2.90	2.00
Process research	0.10	7.72
Analytical research and services	-	8.13
Support	0.80	3.85
TOTAL	4.70	21.70
	26.40	

DISCUSSION

J.D. SHREVE, Jr., United States

As I listened to the two papers this morning, the thought struck me that we might be making a psychological error with all of this work on revegetating tailings piles, and in the legalistic sense "creating an attractive hazard". If we are too successful at revegetation, and create highly attractive areas, we will be inviting human activity on the very spot that we want to limit it. I suggest that the vegetation trials extend to things like cactus, bramble bush, all kinds of hold-off plants that say "go around me" rather than "come on me and have a picnic".

J. HOWIESON, Canada

I could not agree more ; these are pretty attractive meadows up there, but there is not much open ground. Anyone coming along a thousand years from now, this might be a good place to settle.

W.J. SHELLEY, United States

On your grass analysis for radium, was that the whole plant ?

J. HOWIESON, Canada

These were ashed samples.

W.J. SHELLEY, United States

Was it the whole plant or just the cuttings from the top ?

J. HOWIESON, Canada

I did not ask that question, I am afraid I do not know.

W.J. SHELLEY, United States

Other question, did you analyse for radium in your fertilizer ?

J. HOWIESON, Canada

It depends a great deal where the phosphate came from. I doubt if there was any particular attention paid to that.

D.M. LEVINS, Australia

I am surprised that it has taken this long in the conference to talk about sulfides. Because I think that when ore contains sulfides, there is the greatest problem that the uranium industry faces as regards leaching of heavy metals. But I have my doubts about the long term affects of liming. I feel that it is only a short term measure and eventually the acidity is going to overcome the lime and you are going to have a very serious problem.

J. HOWIESON, Canada

        This is apparent from these pits that we are operating. We expect 4 or 5 years for the pits to become acidic and overcome the lime.

D.M. LEVINS, Australia

        Perhaps the long term solution is to try to restrict the oxygen into the pile. This would involve actually putting a covering over the tailings.

J. HOWIESON, Canada

        Well, this is really the main hope for vegetation, that it will reduce the penetration of water and it will reduce the availability of oxygen. That is what we are looking for in the vegetation of the whole piles.

ANALYSIS OF DISPOSAL OF URANIUM MILL TAILINGS
IN A MINED OUT OPEN PIT

W. P. Staub and E. K. Triegel
Oak Ridge National Laboratory
Oak Ridge, Tennessee USA

ABSTRACT

Mined out open pits are presently under consideration
as disposal sites for uranium mill tailings. In this method
of tailings management, the escape of contaminated liquid
into an adjacent aquifer is the principal environmental
concern. The modified Bishop Method was used to analyze the
structural stability of a clay liner along the highwall and
fluid flow models were used to analyze the effect of tailings solutions on ground water under several operating
conditions. The slope stability of a clay liner was analyzed
at three stages of operation: 1) near the beginning of
construction, 2) when the pit is partially filled with
tailings, and 3) at the end of construction. Both clay lined
and unlined pits were considered in the fluid flow modeling.
Finally, the seepage of tailings solutions through the clay
liner was analyzed.

Results of the slope stability analysis showed that it
would be necessary to construct the clay liner as a modified
form of engineered embankment. This embankment would be
similar in construction to that of an earthfill dam. It
could be constructed on a 1:1 slope provided the tailings
slurry were managed properly. It would be necessary to
maintain the freeboard height between the embankment and
tailings at less than 4 m. A partially dewatered sand beach
would have to be located adjacent to the embankment.

Potential leakage and aquifer contamination was modeled
for lined and unlined pits of various designs. Sulfate, and
possibly U and Th, are the most likely contaminants.
Results from the model showed the clay and soil cement lined
pit to be most effective in containing the pollutants.

ANALYSE DU STOCKAGE DES RESIDUS DE TRAITEMENT DE L'URANIUM DANS UNE CARRIERE A CIEL OUVERT DESAFFECTEE

RESUME

Les anciennes carrières à ciel ouvert sont actuellement consi-
dérées comme pouvant se prêter au stockage des résidus de traitement de l'uranium. Selon cette technique de gestion des résidus,
le principal problème qui se pose sur le plan de l'environnement
tient aux fuites de liquide contaminé dans une couche aquifère
adjacente. On a utilisé la méthode de Bishop modifiée pour analy-
ser la stabilité structurale d'un revêtement d'argile le long d'un
mur élevé et des modèles d'écoulement des fluides ont servi à étu-
dier l'effet des solutions de queue de traitement sur la nappe
phréatique dans différentes conditions d'exploitation. La stabilité
de la pente d'un revêtement d'argile a été analysée à trois stades
de l'exploitation : (1) au début de la construction, (2) au moment
où la carrière était partiellement remplie de résidus et (3) à la
fin de la construction. Des carrières avec ou sans revêtement ont
été prises en considération dans l'établissement des modèles d'é-
coulement des fluides. Enfin, on a étudié le suintement des solutions de queue de traitement à travers le revêtement d'argile.

Les résultats de l'analyse de la stabilité de la pente ont
montré qu'il serait nécessaire de construire le revêtement d'argile
suivant la forme modifiée d'une digue. La construction de cette
digue serait identique à celle d'un barrage de terre. Elle pourrait
être construite avec une pente de 1/1, à condition que les boues
résiduelles soient traitées correctement. Il serait nécessaire de
maintenir un dénivelé d'au moins 4 mètres entre le haut du barrage
et les résidus. Une plage de sable partiellement déshydratée
devrait jouxter la digue.

Les fuites potentielles et la contamination de la couche
aquifère ont été modélisées dans le cas de différents schémas de
carrières avec ou sans revêtement. Les produits de contamination
les plus probables sont les sulfates et éventuellement l'uranium
et le thorium. Les résultats obtenus à l'aide du modèle ont montré
que la carrière avec revêtement d'argile et de sol-ciment offre le
moyen le plus efficace de contenir les polluants.

Introduction

Disposal of tailings in depleted ore pits is an appealing
alternative to other tailings management methods of the
recent past. In the 1950s and 60s, tailings were often
piled on the surface and left to dry.[1] Wind erosion and
sheet runoff widely dispersed these tailings with their low
concentrations of radioactive components. Another common
method of containment was the construction of a ring dike
made from the sandy portion of the tailings. Clay slime and
contaminated water were impounded within the ring dikes.
Often the dikes were poorly designed and constructed, they
were located on the flood plains of major streams and their
reservoirs were unlined. Failure of several of these dikes
led to the uncontrolled surface discharge of tailings.
Although dikes have remained intact at most tailings impoundments,

[1] "Phase I Reports on Conditions of Inactive Uranium
Millsites", United States Atomic Energy Commission;
Grand Junction, Colorado; Oct., 1974.

groundwater contamination occurred by seepage through the floor of the reservoir. More recently, high earth fill embankments have been constructed across natural drainage basins to impound slurried tailings within lined reservoirs. While dams provide short-term (tens to perhaps hundreds of years) protection against erosion of tailings, the natural stream course will eventually breach the embankment and cut intricate and progressively deepening channels through the tailings. Even in the short-term there is the risk of catastrophic failure of a poorly constructed or earthquake damaged embankment. It is generally agreed that below grade disposal of tailings in depleted ore pits substantially reduces the impact of wind erosion, eliminates the possibility of catastrophic failure and reduces the possibility of stream erosion.

Little is known about the potential impacts of below grade disposal of tailings on groundwater. The purpose of this study is to analyze the stability of a clay liner on a steep slope and to determine its effectiveness in sealing off contaminated, acidic (pH <2) liquid waste from surrounding aquifers.

Slope Stability Analysis

Oak Ridge National Laboratory (ORNL) staff analyzed two simplified designs for clay lined pits (Figure 1). The design with a 1:2 highwall slope is similar to a tailings management alternative being considered by United Nuclear Corporation [2] for their proposed Morton Ranch uranium mill near Glenrock, Wyoming. ORNL proposed the 1:1 highwall slope which is more stable and may require a smaller volume of material for the clay liner.

Table I lists the engineering properties obtained from triaxial shear tests on undisturbed lithologic units in the Ft. Union Formation and on remolded (Proctor mold) soils. These data were supplied by Dames and Moore, consultants to United Nuclear Corporation. Saturated samples of the Ft. Union were tested under consolidated-drained conditions. The moisture content of remolded soils was on the wet side of optimum and samples were tested under consolidated-undrained conditions. Engineering properties for the tailings were assumed to be like that of Exxon's nearby Highland Mill.

The Modified Bishop method[3] was utilized for slope stability analysis. Table II lists the factors of safety for critical failure arcs under static and maximum probable earthquake loading conditions, (acceleration equal to 8% of gravity), with and without liquefaction of the tailings.

[2] "Environmental Statement Related to Operation of Morton Ranch Uranium Mill, United Nuclear Corporation", (Draft), NUREG-0439, p. 10-4; U. S. Nuclear Regulatory Commission Washington, D. C.; April 1974.

[3] "Slope, User Manual", McDonnell Douglas Automation Company, St. Louis, Missouri, 1974.

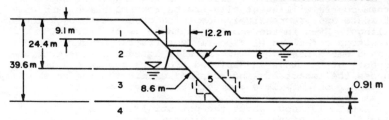

A. PARTIALLY BACKFILLED ORE PIT WITH A
I:I HIGHWALL SLOPE.

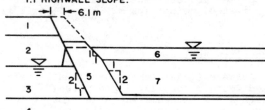

B. PARTIALLY BACKFILLED ORE PIT WITH A
I:2 HIGHWALL SLOPE.

LEGEND

SEMICONSOLIDATED FT. UNION FORMATION (EOCENE)	1	DENSE SANDSTONE
	2	STIFF SHALE
	3	DENSE SANDSTONE
	4	MASSIVE STIFF SHALE

☷ WATER TABLE

BACKFILL	5	CLAY LINER
	6	TAILINGS
	7	UNCLASSIFIED

0   20   40   60
(meters)

FIGURE 1:   CLAY LINER DESIGNS USED IN COMPUTER CODE
FOR SLOPE STABILITY ANALYSIS BY THE
MODIFIED BISHOP METHOD

TABLE I: SHEAR STRENGTH PROPERTIES USED
IN SLOPE STABILITY ANALYSIS

NO.	MATERIAL	TRIAXIAL SHEAR TEST	SPECIFIC WEIGHT		COHESION		FRICTION ANGLE (DEG)
			GM/CM^3	(PCF)	PASCALS X 10^3	(PSF)	
1	DENSE SANDSTONE	CD[B]	2.0	(125)	47.9	(1000)	31
2	STIFF SHALE	CD	2.1	(130)	21.5	(450)	15
3	DENSE SANDSTONE	CD	2.0	(125)	47.9	(1000)	31
4	STIFF SHALE	CD	2.1	(130)	21.5	(450)	15
5	CLAY LINER	CU[C]	2.0	(125)	9.6	(200)	14
6	TAILINGS[A]	NONE	2.0	(125)	7.2	(150)	20
7	UNCLASSIFIED	CU	1.9	(120)	12.0	(250)	17

[A] SHEAR STRENGTH OBTAINED FOR TAILINGS AT EXXON'S NEARBY HIGHLAND MILL

[B] CONSOLIDATED-DRAINED SAMPLE

[C] CONSOLIDATED-UNDRAINED SAMPLE

TABLE II: RESULTS OF SLOPE STABILITY ANALYSIS FOR A
TAILINGS PIT AT VARIOUS STAGES OF OPERATION

A. FAILURE SURFACES PASS THROUGH HIGHWALL AS WELL AS THROUGH LINER AND TAILINGS

| SLOPE STEEPNESS | | STAGE OF DEVELOPMENT | HORIZONTAL DIMENSION OF LINER (M) | | HEIGHT OF FREEBOARD (M) | FACTORS OF SAFETY | | | |
| HIGHWALL | CLAY LINER | | TOP SURFACE OF TAILINGS | AT BASE OF TAILINGS | | STATIC | | EARTHQUAKE[B] | |
						LIQUEFACTION OF TAILINGS NO	LIQUEFACTION OF TAILINGS YES	LIQUEFACTION OF TAILINGS NO	LIQUEFACTION OF TAILINGS YES
1:1	NA[A]	END OF EXCAVATION	NA	NA	NA	0.82	---	---	---
	1:1	START OF TAILINGS EMPLACEMENT	6.1	6.1	3.0	0.88	---	---	---
	1:1	PARTIAL POOL	6.1	6.1	6.1	1.09	---	---	---
	1:1	PARTIAL POOL	6.1	6.1	3.0	1.21	1.14	0.99	0.90
1:2	NA	END OF EXCAVATION	NA	NA	NA	0.91	---	---	---
	1:1	START OF TAILINGS EMPLACEMENT	17.0	18.3	3.0	0.66	---	---	---
	1:1	PARTIAL POOL	12.1	18.3	3.0	1.14	1.07	0.92	0.93
MINIMUM ACCEPTABLE FACTORS OF SAFETY						1.30		1.00	

TABLE II (CONT.):

B. ASSUMED FAILURE SURFACES PASS THROUGH CLAY LINER AND TAILINGS BUT NOT THROUGH THE HIGHWALL.

SLOPE STEEPNESS		STAGE OF DEVELOPMENT	HORIZONTAL DIMENSION OF LINER (M)		HEIGHT OF FREEBOARD (M)	FACTORS OF SAFETY			
						STATIC		EARTHQUAKE	
						LIQUEFACTION OF TAILINGS		LIQUEFACTION OF TAILINGS	
HIGHWALL	CLAY LINER		AT TOP SURFACE	AT BASE OF TAILINGS		NO	YES	NO	YES
1:1[D]	1:1	ALL STAGES	6.1	6.1	3.0	0.86	---	---	---
			12.2	12.2	3.0	1.42	1.05	0.92	0.66
1:2[E]	1:1	FULL POOL	6.1	18.3	3.0	1.30	0.99	0.95	---
		PARTIAL POOL	12.2	18.3	6.1	1.04	---	---	---
		PARTIAL POOL	12.2	18.3	4.6	1.22	---	---	---
		PARTIAL POOL	12.2	18.3	3.7	1.44	1.19	1.22	0.94
		PARTIAL POOL	12.2	18.3	3.0	1.51	1.31	1.31	1.04

(A) NOT APPLICABLE
(B) HORIZONTAL ACCELERATION IS 8% OF GRAVITY (MAXIMUM CREDIBLE EARTHQUAKE FOR THE POWDER RIVER BASIN).
(C) ASSUMED COHESIVE STRENGTHS WERE GREATER BY A FACTOR OF 2 THAN AS INDICATED IN TABLE I.
(D) CRITICAL FAILURE SURFACES INCLUDE SEPARATION AT THE BOUNDARY BETWEEN HIGHWALL AND LINER.
(E) CRITICAL FAILURE SURFACES LIE ENTIRELY WITHIN CLAY LINER AND TAILINGS.

If the shear strengths of rock in the highwall (Table I) are accepted at face value, the slope stability analyses show the designs to be inadequate under all conditions ranging from an empty to partially filled pit based on Nuclear Regulatory Commission's minimum acceptable factors of safety of 1.3 and 1.0 for static and earthquake loading, respectively. [4] However, the results of the analyses conflict with practical experience at the Morton Ranch site where an open pit highwall stands on a 1:2 slope with no apparent evidence of massive slope failure. Murdock [5] states that this perplexing result is not uncommon when standard methods of slope stability analysis are applied to cut slopes. For example, the shear strength of Ft. Union strata in the highwall may have been underestimated. The factor of safety approaches unity for a 1:2 slope at the end of excavation, if cohesive strengths are assumed to be greater by a factor of two than those listed in Table I. Partially saturated strata above the water table may have considerably more strength (due to surface tension) than their saturated counterparts. [6] Such strength is difficult to quantify because it varies with the degree of saturation as well as particle size distribution and clay content. Furthermore, if shale units dry out too much, shrinkage cracks will weaken the strata along the highwall.

Lee, et al.[7] have studied the stability of highwalls in open pit coal mines in the western part of the Powder River Basin. They report much higher shear strengths for the sandstones and shales of the Ft. Union in their study area. They concluded, however, that fracture patterns (whether by drying out, stress release, or pre-existing) play a significant role in reducing the slope stability. Futhermore, they conclude that slope stability is time dependent.

Another possibility is that the analyses are correct but that the presence of massive slope failure is not always readily apparent. Slope failure may be of an insidious nature. Telltale bulges that appear in the floor of a pit may be removed or covered by earthmoving equipment while small escarpments remain unobserved a hundred meters or more behind the highwall.

The results of this analysis and the conclusions of Lee, et al. [7] cast some doubt on the stability of an open pit highwall over a period of years. If the pit can be backfilled over a reasonably short time frame, it is less likely that massive slope failure will become a problem.

---

[4] "Design, Construction, and Inspection of Embankment Retention Systems for Uranium Mills"; Regulatory Guide 3.11; U. S. Nuclear Regulatory Commission, Washington, D. C., March, 1977.

[5] Murdock, L., Geotechnical Engineer, Dames and Moore; Salt Lake City, Utah, "Personal Communication", June, 1978.

[6] Smith, W. K., Geotechnical Engineer, U. S. Geological Survey, Denver, Colorado, "Personal Communication", June, 1978.

[7] Lee, F. T., W. K. Smith and W. Z. Savage, "Stability of Highwalls in Surface Coal Mines, Western Powder River Basin, Wyoming and Montana", U. S. Geological Survey, Open-File Report 76-846; Denver, Colorado; 1976.

Table II also lists the results of slope stability analysis for the clay liner based on the assumption that the critical failure surface may include the liner-highwall interface without passing through the highwall. Because of differential settlement of the clay liner with respect to the adjacent steep highwall, there will be very little shear strength developed at the liner-highwall interface (Robertson[8]). In the interest of producing conservative results, it was assumed that the shearing strength was zero along this interface.

The low shearing strength at the liner-highwall interface is the controlling factor in the design of a clay liner with an outer surface parallel to a 1:1 highwall slope. The minimum practical horizontal width (6.1 m, to allow for freedom of movement of compaction equipment) considered for the clay liner, combined with the minimum practical freeboard (3 m) provides an unacceptably low factor of safety (0.86 under a static load) against slope failure. An acceptable factor of safety can be obtained by doubling the width of the liner to 12.2m. Even than the factor of safety is slightly less than the lower limit for a maximum credible earthquake. The critical failure surface is a combination of plane failure along the liner-highwall interface and circular failure arc through the liner and tailings.

For the clay liner with a 1:1 slope constructed adjacent to a 1:2 highwall slope the critical failure surface does not pass near the liner-highwall interface. The steeper slope angle for the interface together with greater clay liner thickness results in acceptably high factors of safety for combinations of plane and circular arc failure surfaces.

In the 1:2 highwall slope case, the critical failure surfaces are arcs that lie completely within the clay liner and the tailings. Acceptable factors of safety are functions of construction and operational stages of development, the height of the freeboard between liner and tailings, and the physical state of the tailings. A partially filled pool (6.1 m of tailings) with a 3 m freeboard is stable against all failure modes including maximum credible earthquake and liquefaction of tailings. On the otherhand, if the freeboard in the above case is increased to 3.7 m liquefaction of the tailings will induce instability under both static and earthquake loading conditions. Maintenance of a partially dewatered sandy beach around the entire inside perimeter of the tailings impoundment would be a safeguard against liquefaction. At full pool, the clay liner is marginally stable (because of decreasing thickness) under a static load without liquefaction and an earthquake or liquefaction would destabilize the liner. At full pool, however, the critical failure surface would pass through only the upper 1 to 2 meters of tailings. Thus slope failure of the clay liner at full pool would result in the exposure of a small percentage of the tailings fluid along the upper few meters of the highwall.

[8]   Robertson, A. M., Geotechnical Engineer, Steffen, Robertson and Kirsten, Inc., Vancouver, British Colombia, Canada, "Personal Communication", July 1978.

The above slope stability analysis suggests that a clay
liner can be constructed on a steep highwall slope. Design
details would vary depending on the highwall slope angle.
Nevertheless, it is clear that: 1) the slope of the clay
liner should not be greater than 1:1, 2) the thickness of
the liner (or combination of liner and supporting shell with
similar or greater strength) should exceed 6 m at the top of
a 1:2 highwall slope or 12 m for a 1:1 highwall slope, 3)
the height of the freeboard should be less than 4 m, and 4)
a partially dewatered sand beach should encircle the inside
surface of the clay liner. Furthermore, soil cement may be
required along the inside surface of the liner to add strength
along the unconfined slope where it will be difficult to
compact clay to design specifications. Specifically designed equipment may be required for compacting clay against
the steep highwall and it may also be necessary to construct
a filter between the liner and highwall in order to prevent
piping failure (erosion of the clay liner by seepage). From
the standpoint of stability it should not be necessary to
install a drain along the outside of the embankment because
sandstone exposed in the highwall will permit any seepage
through the liner to drain away without building up pore
pressure.

Modeling of Aquifer Contamination

In addition to assessing the structural stability of
various waste pit designs, the effect of such designs on the
possible contamination of an aquifer was analyzed. The
mathematical model of the Morton Ranch waste pit is primarily
based on hydraulic behavior of the pit and aquifer and is
composed of two sections. The first assesses fluid transport
from the waste pit to the aquifer, while the second section
models dispersion of contaminants in the aquifer itself.
The mathematical assumptions correlate with results from
other studies of groundwater contamination. [9]

Waste Pit Drainage

The Morton Ranch waste pit may be approximated as a
long rectangular box, its long axis parallel to the direction
of groundwater flow. In the unlined pit design the wastes
are to be placed above the water table. Drainage is modeled
using a modified Huggins and Monke [9]   equation for surface
infiltration of ponded water:

$$DR = ER - R + F [1 - P/G]^3$$

where
- DR = drainage rate
- ER = average potential evaporation rate
- R = average rainfall rate
- F = steady state infiltration capacity
- P = unsaturated pore volume
- G = maximum gravitational water minus
  field capacity.

[9] Robertson, J. B., "Digital Modeling at Radioactive and
Chemical Waste Transport in the Snake River Plain Aquifer at the
National Reactor Testing Station, Idaho", USGS Open-File Report
IDO-22054, 5/74.

The value of F is assumed to be 1.25 cm/hr (0.5 inches/hr.),
corresponding to a drainage rate typical of sandy soils.[10,11]

The clay lined pit model assumes that the hydraulic
head in the pit is the main driving mechanism for drainage.
The leakage rate decreases exponentially with the decline in
head. The mass flux q(t) from the pit at time t is given
by:

$$q(t) = C_i D_c H_i \exp [-\lambda_c t]$$

where C_i = initial waste concentration

$$D_c = \sum_{j=1}^{n} \frac{K_j A_j}{\ell_j}$$

n = number of sections of clay liner
K_j = hydraulic conductivity of section j
A_j = area of section j
ℓ_j = fluid path length through section j
H_i = initial hydraulic head
λ_c = D_c/(area over which head is dissipated)

If failure occurs in the clay liner, the model predicts that
the level of resulting contamination would be between the
model results for the lined and unlined pits.

Dispersion in the Aquifer

Fluid movement in the aquifer is assumed to conform to
Darcy's law. Vertical averaging of the concentration allows
simplification of the dispersion calculations by modeling
the movement only in the x and y directions. The initial
volume of the waste plume is set equal to the aquifer volume
directly under the waste pit. This plume, assumed to be of
uniform concentration, moves downgradient with time and
disperses horizontally. The amount of dispersion, (calculated in both the x and y directions), is given by:

$$\sigma = (2 \ ay)^{1/2}$$

where a is the dispersivity and y, the distance from the
waste source centerline parallel to flow. The final concentration is calculated considering the amount of dilution
occurring due to the increased plume size.

Several assumptions were used in modeling the drainage
and dispersion of wastes: (1) The aquifer is homogeneous in
its fluid flow properties. (2) Chemical interactions are
not incorporated due to lack of quantitative data on the
chemistry of the aquifer. (3) Seepage takes place only
through 25% of the pit floor, where the sandstone aquifer is
exposed. Numerical assumptions are listed in Table III.

[10] Huggins, L. F. and Monke, E. J., A Mathematical Model
for Simulating the Hydrological Response of a Watershed",
Paper H9, Proc. 48th Annual Meeting A.G.U., Washington, D.C.,
4/17-20/67.

[11] Fleming, George, Computer Simulation Techniques in
Hydrology, Elsevier, N.Y., 1975, 333 pg.

TABLE III: FLUID FLOW MODEL PARAMETERS*

PARAMETER	MAGNITUDE	
AVERAGE HORIZONTAL AREA OF PIT	$239 \times 10^3 M^2$	$(2.57 \times 10^6 \ FT^2)$
INITIAL HEIGHT OF TAILINGS ABOVE WATER TABLE	18 M	(60 FT)
INITIAL WEIGHT % WATER IN TAILINGS	34%	
ESTIMATED POROSITY OF TAILINGS AFTER EVAPORATION	30%	
HYDRAULIC CONDUCTIVITY		
WALL ABOVE WATER TABLE	$.421 \times 10^{-2}$ M/YR	$(.138 \times 10^{-1} \ FT/YR)$
WALL BELOW WATER TABLE	$.2 \times 10^{-2}$ M/YR	$(.7 \times 10^{-2} \ FT/YR)$
FLOOR OF PIT (CLAY ONLY)	$.2 \times 10^{-2}$ M/YR	$(.7 \times 10^{-2} FT/YR)$
LONGITUDINAL DISPERSIVITY	21 M	(70 FT)
TRANSVERSE DISPERSIVITY	4.3 M	(14 FT)
POROSITY OF AQUIFER	23%	
AQUIFER THICKNESS	12 M	(40 FT)
POTENTIAL GROUNDWATER VELOCITY	3 M/YR	(9.8 FT/YR)
EVAPORATION RATE	1.07 M/YR	(42 IN/YR)
RAINFALL RATE	0.30 M/YR	(12 IN/YR)

*
  Viessman, Warren, Jr., John W. Knapp, Gary L. Lewis,
Terence E. Harbaugh, Introduction to Hydrology, IEP-Dun-
Donnelley, New York, 2nd Ed., 1977, 704 pg.

  Davis, Stanley N. and DeWiest, Roger J. M., Hydrogeology,
J. Wiley & Sons, Inc., N.Y., 1966.

In general, the numerical assumptions used in the model
are in the conservative range of values. Since chemical
interactions are not considered quantitatively in the fluid
flow model, the results are conservative and must be considered in terms of the probable behavior of the ions in the
aquifer. Orientation of the pit at some angle to the principal
direction of groundwater flow would shorten the effective
travel path across the pit, producing results similar to
those shown in Figure 3. However, the assumption that the
aquifer is homogeneous is more likely to produce results
which are too low. The presence of structural features,
such as buried stream channels, would act to confine the
flow, reduce the total dispersion and result in less dilution.

Fluid Dispersion Results

Results were calculated for various clay liner thicknesses and permeabilities, and for the more generalized
cases of flow and pit design. Table IV gives the calculated
concentration of those waste constituents which at some
point exceed drinking water standards.

Figures 2 and 3 show the effect of clay liner thickness
on the ionic concentration in the aquifer, under various
conditions of pit orientation and groundwater velocity.
Since the drainage rate decreases exponentially, extending
the draining time by increasing liner thickness has less
effect at long time intervals. At Morton Ranch, the time
span is long due to the low groundwater velocity and large
pit, hence the increase in clay liner thickness has less
effect than in the other cases studied.

The results indicated in the tables and figures must be
considered in terms of the geochemistry of the aquifer-waste
system. The aquifer at Morton Ranch is a clayey sandstone
with low concentrations of carbonates and metal oxides. The
initial pH of the wastes is 1.8, which would increase to an
aquifer pH of approximately 6 to 8. [12,13] Characteristic
values of Eh for shallow aquifers range from +200 to +400
mV [14] (oxidizing environment). Under these conditions,
the following conclusions may be drawn: [15,16] (1) sulfate
is likely to remain mobile due to the high solubility of its
salts, [17] (2) Se will precipitate out as metallic Se^{o},

[12]   Dames & Moore, Report of Investigation and Design,
Tailings Disposal Area, Morton RAnch Mine and Mill, Converse Co.,
Wyoming, for UNC, Salt Lake City, Utah, 1977.

[13]   United Nuclear Corporation, Environmental Report on the
Morton Ranch, Wyoming Uranium Mill, UNC-ER-2, Casper, Wyo., 1976.

[14]   Becking, L., G. M. Baas, I. R. Kaplan, and D. Moore,
"Limits of the Natural Environment in Terms of pH and Oxidiation-
Reduction Potentials", J. of Geology, 68, 3, May, 1969, p. 243-283.

[15]   Garrels, Robert M. and Charles L. Christ, Solutions,
Minerals and Equilibria, Harper & Row, N.Y., 1965, 450 pg.

[16]   Harrison H. Schmitt (ed), Equilibrium Diagrams for Minerals
at Low Temperature and Pressure, Geological Club of Harvard,
Cambridge, Mass., 1962, 199 pg.

[17]   Stephen, H. and Stephen, T. (eds.), Solubilities of
Inorganic and Organic Compounds, MacMillan, NY, 1973.

TABLE IV: CONCENTRATIONS OF SELECTED IONS
UNDER VARIOUS WASTE PIT DESIGNS
ASSUMING ONLY HYDRAULIC DISPERSION

DESIGN	INITIAL WASTE CONCENTRATION	UNLINED PIT	.9M THICK LINER	.9M LINER WITH SOIL CEMENT	9M LINER	EPA DRINKING WATER STANDARDS
DISTANCE FROM PIT	--	10,000 M	10,000 M	10,000 M	1000M	--
GENERAL CASE (C/C_I)	1	.127	.087	.043	.257	---
SO_4 (MG/L)	7350	936	638	316	1889	250
SE (MG/L)	.02	.003	.002	.0004	.005	.01
U (MG/L)	5.1	0.6	0.4	0.2	1.3	4.4
α (PCI/L)	84,300	10,731	7317	3625	21,665	15
β (PCI/L)	73,650	9,376	6393	3167	18,928	1000

*HYDRAULIC CONDUCTIVITY OF CLAY + SOIL CEMENT $= \frac{1}{100}$ CONDUCTIVITY OF CLAY

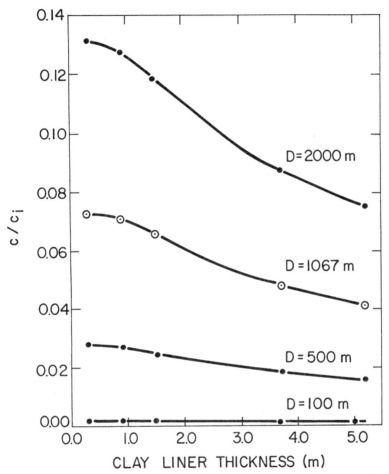

FIGURE 2:   EFFECT OF LINER THICKNESS ON CONCENTRATION, ASSUMING
VARIOUS ORIENTATIONS OF THE WASTE PIT RELATIVE TO
GROUNDWATER FLOW (VOLUME = CONSTANT).   D = LENGTH OF
PIT PARALLEL TO FLOW, GROUNDWATER VELOCITY = 10 M/YR.

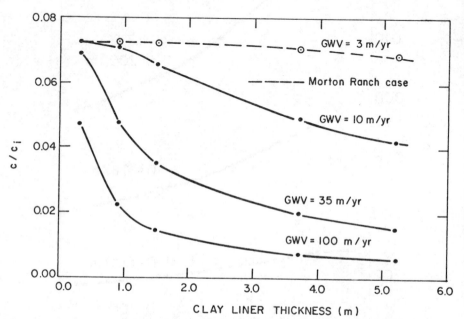

FIGURE 3:   EFFECT OF CLAY LINER THICKNESS ON IONIC CONCENTRATION (C/C_I)
IN THE AQUIFER WITH VARIOUS GROUNDWATER VELOCITIES (GWV). WASTE
PIT = 1067 M X 224 M X 18 M.   DISTANCE FROM PIT = 10000 M.

(3) U will be soluble in the +6 oxidation state, and (4) α and β emissions are likely to be reduced through removal of both ^{226}Ra (by precipitation as a sulfate) and ^{230}Th by precipitation (as hydroxide). Fixation by reaction with clay constituents may also occur. However, the ^{230}Th may not undergo fixation if complex formation occurs. [18] More field data is necessary to determine the extent of ionic migration and the magnitude of the Th problem.

Conclusions

Slope stability of the clay liner along the highwall can be achieved but only through careful design and construction and the use of a large volume of construction material. The liner and its protective shell should be at least 12 m thick, the freeboard should be no greater than 4 m, a filter may be required between the liner and highwall, soil cement may be required on the inside of the liner, and a partially dewatered tailings derived sand beach should completely encircle the inside of the clay liner.

Such a structure would be, in essence, an engineered embankment like an earthfill dam but without a "downstream" shell and a steeper "upstream" shell. The steep upstream shell is made possible only because the freeboard height would be carefully controlled. By contrast, a traditional embankment dam would be built in stages so that the freeboard would vary from 3 to 12 m, thus requiring a gentler slope.

Because the clay liner would completely encircle the tailings, it could be considered as a below grade form of the more traditional "ring-dike" method of tailings containment. The proposed below grade "ring-dike" would be an engineered embankment with a steep sloped inner shell. The rock behind the highwall makes the outer shell unnecessary.

In summary, the proposed construction along the highwall of an open pit is nothing more than an engineered embankment with two uniquely appropriate design features: 1) a steep sloped inner shell made possible by careful control of freeboard, and 2) the absence of an outer shell, its purpose being fulfilled by in-place rock behind the highwall.

In general, tailings disposal design is complicated by the effect of groundwater flow parameters, the large time span involved, the size of the waste pits being considered and the chemical characteristics of the waste. The effectiveness of various design changes, such as clay liner thickness, can only be determined if these factors are considered. In the Morton Ranch case, with large pit areas and low groundwater velocities, large increases in clay liner thicknesses are necessary to reduce the amount of leakage by any substantial amount. The alternative solution of reducing the permeability of the liner by addition of soil cement (assuming it will be acid resistant) allows for thinner liners. Such liners reduce potential leakage much more efficiently than uncemented liners of the same thickness or unlined pits with wastes placed above the water table.

---

[18] D'Applonia Consulting Engineers, Report 3: Environmental Effects of Present and Proposed Tailings of Disposal Practices: Split Rock Mill, Jeffrey City, Wyo., Project No. RM77-419, 1977.

DISCUSSION

R.E. WILLIAMS, United States

Can you go back to the figure where you had that cross-section with the pit, the phreatic surface and the clay liner. The thing I want to point out regarding the slope failure is that when you excavate a pit, you cannot do it instantaneously and I have observed a number of them being excavated and what actually happens is that the drainage caused by pumping off the pit, once the pit hits the water table takes the phreatic surface down, so that you never have this circumstance, where there is a freatic line high on the wall of a pit because it is not excavated instantaneously. The modified Bishop technique that you are using for slope's stability of analysis incorporates the effect of the neutral stress, the poor water pressure, on the effect of the slope's stability on the face. The model probably predicts failure that is a factor of safety less than one because you are modelling the phreatic surfaces as it would exist if you instantaneously excavated the pit instead of how it really exists, so that it is dry because of the slow rate of excavation of the pit. If you put the clay liner in, it becomes a dam essentially and the phreatic surface will build back up to where it would have been, if the pit would have been excavated instantaneously. Now the only thing that will prevent failure of that wall is the clay liner. That assumes that the freatic surface has time to build back up to this elevation before you put the tailings in, and you better watch out, because it will probably fail.

W.P. STAUB, United States

We did consider a gently sloping water table down towards the bottom of the pit. I think that perhaps answers the first part of the question.

R.E. WILLIAMS, United States

The freatic surface will come back up to its normal level once you install the clay liner, in which case the modified Bishop factor of safety probably will be valid ; in which case the only think that will prevent failure is the sheer strength of the clay liner itself. And I do not know what that would be, but my guess is that it would not be very good, because it will be saturated.

W.P. STAUB, United States

I think what we are talking about here, is the return of the water level to its original location. We really are in some sort of a dilemma in determing what the strengths of the characteristics of those units are in the first place. So there are a number of other problems that are also associated with that, particularly analysis.

A.B. GUREGHIAN, United States

In talking about your problem, with the influence of the clay liners which probably seems to be the very end thing at the moment to try to prevent the rate of advance of pollution of ground water. I suppose that the example that you have shown is not really illustrative of clay liner in a real situation where the unsaturated portion of aquafer could be appreciable, the thickness that is. And your problem is more related to a saturated system and probably you

are trying to deal on very steady state conditions. We did not see
any inference off the liner, especially the walls that you have
inside, and the rate of advance of the water for that particular
liner ; what is the retardation effect ?

W.P. STAUB, United States

On the assumption that the clay liner is safely constructed
so that it would not fail, water would begin to appear on the outside of the clay liner after about 2 to 3 years. Using a 3 foot thick
liner of the bottom of the pit, the fluid contained in the pit would
be essentially drained after 1,000 years or so.

A.B. GUREGHIAN, United States

Are you considering a very safe condition right from the
start as far as the water movement is concerned ?

W.P. STAUB, United States

Perhaps I should defer that question to my co-author,
Larry Triegle, who is not here today. I believe that she did assume
steady state. As far as the second part of the problem is concerned,
I suppose that the effect of having a very thin material, like clay,
where absorption could be appreciable will certainly influence it
again, for the concentration of fluids passing.

AN ASSESSMENT OF THE LONG TERM INTERACTION OF URANIUM TAILINGS
WITH THE NATURAL ENVIRONMENT

D. Lush
Beak Consultant, Toronto, Canada

J. Brown
University of Western Ontario, London, Canada

R. Fletcher, J. Goode, T. Jurgens
Kilborn Limited, Toronto, Canada

ABSTRACT

Current investigations of methods for the management of tailings at
uranium mill sites raise pertinent questions as to accuracy of our
identification of potential hazards and their causes. Of particular
relevance are the physical-chemical geomechanisms which may lead to
the dispersion of contaminants into the environment and eventually
into the food chain.

This paper looks at present practices of uranium mill tailings
storage in the light of "Perpetual Care" management concepts and
requirements. It examines the potential geomechanisms of transport
that would be acting on the tailing over the next millenia. Interesting parallels are drawn between these mechanisms and those geomechanisms which led to the formation of many of the original ore
deposits.

Central to the substance of this paper is a study currently being
undertaken for the Atomic Energy Control Board of Canada. Selected
elements to this study will be presented in the paper together with
a discussion of relevant peripheral issues.

EVALUATION DE L'INTERACTION A LONG TERME ENTRE LES RESIDUS
DE TRAITEMENT DE L'URANIUM ET L'ENVIRONNEMENT NATUREL

RESUME

Les études consacrées actuellement aux méthodes de gestion des résidus se trouvant sur le site des installations de traitement de
l'uranium soulèvent des questions fondées quant à l'exactitude de
nos connaissances sur les risques potentiels que comportent les
résidus et leurs causes. Les géomécanismes physico-chimiques qui
peuvent contribuer à la dispersion des produits de contamination
dans l'environnement, voire dans la chaîne alimentaire, présentent
un intérêt particulier à cet égard.

Cette communication traite des méthodes actuelles de stockage des résidus de traitement de l'uranium, compte tenu des principes d'une gestion à perpétuité et des exigences y afférentes. On examine les géomécanismes de transport susceptibles d'agir sur les résidus de traitement au cours des prochains millénaires. Des parallèles intéressants sont établis entre ces mécanismes et les géomécanismes qui ont entraîné la formation de la plupart des dépôts de minerai originaux.

Cette communication est principalement axée sur une étude qui a été entreprise pour le compte de la Commission de contrôle de l'énergie atomique du Canada. On y trouvera certains extraits de cette étude, ainsi qu'un examen des questions connexes pertinentes.

INTRODUCTION

During the past decade there has been an increasing awareness and concern in western society about the environment and our impact on it. In many instances, these concerns have been well founded and have lead to very positive action against pollution and wastefulness. However, it has not been uncommon for there to be perceived concerns, with no basis in fact.

Ever since the explosion of the first nuclear device in 1945, there has been a general distrust of nuclear energy, derived principally from a lack of understanding. Unlike other major energy sources such as coal, oil and hydroelectric power, people cannot see, taste or touch nuclear power and are aware of it only in its most destructive form. Nuclear power installations are housed in huge concrete structures behind tall fences and guarded gates, and make popular subject matter for the press, where radioactive leaks, safety standards and waste handling get front page coverage. Much of this occurs because of a general lack of knowledge regarding the unseen dangers of nuclear power.

Recently in the western world, through public hearings associated with environmental impact assessment studies, the entire nuclear fuel cycle has come under considerable scrutiny, from mining and milling, through refining, fuel fabrication, and fuel burnup to reprocessing and waste disposal. In terms of sheer volume of waste and long term potential for radiation dosage committment, uranium mining wastes have been identified as a major cause for concern. Unlike spent fuel wastes, which have very high specific activities but decay rapidly to lower levels, uranium mill tailings and mining wastes have a low specific activity but very long half lives which present a long term public health concern potentially greater than that from high level wastes.

Accentuating the potential long term effects are past examples of waste mismanagement, both in the United States and Canada, which have aroused the public's concern further. Major research is underway to isolate and provide "perpetual care" of high level wastes at the end of the nuclear fuel cycle and the question of wastes at the beginning of the cycle is also being addressed. The low level waste generated at the uranium mill site is much too massive in volume to isolate in the manner proposed for high level wastes. This begs the question of management for the long term both in terms of its philosophy and implementation.

The examples, as discussed in this paper, are specific to the Canadian temperate environment, although the approach has more general application.

PHILOSOPHY OF MANAGEMENT

Technical Considerations

The development of a long term waste management philosophy requires the acceptance of a basic set of management criteria. Our societies' approach has, as its basic tenets, that the present generation of waste managers should leave the wastes in such a manner that there is no foreseeable threat to future generations and future generations will not have to be involved in the care of the wastes. Implied is that the future bleed rate of contaminants from waste management sites should not exceed present regulatory levels, and not rely on continued monitoring to demonstrate that fact.

At present, surface water bleed rates of radionuclides from uranium
tailings management areas are maintained within regulatory levels by
treatment before final discharge to the open environment. This
requires a long term management committment to the waste site which
our basic philosophy and past history of society suggests is unacceptable. Any acceptable long term solutions must not rely solely on a
social management system, but rather on a management controlled by
those basic geochemical and biochemical cycles which have been
determining the flux rates of radionuclides through the open environment during the evolution of life on earth.

In order to develop a management system which meets these criteria,
a basic knowledge is needed of (a) those geochemical and biochemical
reactions affecting the mobility of radionuclides (b) the form into
which those radionuclides are transformed during the milling and
waste treatment operation and (c) the consequences of this transformation as related to their mobility within and the flux rate from
the waste management area over geological time. Only through an
understanding of these processes and their interactions with the
waste management environment can we determine whether present or
future management areas will transcend smoothly the cessation of
active management and determine whether future passively managed
bleed rates acceptable under present day and hopefully future standards
are achieved.

Before discussing the geochemical and biochemical cycles controlling
the flux rates of radionuclides to the open environment, it is
important also to have a perspective in philosophical terms of
social man and management.

Social Man and Environmental Management

The present notion of man's ability to manage or control his surroundings beyond the confines of our cities is not recent. This was
evident in the classical or Roman civilization in the development of
formal gardens and artificial irrigation schemes. Our ability today
to achieve control of our environment has never been greater.
Similarly, the need to control man's activity within this environment
has never before risen to such an intensity. Nevertheless, our
ability to control applies almost exclusively to human action and
not to natural phenomenon. Also in the time frame of our present
concern for the management and control of uranium mill tailings,
there is insufficient evidence to show that mankind can exercise a
sustained control of sufficient duration.

As shown in Figure 1, man's presence on earth was quite modest until
the arrival of Homo sapiens less than 60,000 years ago. Even
though Neanderthal man was considered to have a rudimentary social
organization, his small numbers, (five million) afforded only the
puniest of efforts in environmental control.* In contrast, the
recurrent glaciations of the world, each of roughly 70,000 years
duration were mammoth in extent. These glaciations with their
associated natural activities are the outstanding geo-forces that
occurred during man's presence on earth. The ability to effect a
sustained control over the environment is directly linked to man's
ability to effect a sustained control over his own activities. This
we identify in the highest state of achievement as civilization.
Throughout man's history, we can detect no more than 2 dozen civilizations, all of which occurred during the last 10,000 years. Furthermore, from Table I, we can see that the greatest duration of any
human civilization was less than 6,000 years and nearly one-half of
these had less than 2,000 years duration.

* Our world population today is 4.1 billion people.

TABLE I DETECTED CIVILIZATIONS

Description	Era Starts	Approximate Duration	Final Empire	Invader
1. Mesopotamian/Sumerian	BC 6000	5700 years	Persian	European
2. Egyptian	BC 5500	5200 years	Egyptian	Greek
3. Indus	BC 3500	2000 years	Harappa	Aryan
4. Cretan	BC 3000	1900 years	Minoan	Dorian
5. Canaanite	BC 2200	2100 years	Punic	Roman
6. Sinic	BC 2000	2400 years	Han	Huns
7. Hittite	BC 1900	900 years	Hittite	Phygians
8. Andean	BC 1500	3100 years	Inca	European
9. Hindu	BC 1500	3400 years	Mogul	European
10. Classical	BC 1100	1600 years	Roman	German
11. Mesoamerican	BC 1000	2500 years	Aztec	European
12. Islamic	AD 600	1300 years	Ottoman	European
13. Orthodox	AD 600	?		
14. Western	AD 500	?		
15. Chinese	AD 400	1500 years	Manchu	European
16. Japanese	AD 100	1800 years	Tokugawa	European

Reference: Carroll Quigley, Evolution of Civilizations.

The earth's climate during man's civilized era has been mainly
Boreal or the initial post glacial period following the Wisconsin
(Würm in Europe). The features of this climate are relatively dry
and cool. If we were to presume a continuance of the glaciation
events, the next glaciation could possibly begin within a few thousand
years. During this period, there conceivably would be an increase
in precipitation within the northern hemisphere with dropping sea
levels. That is to say, it may well be within man's ability and
need to adjust to the natural climate changes through adjustment of
city levels at the coast and to expand irrigated lands beyond river
valleys.

Management for the Long Term

Management in popular usage has the connotation of success in reaching
one's objectives. In this context, management over a term of 100,000
years has no precedence in human experience. As noted earlier, no
civilization has succeeded in controlling human activity longer than
6000 years and indeed no evidence exists for man's ability to control
immense natural events such as glaciations with their global consequences. Nevertheless, it can be said that some presumed objectives
of ancient man have been achieved over a long term. The burial of
human remains of Neanderthal Man for example, in a nearly intact
form, has been achieved, in the instances of our discoveries over
115,000 years. The excavation at other burial chambers in Egypt,
China and Greece show that for those people their long term objectives
were won for periods up to 2500 years. Even with these achievements
in isolated instances, we have no way of knowing what success rate
has been achieved. How many Neanderthal burials were there? How
many burial tombs of ancient emperors go undetected?

From the above examples, we see an achievement of management objectives by chance rather than by control since modern man rather
ammorally has no acceptance of the ancient man's desire for the
sustained burial of his remains. This facet leads to the question
of management for whom? To what extent are we to carry forward the
mores and beliefs of this era into another era 100,000 years later.
While proof "exists" that waste dispersed into the environment is
undesirable and hazardous, what can we use as a basis for constituting

the notion of harm for a society 100,000 years hence? To what
extent will our social and scientific dogma be acceptable in 100,000
years or even 10,000 years? To what extent will our presence on the
earth in population scale and distribution be a significant factor
in achieving management as we know it today?

Prior to the early 1900's, management as a science or theme of human
activity did not exist. Up until that time, attemps were made by
Adam Smith and others in conditioning social thinking towards the
concept of management for the factors of production in an embryo
industrial society. It took about 200 years to effect the changes
he propounded. Thus, it can be presumed that our present concept of
management has a life no longer than, say, 200 years. In this
regard, the probable limits to our management (as presently conceived)
of uranium mine tailings would not exceed 200 years. A detention or
control beyond that period would need to rely on chance or on natural
geophysical/geochemical forces.

GEOCHEMICAL CYCLES

The management philosophy discussed previously, essentially suggests
a smooth phasing into, and return to a natural geochemical cycle.
In order to pursue this philosophy, a basic knowledge of geochemical
behaviour and the geochemical cycle of ore deposition is needed.
Figure 2 illustrates in a simplified manner the geochemical deposition
cycle for ore at two mining camps in Canada.

In Northern Saskatchewan, the mechanism of transport of the low
thorium ores has been hydrothermal in nature with a resulting deposition by biological and chemical reduction, chelation and exchange
processes of vein type deposits. In contrast, ores in the Elliot
Lake area of Ontario have moderate to high thorium content and the
transport mechanism has been detrital to hydrothermal. This has
resulted in quartz-pebble conglomerate types of deposits with high
levels of associated sulphide mineralogy. The third type, which is
for the most part the original ore, is that characterized by the
pegmatite deposits of the Bancroft area which are high in both
thorium and uranium. These pegmatites are classed as primary, the
ore mineral having crystallized from a felsic melt. Uranium,
thorium, and the rare earth elements, concentrate in these late
stage mineral deposits because of their large ionic size and charge.
Radium, because of its relatively short half life, is not known to
form any natural minerals but rather is probably found in a substituted
lattice position in the uranium minerals, or possibly to a lesser
degree within recent thorium mineralization. This has important
consequences for mobilization of both Radium and Thorium-230 within
the milling process.

The conglomerate and vein deposits are classed as secondary because
the uranium is derived from earlier rock, transported, and concentrated
by various sedimentary processes (fluvial transport, sorption,
and/or reduction of uranium ions from aqueous solution.) Both U^{+4}
and Th^{+4} are relatively insoluble in aqueous solution. However, $+2$
unlike thorium, uranium oxidizes readily to the 6^+ species $(UO_2)^{+2}$
which is relatively soluble. This basic difference in chemistry is
what has resulted in the differential mobilities of thorium and
uranium, and their geochemical segregation. The mobile uranyl ion
$(UO_2)^{+2}$ is transported in solution as soluble hydroxide, carbonate,
sulphate and organic complexes and accordingly is soluble in either
acidic or basic media. Soluble UO_2^{+2} complexes can be precipitated
by four major processes. These processes have been acting throughout
geological time and are still active today within the tailings pile

(3) reaction with pre-existing sulphide minerals with or without the presence of bacteria and (4) reaction with or absorption onto mineral surfaces such as silica SiO_2.

The information available on radium, a radioactive decay product of thorium, indicates a relatively simple geochemistry. Only the divalent state is known and as an alkaline earth element, it most closely follows the aqueous chemistry of barium. As mentioned previously, no radium minerals are known, probably because of its relatively short half life in a geochemical sense. However, radium readily substitues into many inorganic compounds and mineral lattice sites. At radioactive equilibrium, the theoretical ratio of radium-226 to uranium-238 is approximately 3.5×10^{-7}. The solubility of radium and uranium salts are quite similar and show a much higher solubility and consequent mobility than the salts of Thorium. The one very important exception to this is that the sulphate of radium, is highly insoluble in contrast to thorium sulphate which is quite soluble under acidic conditions.

These differential mobilities consequently lend a high degree of potential variation in radionuclide content to natural ore bodies. These variations are very closely reflected in the milling and extraction processes employed with alkaline leach methods used for carbonate rich ores, acid leach processes used for low carbonate ores, and other processes used for removal of associated materials such as sulphide minerals.

MILLING CYCLES

The uranium mill is primarily designed to maximize the recovery of the uranium present in the ore, and to ensure that the purity of the uranium concentrate is high in order to be marketable. Consequently, uranium concentrate must contain only low levels of most other elements including thorium and radium. Most uranium mills satisfact-orily achieve their design objectives and generally recover greater than 90% of the uranium in the original ore as a uranium concentrate containing at least 80% U_3O_8 and reject virtually all thorium, radium and other radionuclides to the tailings.

With very few exceptions, the current approach to the recovery of uranium from an ore body is to mine with maximum selectivity con-sistent with the use of modern mining equipment and to then recover uranium from the ore, utilizing hydro-metallurgical milling techniques.

During the milling process, uranium minerals are solubilized using either dilute sulphuric acid or carbonate solutions.

The fate of uranium daughter products during the dissolution process has not been clearly defined. However, it is suspected that the chemical extraction of uranium under acid leach conditions also releases associated thorium-230, radium, and some other heavy metals. The thorium present as monazite is likely not significantly attacked. The radium likely immediately reacts with free sulphate and precipitates as radium sulphate. Thorium liberated from the uranium minerals likely stays in solution until the acid leach solutions are neutralized, at which point, it precipitates as a sulphate ($Th(SO_4)_2$) or hydroxide ($Th(OH)_4$) before discharge to the tailings pond. Conversely the solubility of thorium in carbonate solutions is quite low. Radium has a low solubility in both carbonate and sulphate solutions. Despite the low solubility of radium species and thorium carbonate complexes, some solutional mobilization and reprecipitation of these elements appears to take place during the uranium dissolution process.

Some general observations can be made at this stage with respect to advance techniques of radionuclide removal. Firstly, the prospect of chemically isolating radium in the mill is very low. Secondly, thorium could probably be isolated in the sulphuric acid milling process but not in the carbonate milling process. Thirdly, because of the mobilization and reprecipitation processes occurring in the mill, the radium is predominantly found associated with the finer tailings particles.

In some cases, such as at Elliot Lake, the conditions for optimum uranium dissolution are not conducive to a high degree of thorium solubilization since the thorium occurs partly as the refractory mineral monazite. Thus, natural thorium (Th^{232}) is not completely solubilized, and depending upon its mineralization, complete isolation may not be possible in the mill without radical modification to the milling process.

It is clear that the question of isolating radium and thorium in the mill is complex and there is a low probability of successful simple modification at existing mills to enable them to produce tailings of lower radioactivity.

A technique that appears to offer some potential for isolation of radium and thorium is preconcentration. Preconcentration methods available to the metallurgist for the separation of uranium ore into high activity and low activity fractions, include handsorting, radiometric sorting, photometric sorting, conductivity sorting, heavy media separation, magnetic separation and flotation.

Depending on the geology and mineralization of an orebody, the listed techniques can result in uranium recoveries of 95% in concentrates of between 20% to 60% of the mass of the as-mined ore. If the ore is at/or near secular equilibrium, then all uranium daughter products will be concentrated with the uranium. Because of the general physical and physio-chemical similarities between thorium and uranium minerals, thorium will generally tend to concentrate with uranium in primary and detrital deposits. Thorium-230 may be expected to be found associated with uranium in hydrothermal deposits.

Preconcentration of uranium could therefore yield a primary tailings with a low content of radionuclides which might be similar to a base metal tailings and be disposed similarly. The uranifereous product from preconcentration would be processed by slightly modified versions of conventional mills to yield a tailings some 2 to 5 times more active than the run-of-mine ore but of one half to one fifth the mass. The high activity tailings would be more readily and economically processed or disposed of due to the smaller volumes involved.

Preconcentration as a means of reducing the tailings problem appears to be a promising technique. It is suggested that preconcentration tests should be featured in any evaluation program for new uranium deposits. In general, preconcentration of radionuclides is believed to be more readily accomplished than isolation of such elements within an existing mill through modification of the circuit. Preliminary studies indicate that preconcentration might be more effective than the reprocessing of tailings for radionuclide removal.

Figure 3 illustrates the tailings generation from the milling cycle in simplistic terms by separating the solid phase, or initial gangue materials from the hydro-metallurgical liquid phase residue from the extraction process. As mentioned previously, preconcentration offers a possible technique for advanced thorium and radium isolation in the mill which conceivably would alter flux rates of radionuclides from the resultant tailings.

GENERIC MANAGEMENT OPTIONS

Before discussing the tailings mobilization cycle, it is important to briefly highlight the present practise of surface tailings management in Canada, and comment briefly on the less widely used underground storage, and, presently being evaluated underwater management option.

While some underground, and to a lesser extent, underwater placement of a portion of the uranium mill tailings is, or has been practiced in Canada, almost all mills convey at least a major portion of their tailings as a slurry to an above ground containment area. The volume of tails material produced exceeds the volume of ore removed and consequently requires a large containment area. The tailings are discharged behind retention dam where the solids and liquids separate, and the transport water is treated and recycled to the mill. Barium chloride constitutes the principal treatment and is added to precipitate radium from any water discharge to the natural watercourse. In all leach operations, the addition of lime is undertaken for pH adjustment. A typical tailings storage system is shown in Figure 4.

The tailings containment dam must be impervious and stable and can be constructed in a variety of ways, two of which are shown in Figure 5. At some tailings containment operations, attempts have been made to install an impermeable or semi-permeable lining at the bottom of the containment area either with natural materials or synthetic liners. Modern containment systems usually provide for the diversion of upstream drainage around the tailings area and hydraulic facilities to pass the rainfall that falls on the tailings area itself. Figure 4 shows one method of providing stability at a tailing containment for the final abandonment period. The monitoring and/or treatment of the drainage water is readily handled with this method.

Following final abandonment, several pathways are available for the transport of radionuclides to the open environment. These routes are indicated on Figure 6 in terms of short term and long term transport. In the short term, man has the ability to provide good management practice and can recover radionuclides for further treatment and immobilization.

As mentioned above, underground placement of tailings is practiced in Canada on a small scale. The generic approach appears to offer a good chance of regulating bleed rates of radionuclides to low levels. This is especially if practiced in a hydrogeologically suitable environment. It could be used principally on the high specific activity tailings originating from a preconcentration circuit, and those sludges produced from water treatment operations.

Underwater placement of tailings at present is not a sanctioned method of tailings management in Canada. It would appear, however, that under the proper set of environmental constraints, it could be a viable management option. Recent studies have shown that certain types of bentonitic clays have very high geochemical retardation coefficients for radionuclides of concern such as Radium-226. Coating the substrate of a tailings lake with a layer of bentonite could potentially act as a primary geochemical barrier, lowering radium diffusion coefficients into the overlying water several

orders of magnitude. A secondary barrier to transfer to the biosphere
would, in the proper environment, build with time, in the form of a
humic and fulvic rich organic detrital blanket overlying the clay.
These complex organic acids with molecular weights as high as several
hundred thousand are very reactive, and bind readily with heavy
metals. Their biological and chemical oxidative half lives are in
the order of several hundred years, resulting in a continual accumulation throughout the geomorphic life of the lake. It is this
process which has resulted in the high levels of uranium, thorium
and radium along with other heavy metals which are found in coal
deposits.

PREDICTIVE APPROACH TO TAILINGS MANAGEMENT

Obviously, each of the three generic management options, dry land,
underground, and underwater, have environmental and economic costs
and benefits associated with them. The economic costs and benefits
can be handled by what are already mostly standardized procedures.
The environmental costs, however, must be calculated far into
the future. In order to attempt this, it is necessary to predict
the behavior of radionuclides within the tailings under a given
management system, and equate the environmental costs with that
behavior.

For purposes of illustration, the predicted behaviour of a hypothetical
dry land tailings pile is shown in Figure 8. In order to develop
these predictions, the same basic chemical reactions identified as
operative under natural geochemical cycles have been employed. In
this example, several assumptions are made. The most important are
that (1) the area has a positive net water balance and that water
tables are dropping; (2) the tailings contain a significant amount
of sulphides and have been limed prior to discharge; (3) the surface
tailings have been over limed and fertilized so that a stable floral
community exists, and (4) all sorption and biochemical reactions are
ignored as they will principally reflect flux rates and not the
overall evolution of the system. It should also be emphasized that
a real tailings pile will not be of a homogeneous nature, but rather
because of gravitational separation of fine and coarse fractions and
intermittent movement of the point of discharge, the stratigraphy of
the system will be quite complex. Compounding this is the fact that
the various stratigraphic units will be characterized by differing
permeabilities, porosities and mineralogies. Any field experiments
undertaken to verify this first order model must take these factors
into account.

One of the major controlling factors regulating the internal evolution
of the tailings will be the flux rate of oxygen into the system.
Since a gas such as oxygen can diffuse more readily through air than
water, the depth of the water table and its seasonal fluctuations
through time will greatly affect what occurs to all parameters of
concern.

Water table levels will likely control the shape of the Eh profile.
Slowly dropping water levels with a narrow water table fluctuation
will likely result in a compression of the redox potential band
whereas more rapidly dropping levels, with wide fluctuations will
likely result in its spread.

As discussed previously, the presence of free oxygen will likely
result in oxidation of sulphides and a drop in pH will occur to a
level of 1 - 2. This will result in an increase in soluble sulphate
and heavy metal levels. The reaction responsible for this change
can be expressed as $15/2\ O_2 + 4H_2O + 2Fe\ S_2 \rightarrow Fe_2O_3 + 8H^+ + 4SO_4^=$.

These reaction products will be carried downward, neutralizing lime and precipitating as calcium sulphate and other metal sulphates. As these sulphates migrate into a region of negative redox potential, bacterial populations may reduce the free sulphate resulting in the precipitation of metal oxides and the generation of H_2S. Although our model does not consider this mechanism, it is mentioned because of its potential importance in altering migration rates and because the characteristic rotten egg odor of hydrogen sulphide can be easily recognized.

As the sulphates become depleted, radium sulphate will become mobilized and begin to follow the trailing edge of the sulphate and thorium plume. Uranium being mostly present as residual mineral, will leach under both basic and acidic conditions at a very low rate. It is suspected that much of the uranium that leaves the milling circuit in solution will not precipitate in the tailings, but rather be carried on through the system during the operation of the mine.

Some field data and bench scale leaching studies do support the hypothesis. If these predictions are meaningful, or indeed others are formulated which suggest a definite internal evolution under a prescribed set of environmental conditions, the implications for management in the long term are clear.

They suggest that present leachate systems are treating those materials mobilized within the zone of positive redox potential, and which stay within that zone for the duration of their passage through the tailings to the treatment system. It also suggests that the present dropping levels of contaminants may simply be reflecting a short term shrinkage of that mobilization zone, and that if a set of suitable conditions were to exist at some time in the future, there may be a rapid initial drop in pH followed by a sulphate and thorium slug followed by a radium pulse.

On the positive side, release of radon gas from the surface and potential for plant uptake of metals along with surface gamma dosage appears to be decreasing with time. The rate will depend upon the rate of water and oxygen infiltration, which in turn depends upon sulphide levels, permeability, etc. Only through a series of well planned and executed field measurements can we begin to collect the data which will enable us to confirm or refute this model and place the appropriate scales on the graph.

The evolution of a tailings waste management area and its interaction with the surrounding environment is a very site specific problem. A generic management model monitoring approach can serve to pinpoint those areas where we should focus our research efforts. However, the design of any particular program must, of necessity, be quite site specific in approachtaking into account the nature of the host ore, the milling process employed and the site options available for management both as they exist today and as they may be expected to change in the future.

SUMMARY

This paper has endeavored to come to grips with the very thought provoking subject of the long term interaction of uranium tailings with the natural environment and has addressed the subject of the philosophy of management related to "perpetual care" of such tailings confinement areas.

Arising out of the study is the parallelism which can be drawn
between the potential biochemical and geochemical mechamisms which
will be acting on the tailings over the long term, to the geomechanisms
which led to the formation of the ore deposits in the first place.
A basic understanding of the interaction of biochemical and geochemical
cycles is deemed to be an important prerequisite to tailings system
design.

The examination of generic tailings management techniques and example
scenarios, convincingly points out the need for site specificity in
any long term assessment approach. The host rock genesis and the
special features of the geochemical and milling cycles at each site
uniquely determine the special treatment systems required and the
best management techniques and options available for final abandonment.

The reader is referred to the forthcoming AECB study for references
and background data to this paper.

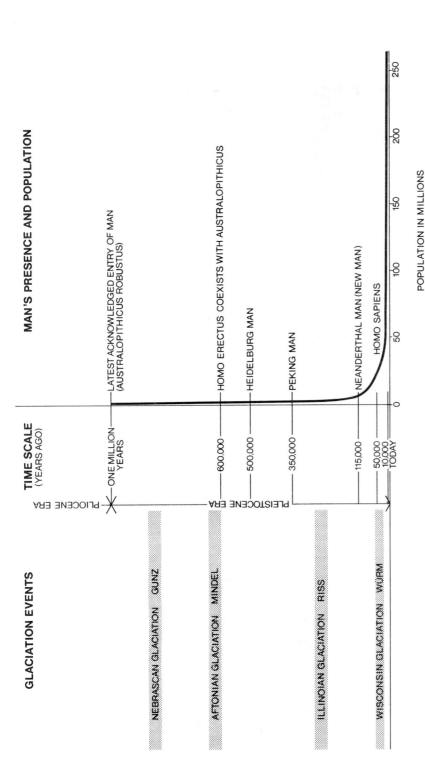

MAN'S PRESENCE ON EARTH
FIGURE 1

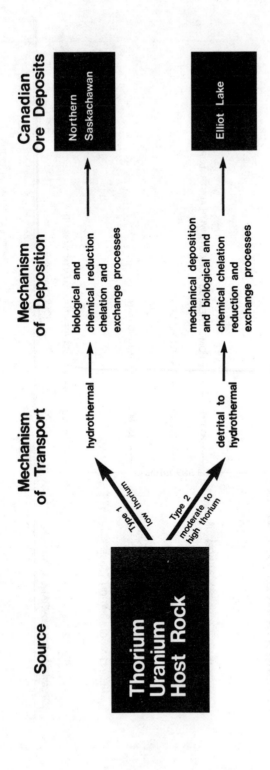

Geochemical Deposition Cycle

Source

Mechanism of Transport

Mechanism of Deposition

Canadian Ore Deposits

Thorium Uranium Host Rock

Type 1
low thorium

Type 2
moderate to high thorium

hydrothermal

detrital to hydrothermal

biological and chemical reduction chelation and exchange processes

mechanical deposition and biological and chemical chelation reduction and exchange processes

Northern Saskachewan

Elliot Lake

FIGURE 2

- 386 -

Tailings Generation

| Source | Mechanism of Transport | Mechanism of Deposition | Ore Deposits |

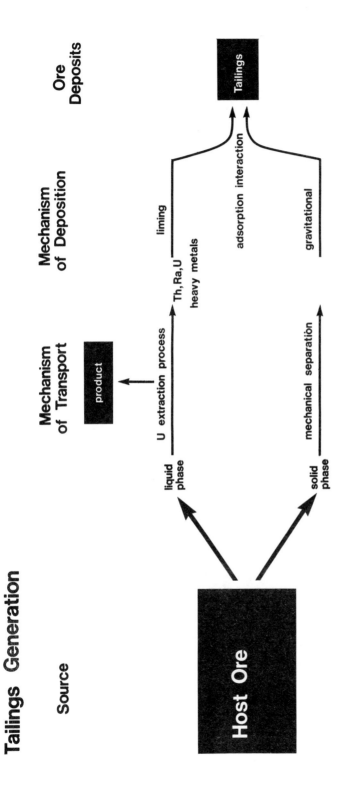

FIGURE 3

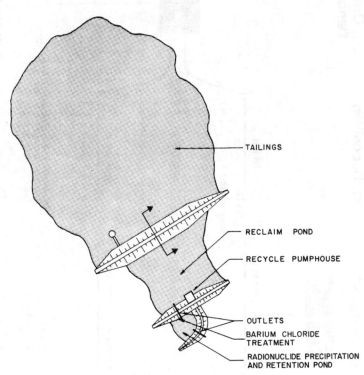

TAILINGS

RECLAIM POND

RECYCLE PUMPHOUSE

OUTLETS

BARIUM CHLORIDE
TREATMENT

RADIONUCLIDE PRECIPITATION
AND RETENTION POND

TYPICAL TAILINGS CONTAINMENT AREA

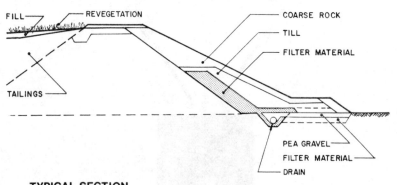

FILL

REVEGETATION

COARSE ROCK

TILL

FILTER MATERIAL

TAILINGS

PEA GRAVEL

FILTER MATERIAL

DRAIN

TYPICAL SECTION
FINAL ABANDONMENT PROVISION FOR TAILINGS AREA

FIGURE 4

- 388 -

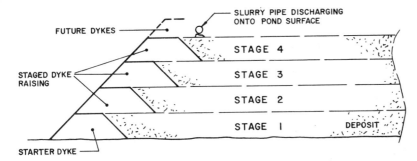

UPSTREAM CONSTRUCTION

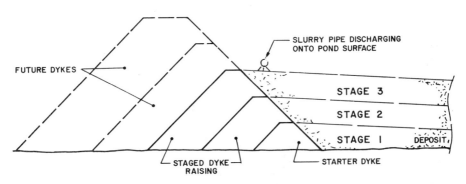

DOWNSTREAM CONSTRUCTION

TYPICAL TAILINGS RETENTION DAM
CONSTRUCTION METHODS
FIGURE 5

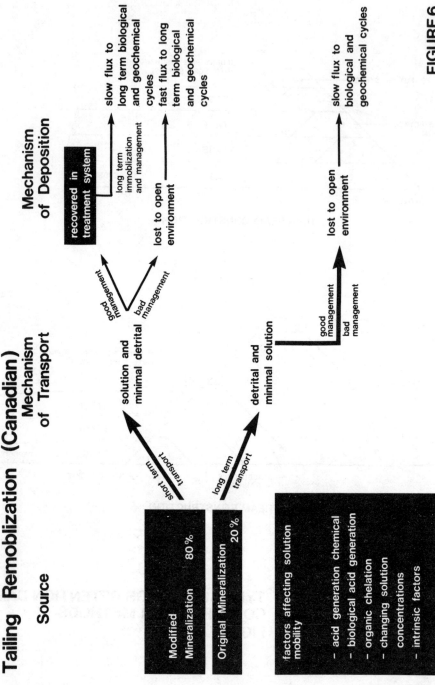

Tailing Remobilization (Canadian)

Source   **Mechanism of Transport**   **Mechanism of Deposition**

Modified Mineralization 80%

Original Mineralization 20%

factors affecting solution mobility
- acid generation chemical
- biological acid generation
- organic chelation
- changing solution concentrations
- intrinsic factors

short term transport

long term transport

solution and minimal detrital

detrital and minimal solution

good management

bad management

good management

bad management

recovered in treatment system

lost to open environment

lost to open environment

long term immobilization and management

slow flux to long term biological and geochemical cycles

fast flux to long term biological and geochemical cycles

slow flux to biological and geochemical cycles

FIGURE 6

Land Disposal Revegetated

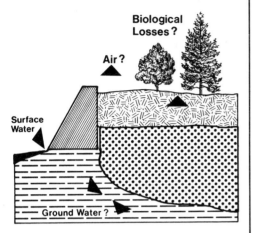

Under Water Disposal

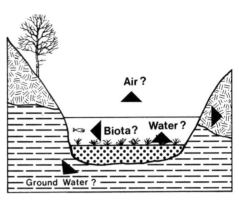

Underground Disposal

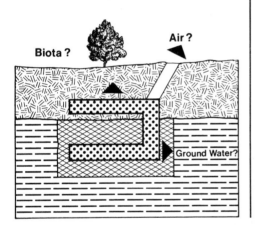

Legend

:::: Tailings

Unconsolidated material

== Consolidated material

Three Management Scenarios Under Evaluation
FIGURE 7

Hypothetical Tailings Pile Leaching Scenario

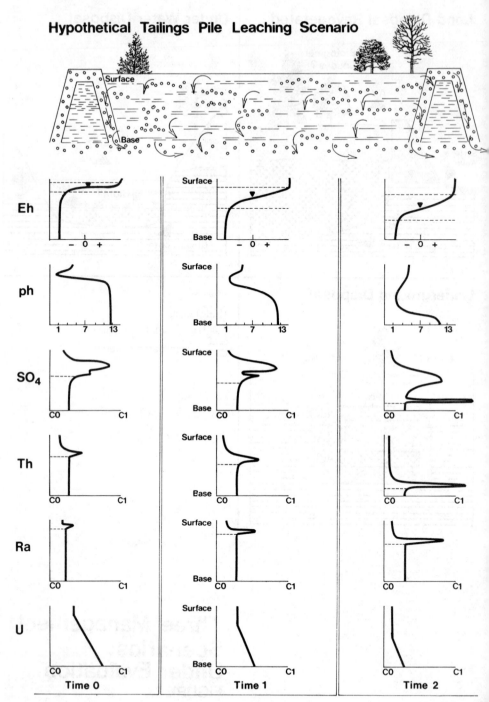

FIGURE 8

DISCUSSION

A.B. GUREGHIAN, United States

I could not very well interpret your text when you refer to pH. I mean were you trying to show a sort of a pressure distribution and tailings ponds and the time from where things are stopped ?

D.L. LUSH, Canada

We are looking at post abandonment many years after the surface is dried up and I am referring to pH there as the availability of free oxygen down into the system.

A.B. GUREGHIAN, United States

Are you planning on doing any modeling in the unsaturated region there ?

D.L. LUSH, Canada

Not in terms of this study. It was mentioned earlier today, for instance in some of the tailings piles that Canada is doing work on up in the nortic tailings areas. I think they are planning a drilling program into the tailings and these are some of the things that will probably be looked at. What sort of development we are getting in terms of pH down into the tailings and how this correlates with the sulfite levels, how this correlates with grain size. In other words, a lot of your radium, a lot of your sulfate minerals are tied up with the fines, and slimes, as opposed to the sands. When you look down into the tailings pile you will not get the sort of profiles that I have illustrated here. If this model was 100 % accurate, you would get a whole series of mini-profiles developing into both the coarse and fine materials with the extent of the profile relating almost directly to the hydraulic conductivity of the material and the ability of free oxygen to flex down into it.

A.B. GUREGHIAN, United States

You brought up a very good point about the hydraulic conductivity. Given the difficulties in assessing the hydraulic conductivity, and initially you mentioned also about the homogeneity of the tailing ponds. What is your assumption ?

D.L. LUSH, Canada

It is a completely invalid assumption.

A.B. GUREGHIAN, United States

How are you planning to determine the hydraulic conductivity of tailings ?

D.L. LUSH, Canada

In terms of this study, we are not. For determining the bulk hydraulic conductivity of a tailing pond, I could advocate here or ask Mr. Gillham to reply.

R.W. GILLHAM, Canada

What Mr. Lush has presented is a conceptual plan and I think a plan or a point of departure for discussion. In terms of our work in the flowing and consolidated material the actual analysis we are performing is below the bottom of the actual tailings. Now in terms of determining the hydraulic properties of the saturated tailings : a person can go up to the field and get some numbers which may have some degree of confidence with reference to the saturated tailings. In unsaturated tailings, there is no data available that I am aware of. The only study that I am actually aware of is in the U.S. at Colorado State ; there the absorption curves and the hydraulic conductivity water content relations are being measured and are going to be incorporated into the numberical models of water transport. In that particular study, one of the objectives as I understand it, is to build or design or engineer the tailings themselves, using the actual tailings from the site distributions in order to create an advantageous hydraulic environ-ment both to prevent the exhalation of radon and also for crop production. I suppose in going back to the point in measuring properties in the actual tailings, you very rapidly get into a problem scale and whether you are going to do a pump test and get a bulk number and what you do with it after that, I do not know. You can go the other extreme and try to measure that conductivity in every stratographic unit, again within a tailings zone that becomes extremely complicated and time consuming. You end up with so many numbers, again I do not know what to do with them, but I think it is a very difficult problem. In terms of Mr. Lush's presentation and his analysis, it is certainly not a type of complexity that is built into this analysis at this point.

A.B. GUREGHIAN, United States

I would certainly be delighted to see the results once you will get them. It would be enlightening. We do not know much about what goes on there. It will help you tremendously, I suppose, if you are considering what route it takes, I mean you should have a fair idea of those relationships.

W.P. STAUB, United States

There is a lot of conflicting information when you look at biological concentration factors in water, for instance where you get concentration factors ranging in there between 2 and 6,000 and you start to look into the data and a lot of things start to become obvious that you have got to take into account. If you are not looking at all the populations, for instance you have got to consider what phase of growth you are in. You have got to consider the production of what we call extracellular fibrillo material which has very high bionic exchange properties and that can concentrate a lot of these heavy metals and radionuclides around the cell itself, in a very strong manner and act as a source for a trace metal nutrient and for a potential uptake by individual cells. These same organics that are produced by unicellular algae in the water are also responsible for the precipitation for a lot of materials to the bottom of lakes and your complex organics produced in that way have very long biological half life, we are talking about several hundred years. They can lie in those sediments for several hundred years and build up concentra-tions in the sediments that are biologically unavailable through time. Again this is the same sort of process that relates to heavy metals and coal, radionuclides and coal. A lot of your Canadian lakes have substrates that have organic contents as high as 40 or 50 %. About 50 or 60 % of that which will be humical and full of the acid complexes which have a very high level of heavy metal. We get manganese in some of the sediments that are up a couple of percent.

Iron is up around 10 %. There is an extremely efficient complex and concentrating mechanism that operates in Canadian lakes. The water chemistry, and again this was brought up in several other talks, they were considering both in terms of leaching the tailings and in the behavior of the materials once it has been leached from the tailings, is extremely important and gets down to a site specific type of thing. You cannot look at the water chemistry in the Southwestern U.S. and say that materials behave in that type of water in a certain way and extrapolate that to the Canadian environment ; it just would not work.

D. MARKLEY, United States

We have looked at this study in terms of environmental movement on various parameters, I guess the question that comes to my mind, not only in this paper, but in some of the others. Do you have any intentions of linking your environmental movement assumptions to environmental or health effects ?

D.L. LUSH, Canada

We are addressing that in a chapter on Radiological and Health Effects that are a part of this study. Again, it is an area that is open for a lot of debate and it is one that is not the primary purpose of this study. In terms of regulation, what are safe levels going to be through time and safe levels through the open environment that is a whole other area that is in a state of flux right now. It is not the objective of this study to try and clarify that.

D.M. LEVINS, Australia

According to your theory, when you leach a uranium ore the thorium-230 should be leached to a greater extent that the thorium-232. Do you in fact find that this is the case ?

D.L. LUSH, Canada

We are hypothesizing that and I have not got any good numbers to back it up. It is, I would say, site specific, depending upon the minerology that the thorium-232 is in and depending upon the edge of the deposit. If the deposit is very old and there is a chance for a 230 segregation from the actual uranium itself, it is possible that you can get a certain type of ore where you would not get a very high percentage of extraction. If there has not been that type of segregation, if there has been a stable deposit over the last couple of million years, you probably would get a very high level of extraction, so I would have to waffle on that one and say that it depends on the mineralogy of the ore that you are talking about.

D.M. LEVINS, Australia

Your observations about the actual mobility of radium during leaching is interesting. I think it generally conflicts with what people believe, but it is supported by work at Oak Ridge on analyzing the radium concentration, or concentrations of particles size in both ore and tailings and they found that the radium actually moved. In other words, in the tailings the radium had moved from the coarse material to the slime material. So this suggests that radium is quite mobile and moves to the more accessible absorption sites.

D.L. LUSH, Canada

        I fully agree with you on that and I think that there is
another little interesting piece of data here. If you take your
tailing finds and you extract them with EDTA or any other kealating
agent, you find that about 80 % of the radium is stripped off. Now
if the radium is bound up inside any significantly large mineral
structure, you would not expect a surface reacting agent, such as
EDTA to strip it off, yet it does. If you have then leached the
material, or if you leached that s me material with nitric acid,
you find that about 90 % of the radium is taken off. So it appears
that a very high percentage of the radium is surface associated or
as a very fine colloidal material that can be easily stripped by
kealating agents, either natural ones such as humec and folic acids,
leaching down through your soil profile or man made ones, such as
EDTA, and that a small percentage of it is locked up in terms of
the mineral in which you need a nitric acid to strip it.

M.D. HOOVER, United States

        As I understand your long term solution is dispersal. Have
you looked at methods of controlled dispersal ?

D.L. LUSH, Canada.

        That is the whole objective of this study. If we adopt the
premise, and again, this is a personal one, there is a lot of feeling
that we are looking at containment in the long term when you are
looking at geological time, I think containment is unrealistic here.
You are looking at eventual dispersion. Then the whole question of
flux rate of the material comes up and the ideal management system
would be to design a system in such a way that the material is
dispersed to the open environment and put back into the geo-chemical
cycles which it came from at a flux rate which is below regulatory
standards, below those levels at which we may expect any type of a
detrimental effect to the biosphere. How you do that is the question
that we are all here to address, and there is no simple easy answer
to it.

D.B. CURTIS, United States

        Your ideas are admirable with regards to dispersal, could
I just ask your opinion about unnatural things like plutonium,
fission products, what you are feeling about containment vs.
dispersal ?

D.L. LUSH, Canada

        Again, I think we get to a philosophical discussion there.
I think the official policy and everything else is containment
through geological time and whether we certainly will not be around
in this type of moral atmosphere or society to worry about the
consequences of it. But, again you are probably just looking at
dispersal in the metal of the earth through geological time. But,
in terms of a management philosophy I think we are dealing with
something completely different in our high level or reactive waste
situation. There we do not have the volumes that we are talking
about here, we can handle them in a very specialized manner. I guess
the Swedes are putting on stream now, where the objective of management is containment. And containment for a very long period of time
until radioactive decay solves most of the problem for us. As I
mentioned at the outset, many other people have mentioned here that
tailings are a different bag of rocks. We are looking at a lot of

stable isotopes that are potentially just as dangerous to the biosphere as the radioactive ones in tailings. That is not the case to the same extent with the reactor wastes, so there is a different philosophy of approach that is needed there.

D.B. CURTIS, United States

If I may say so, I would not like to push that distinction between the two types of wastes. I think that when you look very closely at either of the wastes there are many long lived isotopes there which approximate the stay of elements, and also the end product of all the decay of all the radioactive elements is essentially lead. I would like to say one thing about this question of dispersal. I think a little clarification is necessary that the dispersion that you are talking about is an acceptance of the fact that on a very long time scale there is very little we can do to completely contain the material and it will eventually end up in the environment even with the high level wastes. That is a tectonic time scale and I think the time scale for the dispersion that you are mentioning is of the same kind. It is not that we are presently engaged in an acute dispersion of tailings in the environment as a method of disposal.

D.L. LUSH, Canada

I did not mean to leave that impression. We are looking at, and again if you want to do some calculations where you look at the amount of radium in tailings piles and you look at an acceptable bleed rate at say below 3 picocuries per liter, and the amount of leaching that is going on in tailings piles works out to quite a few thousand years to leach that material out. As I mentioned before, it is not only the dispersal that we are concerned about in this sense ; it is the reconcentration process. i.e. What happens to that radium, thorium, or uranium as it washes into that stream. It goes downstream into that swamp. Are we pulling it all out and are we going to end up in another thousand years from now with another uranium deposit in that swamp that is downstream of 50 % uranium with the associated heavy metals and other radioisotopes. Those are the secondary cycles that we are also addressing, at least in a conceptual sense in the study.

J. HOWIESON, Canada

I would like to congratulate you very much on the approach you have taken. It seems to me, that the idea of dealing with this is on a geological time scale and this is what we have to do. I would like to ask a specific question if you do intend in your study to develop the same sort of profiles, speculative that they may be, for underwater or underground storage as you have for the onland storage ?

D.L. LUSH, Canada

Yes. We are going to be addressing the same sort of issues, and we are going to be trying to, and again we are looking at a function of time. If we are talking underwater storage, we are talking underwater storage as long as it remains underwater. When the water disappears, as in Canada, where we have isostatic rebound which alters drainage patterns, which drains lakes and things like that, we may transform into a land disposal situation, 10,000 years down the road. If we are talking underground disposal, we may be looking at, as I said before, tectonic movement, the development of fracturing systems, the widening of microfractures through possibly acid leaching if we get oxygen into the system. But, I think the advantages of underwater and underground disposal are that we are looking at differences in time scale. We are extending the T-0, T-1, T-2 numbers that I had on the graphs, much further into the future. But in terms of instantaneous bleed rate, 10,000 years down the road, that is going to be a hard one to adjust.

D. JACOBS, United States

I think that you have pointed out something that is very important that these are transient conditions that we have imposed in the environment and they are likely to change drastically over long time periods. However, I am not really ready yet to buy your concept that these things are going to be completely dispersed ; because we know very well from the OKLO formation that those materials have been in containment for over 800 million years and that is 10 half lives of thorium. So, I am not so sure about what even with the tailings, the radioactive half life of important precursers of radium are going to be fairly well contained within the geologic formation.

D.L. LUSH, Canada

Sure, if we wanted to mimick an OKLO formation here, such as burrowing down in an open pit and putting it that far underneath a sandstone deposit, we are buying an awful lot fo time, like at least in an excess of 800 million years. Most of the problems that we are faced with, at least in the Canadian environment are surface management situations. The economics of digging open pits and moving all this material into open pits and everything else, makes it very questionable and it is a site specific type of problem. In northern Saskatchewan, for instance, where we have a lot of deposits that are located right at the interface between the alpha beta sandstone and the underlying basement rock, which is a very highly fractured and weathered system, there we have our major water fluxes are along that contact interface and if we were backfilling into an open pit there, we would be putting the material back into a very hydroanato-mically open environment and that raises all sorts of questions. The one thing that obviously comes to mind is O.K. well you have got all those things there anyway in that system ; why are not they moving ? That gets us back to the point I was bringing up before that we are not putting them back into the original form. We are solubilizing them, we are precipitating them as a flock, we are increasing the surface area and we are potentially allowing things like oxygen to get in there to start to demobilize things out of that reduced environment. There are a lot of site specific questions that have to be addressed.

Session 4

POLICIES AND REGULATORY ASPECTS

Chairman — Président
Mr. J.B. MARTIN
United States

Séance 4

PRATIQUES REGLEMENTAIRES

PUTTING TAILINGS IN PERSPECTIVE
RADIOLOGICALLY, ENVIRONMENTALLY AND MANAGERIALLY

J. D. Shreve, Jr.
Kerr-McGee Corporation
Oklahoma City, Oklahoma
U. S. A.

ABSTRACT

Present management of uranium mill tailings is sophisticated compared with recounted practises of 15 to 20 years ago. Further improvements while possible and wanted should not be rushed. Many factors enter. The paper suggests routes to selecting radiological protective standards, elucidates the site-specific and process-specific nature of tailings control, provides comparative references on source terms of radon, uranium, thorium and radium, defends the need for multiple control options, and finally argues for preoperational plans for tailings management and for plan endorsement in perpetuity once approval is granted, except as improved control at lesser cost proves doable.

One idea is proffered for reducing tailings mass to one-third that accumulated by current practises.

LES RESIDUS DE TRAITEMENT CONSIDERES SOUS L'ANGLE
DE LA RADIOLOGIE, DE L'ENVIRONNEMENT ET DE LA GESTION

RESUME

Les pratiques actuelles en matière de gestion des résidus de traitement de l'uranium sont très élaborées par rapport à celles observées il y a quinze à vingt ans. Il ne faudrait pas leur apporter précipitamment de nouveaux perfectionnements, même si ceux-ci sont possibles et souhaités. De nombreux facteurs doivent être pris en considération. Cette communication suggère certaines voies permettant de sélectionner des normes de radioprotection, met en lumière les aspects du contrôle des résidus de traitement qui se rapportent spécifiquement au site et au procédé utilisé, fournit des références comparatives sur les termes-sources du radon, de l'uranium, du thorium et du radium, démontre la nécessité de disposer de multiples options de contrôle et enfin avance des arguments en faveur de plans préopérationnels de gestion des déchets et de l'adoption de plans de gestion à perpétuité une fois l'autorisation obtenue, à moins qu'un contrôle amélioré à moindres coûts ne s'avère réalisable.

On présente une suggestion visant à ramener la quantité des résidus de traitement à un tiers de celle accumulée du fait des pratiques actuelles.

Outlook

It is important for industry, regulatory authorities and the public at large that uranium mill tailings have become a topic of concentrated attention; [1, 2, 3]; not so much to immediately tighten control of tailings accumulations and their treatment; but rather to characterize tailings, quantify associated effluents and define their paths and mode of translocation. Only by a fuller understanding of tailings and their interaction with the specific environs - an understanding not yet achieved certainly - can realistic methodology for tailings management be invoked, and safe but practical standards of control be stipulated. Previous presentations in the seminar attest the intensity of recent and current investigations and the need for continuation of study. The talks also lend accent to the variability among tailings systems. By a "tailings system", I mean the tailings plus the potential reach and impact of their constituents - the air mass receiving radon and wind blown particles, the surface area affected thereby and the ground mass and waters, surface and subsurface, to which influence may extend. Potential recipients, animals, vegetation and people, and their rate of receipt are the ultimate system elements. Above all, the role of background must be recognized as well.

Uniqueness and Site-Specifics

Every ore deposit has its peculiarities of mineral suite, depth of occurrence, range of uranium grade and extractability, and relation to water table and aquifers. Likewise, every tailings pile has a unique underpinning. The substrates range in depth-to-shale, contours of interfaces, permeability, sorption capacity and chemical and physical nature. Tailings characteristics differ too.

Most mills are one of two distinct types, alkaline or acid leach [4, 5]. Since the alkaline process is more subtle, it requires greater mechanical comminution of the ore. Hence, tailings contain more fines but deviate less from chemical neutrality (pH 10-11) and contain fewer mobile constituents. The acid treatment, more brute force in action, chemically intrudes uranium-rich particles so less grinding is required; hence, the tailings are left coarser and more permeable to radon and highly acid with pH 1.7-2.0. As a result, many elements well fixed in the original ore are made mobile or labile, and so they remain until by interaction with soil or fresh water the acidity is sweetened and precipitation or chemical fixation can lock them up again.

Historical Footings and Earliest Pollution Control

In the early days, situation of a mill and the tailings pile were choices of convenience. Only quite recently have the hydrology and substrate details been elucidated to estimate the consequences of site selection. Prior practice was not as cavalier as it might sound. In all but a few cases, the mill and tailings sites were well removed from nearest populations and guarded by poor roads and unattractive surroundings, albeit the exceptions are notable as we'll see later. To make matters worse, adjoining land was seldom purchased as a buffer zone for long term isolation. Of course, there was no pressure to do so, either by edict or out of common practice. These were the days when the exclusive purchaser of yellowcake was the U.S. Government, when maximum production was urged with few questions raised as to methodology. Essentially all abandoned tailings piles in the USA - some 23 million tons - were generated during that period as was a significant fraction of many piles still active - another 50 million tons in fact. In short, a different set of values guided the U.S. Government and the uranium mining and milling operations in that time interval [4,5,6]. At least two mills and their tailings were perched on river banks. Before 1960, direct discharges to rivers are estimated to have been as high as 11 tons/day of dissolved solids from a single mill. The Federal Water Pollution Control Administration, created in April, 1958, took formal action in 1960 with the cooperation of Colorado and adjoining states to eliminate direct discharges and to monitor the Animas and Colorado Rivers. But that is behind us. Industry and Government agree it is right to be concerned, right to rectify existing ills, and fundamental to plan ahead on new facilities so that uranium mill tailings will have their effluents and emanations controlled. However, with the history just cited, it is human nature to overreact.

This must be resisted. Else we indulge foolhardiness of the opposite sign. I repeat my earlier message for emphasis: Understanding needs precede regulatory action.

Other Radium Sources

It is becoming more and more apparent that the uranium industry is not alone as a generator of radon and as a concentrator of radium. The U.S. EPA has plans to look at the processing of copper, fireclay, zinc, limestone and iron. England is already worrying with radon in the Cornwall tin mines as South Africa has for many years in their gold mines. And some chambers in the Carlsbad Caverns in New Mexico exhibit radon concentrations that exceed the level permitted in a working uranium mine. The phosphate industry has been studied for the last several years. We all remember Joachimsthal and Schneeberg.

Uranium and Thorium are ubiquitous to the extreme of being everywhere at a concentration of 1 to 4 ppm in surface soil. Robley Evans has calculated that a square mile of ordinary backyard dirt, 5 feet deep, would yield 30 tons of elemental uranium and 10g of radium [7]. Each cubic yard of common soil or rock has about $2\mu Ci$ (2 million pCi) of radium. Your backyard is no different.

Radon Source Equivalences and Radiological Impact

Various estimates of the total radon from natural and man-aided radon sources place the latter at less than 1% of the former - the natural radon release is about 100 times the total from all uranium tailings. Moreover, regions surrounding workable ore deposits generally have abnormal amounts of U and Ra. It follows that the radon release from every tailings pile is matched or exceeded by that from a contiguous area about 20 to 100 times the pile area, i.e. ten to a few tens of square miles. This means that at the edge of a six-mile square, there is essentially no evidence of a tailings pile at its center. For most seasons and times of day, there is little or no evidence at the boundary of a 2 or 3-mile square containing a central tailings pile [8]. To wit, non-occupational exposures to radon and wind blown tailings are quite small from milling operations remote from population clusters. The man-rem totals are disappearingly small and we are left with the need to protect only nearby isolated inhabitants for which maximum permissible dose and dose rate become the proper limiting criteria of protection as set forth in the long established guidance of the ICRP, and, in the USA, the NCRP and the Code of Federal Regulations, Title 10 part 20. Only against this backdrop can the subjectivity of ALARA be judged and genuine gains in control appraised. Quickly, we see that the day to day radiological and environmental impacts are quite small. Long term implications are another subject.

Tailings Development and Character

As mining and ore processing proceeds, tailings accumulations grow in depth and area. The increasing depth keeps radon effusion about the same, but the advance in area enlarges it. The real drama is enacted under the pile since most piles and ponds are unlined. This may well be the preferred situation since water and dissolved solids seep downward where sorption and ultimately chemical fixation, via the neutralizing tendencies of underlying soil, lead to considerable if not complete impoundment of radioactivity. Given enough vadose soil mass with which to interact, all values can be contained. Part of good management of tailings is the promotion of interaction to the degree required and to accomplish it well within the confines of the company property. Admittedly this is sometimes difficult because of the elements mobilized as chlorides (and sulfides to a lesser extent) by acid treatment. For all its merits as the most effective process for winning uranium from most ores, acid leaches can generate a more intractable set of products as far as confinement goes.

Some Ideas For Bettering Control

It may well be that intervention should occur earlier in waste stream handling, namely, while slime and sand fractions are still separate. In that more than 90% of the radium reports in the slimes, the slime should be kept to itself; it may eventually comprise the only waste of concern. Once dewatered, its small mass compared with what now constitutes tailings (<1/3), suggests numerous control measures of novel nature. Moreover, just about all the chlorides and sulfides are in the slime branch.

Direct neutralization of acidity could prove economically feasible as may further treatment particularly if by some new technique the traces of uranium can be extracted. The sand stream, much lower in radioactivity (30 to 80 pCi/g) would, if cleanable to <20 pCi/g, allow uses beyond the obvious one of sand-fill for old underground stopes. The standard technique in contrast uses the sand to form dikes or pond boundaries and the slime discharge is moved along the inside edge near the top of these sand berms. Much penetrates the sand as it flows downward to enter the pond. In essence, rather clean sand is recontaminated in an effort to coax solids from the slime and hasten dewatering. In a real sense, this procedure of such long standing emerges as counterproductive. To be sure this alternate scheme has surfaced so recently that only limited tests of it have been made. Even with multiple water washing, the sand retains about 2/3 of its original radium.

Conclusions

With the impressive number of new mill sites proposed, the time is ripe for innovative thinking and unprecedented, preoperational planning. Memory of the brash practices of the past should serve to stimulate cerebration. Improvements to tailings confinement in the last 5 years have been praiseworthy indeed. Yet, piloted by the ever better measurements of airborne and subsurface transport of tailings-derived material, fuller attention to meteorological and hydrological correlates and sample analysis, large refinements in tailings management are promised. The problems are complex however, and will require a mix of disciplines, time, and agency/industry cooperation for solutions.

Selenium earns special mention as a topic demanding attention. A thing of immense chemical ambivalence literally and figuratively, it is a most difficult element to control. Concerted effort will be required to establish workable means for handling it at the tailings scale wheresoever it occurs.

The real test of our ingenuity will be the economics. Pretty surely we'll come upon landmarks of improvement that make premature decisions inviting. Therein lies the danger if costs are unbearable. The key will be patience and confidence that the proper, balanced approaches are just down the road. This will make it an exciting and productive endeavor. Somehow the regulators must bridle that great human urge to appear wise, and instead, exercise patience until they can be wise, until they can set standards based on comprehensive information and then leave them open to adjustment as added knowledge evolves.

To study too little and intuit too much could lead to industrial paralysis by presumption, a case of putative purism perpetrating protection by preying on persons processing piles for power in peace .... PERIOD!

REFERENCES

1. Sears, M.B., Blanco, R.E., Dahlman, R.C., Hill, G.S., Ryon, A.D., and Witherspoon, J.P.: "Correlation of Radioactive Waste Treatment Costs and the Environmental Impact of Waste Effluents in the Nuclear Fuel Cycle for Use in Establishing 'As Low As Practible' Guides - Milling of Uranium Ores," ORNL - TM - 4903, Vol. 1, Oak Ridge National Laboratory, Oak Ridge, Tennessee (May, 1975).

2. Bernhardt, D.E., Johns, F.B., and Kaufman, R.F.: "Radon Exhalation from Uranium Mill Tailings Piles - Description and Verification of the Measurement Method," Technical Note ORP/LV-75-7(A), U.S. Environmental Protection Agency, Las Vegas, Nevada (November, 1975).

3. Eadie, G.E., and Kaufman, R.F.: "Radiological Evaluation of Uranium Mining and Milling Operations on Selected Ground Water Supplies in the Grants Mineral Bell, New Mexico," Health Physics, Vol. 32, pp231-241, (1977).

4. Tsivoglou, E.C. and O'Connell, R.L., "Waste Guide for the Uranium Milling Industry," Technical Report W62-12, Robert A. Taft Sanitary Engineering Center, US Public Health Service, Cincinnati, Ohio (1962).

5. "Process and Waste Characteristics at Selected Uranium Mills," Technical Report W62-17, Robert A. Taft Sanitary Engineering Center, Cincinnati, Ohio (1962).

6. "Uranium Wastes and Colorado's Environment," Colorado Department of Health, Denver, Colorado (August, 1970).

7. Evans, R.D., "Engineers' Guide to the Elementary Behaviour of Radon Daughters," Health Physics, Vol. 17, pp229-252 (1969).

8. Shearer, S.D., Jr. and Sill, C.W., "Evaluation of Atmospheric Radon in the Vicinity of Uranium Mill Tailings," Health Physics, Vol. 17, pp77-88 (1969).

DISCUSSION

D.M. LEVINS, Australia

       With the figure of 90 % of radium in the slimes, I think is a very unusually high figure. The figures that are generally seen in the Australian tailings is the figure of something like 75-90 % of the radium in the slimes and the concentration in the sands run about 200 picocuries per gram.

J.D. SHREVE Jr., United States

       This is a variable number. It may have to do with how well the classification is done by size in the specific mill and we commonly run up to 90 % in our mill and I would allow for the fact that it could be less than this.

J. HOWIESON, Canada

       You suggest that the separation of slimes is a good thing and that there are many ways of dealing with it, but you do not specify what these ways might be. Could you give us some more of your suggestions in that regard ?

J.D. SHREVE Jr., United States

       I think it is simple. In my mind looking at the problem as long term, I personally think we are wrong to worry about five to ten years in the light of the last paper this morning. So, what I feel you can do is to discharge slimes directly on to a well chosen site now based on doing the hydrology up front and studying the substrate and looking for sand lenses and what have you. Put the slime right down by itself. I think when this finally consolidates, either by simple evaporation or by using the techniques that some of the phosphate people are using to hasten a consolidation, you now have a much tighter site spectrum in that you avoid what we do now by mixing the sands back to recreate a medium with the same pore volume. We end up with a more permeable mix. To use the slimes alone, which have all the fines. I think that over a wide range of receiving base and characteristics that these will tend to move down rapidly at first, but very quickly with a little neutralization either applied from the surface or with the slime. They will mechanically tighten that structure and form almost a natural liner of much lower permeability than the mix of sands plus tailings.

CURRENT U.S. NUCLEAR REGULATORY COMMISSION LICENSING REVIEW PROCESS: URANIUM MILL TAILINGS MANAGEMENT

R. A. Scarano and J. J. Linehan

Uranium Mill Licensing Section

U.S. Nuclear Regulatory Commission

Washington, D.C.

ABSTRACT

It is incumbent on the NRC to assure that authorized tailings management programs provide for isolation of these long-lived radioactive materials over the long term. The review of a tailings management program encompasses three basic issues: siting, design stability, and final reclamation plan. The NRC has provided guidance to the industry for the development of acceptable tailings management programs. The industry has responded by proposing innovative schemes keyed to the specific geohydrological characteristics of the proposed milling operation site. Current reviews have indicated that the preferred disposal method is below the natural grade or above grade in an area where the final reclamation configuration would result in the same erosion resistant characteristics as below grade.

PROCEDURE D'EXAMEN ACTUELLEMENT APPLIQUEE A LA DELIVRANCE D'AUTORISATIONS PAR LA COMMISSION DE LA REGLEMENTATION NUCLEAIRE DES ETATS-UNIS : GESTION DES RESIDUS DE TRAITEMENT DE L'URANIUM

RESUME

Il incombe à la Commission de la réglementation nucléaire des Etats-Unis de s'assurer que les programmes autorisés de gestion des résidus de traitement prévoient l'isolement à long terme de ces matières radioactives de longue période. L'examen du programme de gestion des résidus de traitement fait intervenir trois questions fondamentales : choix du site, stabilité de la conception et plan final de remise en état du terrain. La Commission de la réglementation nucléaire a formulé des directives à l'intention de l'industrie pour l'élaboration de programmes acceptables de gestion des résidus. L'industrie y a répondu en proposant des systèmes novateurs axés sur les caractéristiques géohydrologiques spécifiques du site proposé pour les opérations de traitement du minerai. Il ressort des examens en cours que la méthode d'évacuation préférée est celle qui consiste à placer les résidus au-dessous du niveau naturel du sol ou au-dessus dans une zone où la configuration finale du terrain remis en état se traduirait par les mêmes caractéristiques de résistance à l'érosion qu'en-dessous du sol.

INTRODUCTION

A detailed review of the environmental impacts related to a uranium mill project was not included in a licensing review until the passage of the National Environmental Policy Act of 1969 (NEPA). The act requires the preparation of an environmental impact statement (EIS) for major federal actions, including licensing of a uranium mill. From the passage of the act until 1976, the Atomic Energy Commission prepared two EIS's related to new uranium mills. While the environmental impacts and mitigating measures associated with the milling operation were fully addressed in these statements, the development of an adequate stabilization program for the tailings was deferred until termination of milling operations. It was not until late 1976 following the U.S. Department of Energy studies of the inactive tailings sites and the increasing public interest in the environmental impacts associated with mill tailings that the NRC decided that the tailings management issue must be met head on. An acceptable tailings management program must be developed and surety arrangements be in place to insure its execution prior to issuance of an operating license.

CURRENT LICENSING PRACTICES

Since it was evident that the results of a generic environmental impact statement related to uranium milling which the NRC is preparing were a long way off, the licensing staff issued interim guidelines in May 1977 for the industry in the form of the following performance objectives for tailings management:

Siting and Design

1. Locate the tailings isolation area remote from people such that population exposures would be reduced to the maximum extent reasonably achievable.

2. Locate the tailings isolation area such that disruption and dispersion by natural forces is eliminated or reduced to the maximum extent reasonably achievable.

3. Design the isolation area such that seepage of toxic materials into the groundwater system would be eliminated or reduced to the maximum extent reasonably achievable.

During Operations

4. Eliminate the blowing of tailings to unrestricted areas during normal operating conditions.

Post Reclamation

5. Reduce direct gamma radiation from the impoundment area to essentially background.

6. Reduce the radon emanation rate from the impoundment area to about twice the emanation rate in the surrounding environs.

7. Eliminate the need for an ongoing monitoring and maintenance program following successful reclamation.

8. Provide surety arrangments to assure that sufficient funds are available to complete the full reclamation plan.

As can be seen, these objectives are tailored to allow industry flexibility in developing tailings management alternatives for specific sites.

In support of a license application the applicant is required to perform and submit to the NRC an evaluation of viable tailings management alternatives for the proposed project. In reviewing the applicant's evaluation, the NRC staff places special emphasis on the relationship between proposed tailings management alternatives and the possible mill site alternatives. It is not unlikely that some sites will not be suitable for disposal of tailings. In short, there may be some sites that should not have a milling operation. In the year since we issued the performance objectives, the industry has responded with innovative schemes which meet the objectives. I will discuss four of these methods in detail.

The first plan (Figure 1) involves disposing of slurried tailings into a mined out pit. The floor of the pit will be lined with three-foot minimum compacted clay. The walls of the pit will be lined with as much as twenty feet of compacted clay, not that twenty feet is needed; but it is laid as a road around the pit, and that width is necessary for equipment. Compacted overburden materials will be placed in the pit to a point at least ten feet above the groundwater table. During operations, the tailings are slurried into the pit and excess water is decanted from the pit and transferred to a lined evaporation pond located on the surface. Following a drying out period, the tailings will be covered with the following combination of materials; four-foot overburden, two-foot compacted clay, four-foot overburden, and six inches of topsoil. The reclaimed area will be contoured and vegetated to blend with the natural contours of the surrounding land.

Because the open pit mine to be used covers 130 acres and will take four years to mine, impoundment of tailings will take place in stages (Figure 2). This scheme has the advantage of allowing for a staged reclamation program. Reclamation of the early impoundment areas will take place during the operating life of the mill. Following operations, the dried material and liner from the evaporation pond will be transferred to the last stage of the tailings impoundment area.

The second plan (Figure 3) is just a variation of the first in that dewatered tailings, i.e., 20-30 w/o moisture, will be impounded in the mined out pits. A three-foot compacted clay liner will be placed on the bottom of the pit, and compacted fill will be added to a minimum of ten feet above the water table; but a clay liner on the side walls is not necessary. Because of the reduction in the amount of solution that could migrate, which in turn inhibits the mobility of the toxic materials left in the tailings, impermeable side walls are not considered necessary.

The advantages of the second plan are the elimination of the expense involved in laying a clay liner on the side walls and the increased capacity for the tailings.

The staged tailings impoundment and reclamation and covering are the same as discussed in the first plan.

The third plan (Figure 4) involves discharge of tailings slurry into a below-grade impoundment consisting of four individually constructed cells excavated to a depth of 40 to 50 feet below the existing grade.

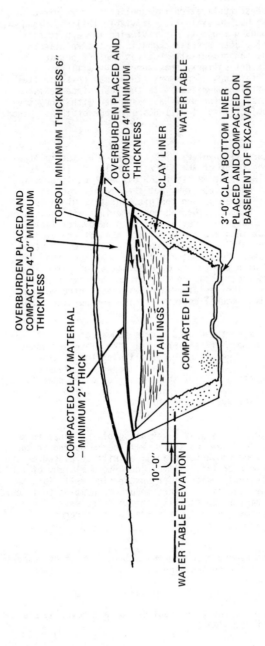

OVERBURDEN PLACED AND
COMPACTED 4'-0" MINIMUM
THICKNESS

TOPSOIL MINIMUM THICKNESS 6"

OVERBURDEN PLACED AND
CROWNED 4' MINIMUM
THICKNESS

CLAY LINER

WATER TABLE

COMPACTED CLAY MATERIAL
– MINIMUM 2' THICK

3'-0" CLAY BOTTOM LINER
PLACED AND COMPACTED ON
BASEMENT OF EXCAVATION

TAILINGS

COMPACTED FILL

10'-0"

WATER TABLE ELEVATION

FIGURE 1. MINED-OUT PIT BURIAL-SLURRIED TAILINGS

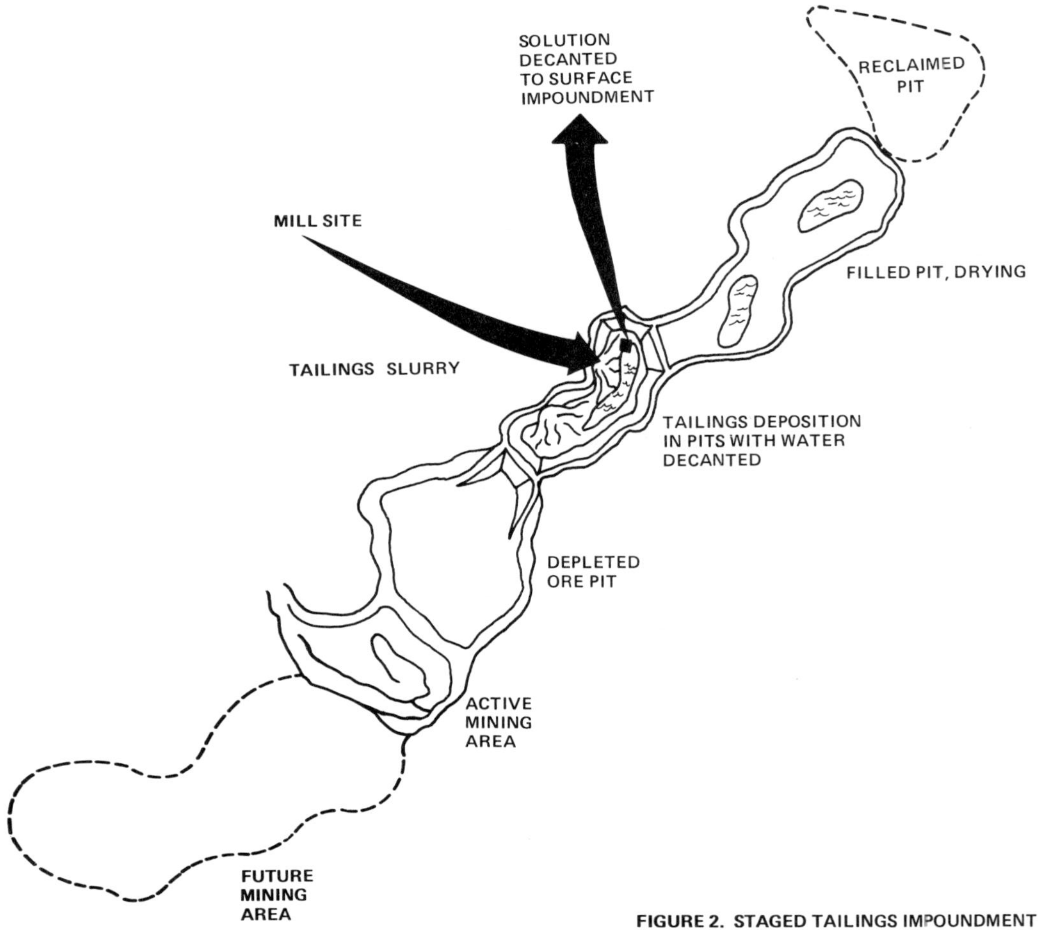

FIGURE 2. STAGED TAILINGS IMPOUNDMENT IN AN ACTIVE OPEN PIT MINE

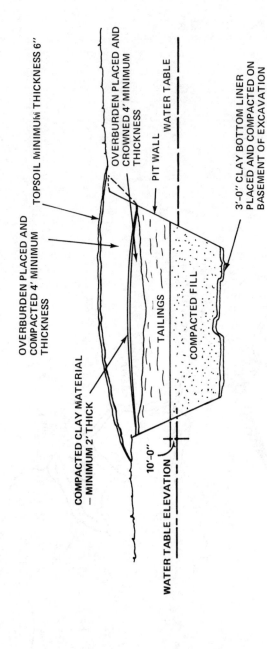

OVERBURDEN PLACED AND COMPACTED 4' MINIMUM THICKNESS

TOPSOIL MINIMUM THICKNESS 6"

OVERBURDEN PLACED AND CROWNED 4' MINIMUM THICKNESS

WATER TABLE

PIT WALL

3'-0" CLAY BOTTOM LINER PLACED AND COMPACTED ON BASEMENT OF EXCAVATION

COMPACTED CLAY MATERIAL — MINIMUM 2' THICK

TAILINGS

COMPACTED FILL

10'-0"

WATER TABLE ELEVATION

FIGURE 3. MINED-OUT PIT BURIAL–DEWATERED TAILINGS

TYPICAL CELL CONSTRUCTION

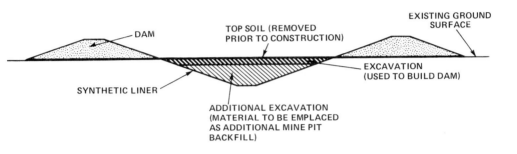

POST OPERATION/PRE-RECLAMATION

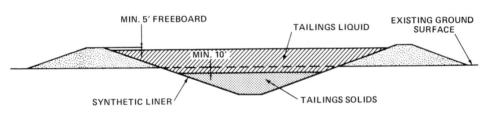

ABANDONMENT

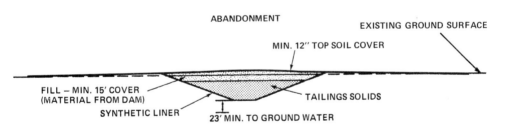

FIGURE 4. SPECIALLY EXCAVATED PITS FOR TAILINGS BURIAL

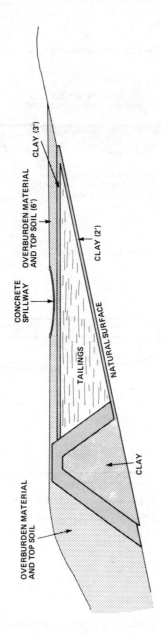

CONCRETE SPILLWAY

OVERBURDEN MATERIAL AND TOP SOIL (6')

CLAY (3')

CLAY (2')

TAILINGS

NATURAL SURFACE

CLAY

OVERBURDEN MATERIAL AND TOP SOIL

FIGURE 5
TAILINGS IMPOUNDMENT AT THE
HEAD-END OF A NATURAL VALLEY

The sides and bottom of each cell will be lined with a synthetic liner:
a 30-mil reinforced Hypalon liner on the sides and a 30-mil PVC liner
on the cell bottoms. Each cell will be surrounded by an above-grade
dam (40' high) that will provide an evaporation pond for the liquid
portion of the tailings and prevent any surface runoff from flowing
into the tailings impoundment. During operations, tailings will be
deposited sufficiently below the natural grade to allow for the placing
of a 15' cover of overburden and topsoil over the tailings cells without
creating an above-ground mound. At the time of reclamation a portion
of the material from the dams surrounding the cells will be used as
cover over the tailings and the remainder will be hauled to the mine
waste dump or used for reclamation of mining areas. Following completion
of reclamation, the cover over the tailings will be contoured to the
natural levels present prior to cell excavation. The use of four
individual cells constructed sequentially at three to four-year intervals
allows the applicant to make improvements and refinements resulting
from experience with the construction and operation of the first cell
and provides for a staged reclamation of the tailings impoundment areas.

The fourth scheme (Figure 5) I will discuss today involves slurrying
tailings into a surface impoundment at the head end of a natural valley.
The area is surrounded on three sides by natural hills, and a dam will be
built on the lower fourth side. The floor of the basin will be lined
with two feet of compacted clay which will be keyed into the clay core of
the dam. Following operation and a drying out period, a covering of
three-foot compacted clay, five-foot overburden, and one-foot topsoil
will be put in place. The area will be contoured and revegetated with
appropriate natural species. A very important feature of this program
is that final contouring will provide for sloping the area towards a
concrete spillway located on the side of the area. It is designed to
divert water runoff from topping the dam and thereby minimize water
erosion of the downstream side of the dam over the long term.

By examining elements of these schemes, we can see examples of how
each of the performance objectives can be met.

Since we have some currently operating mills that are good illustra-
tions of how not to site mills, such as next to a town or on the banks of
a river, our first two objectives address appropriate siting. None of
the proposals under consideration are sited in conflict with these objectives. In fact, the proposed sites are in areas with average population
densities less than two persons per square mile. Therefore, the objective
of minimizing population radiation dose is met.

Seepage is minimized in all these schemes by providing for a liner
or, in the case of plan 2, by reducing the amount of solution that could
migrate which in turn inhibits the mobility of the toxic materials left
in the tailings.

It is not shown in the figures; but in each case, elimination of
blowing tailings during operation will be achieved by the implementation
of an interim stabilization program which will be required by license
condition. This program may include the use of chemical crusting agents,
water sprays, or physically covering the tailings with soil.

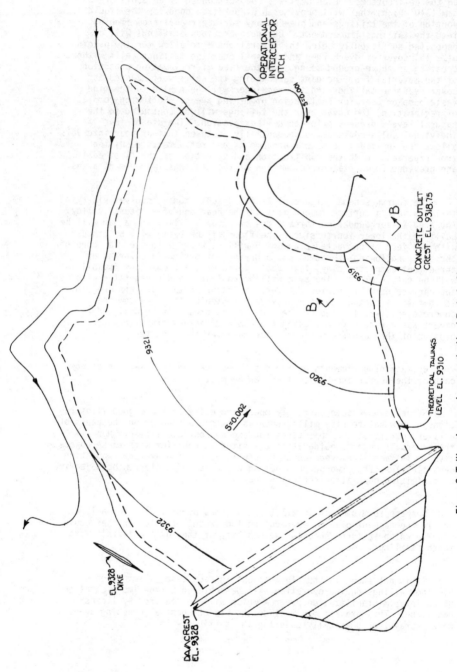

Figure 6 Tailings Impoundment at the Head End of a Natural Valley-Plan View.

Following operations, the reclamation plans for all schemes provide for the reduction of direct gamma radiation to essentially background and radon emanation to about twice background. This will be achieved by covering with various combinations of clay, overburden, and topsoil as discussed earlier. The depth of the proposed covers, which is dependent on the radon diffusion coefficients and gamma attenuation coefficients of each component, ranges from nine feet of cover for the fourth plan to 15 feet of cover for the third plan.

The reclaimed areas will be contoured, and vegetative or riprap cover will be placed to provide protection against wind and water erosion over the long term. Although it is recognized that some finite period (five to ten years) monitoring program will have to be implemented to assure that the reclamation specifications have been met, the programs have been designed to eliminate the need for an ongoing monitoring and maintenance program. Since plans 1, 2, and 3 utilize below-grade disposal of tailings, the reclaimed area will blend in with the natural contours of the land. Plan 4, while it is a surface impoundment plan, is located and designed such that the reclaimed area will be contoured to blend with the natural hills and will have erosion resistant characteristics comparable to a below-grade scheme in the same area.

Finally, for all four plans, the applicant will be required by license condition to provide surety arrangements to assure that sufficient funds are available to complete the authorized reclamation plan.

As demonstrated by the plans presented, the performance objectives were designed to allow industry some flexibility in proposing various engineering solutions for disposal of tailings. We look to the uranium industry to take the lead in developing additional methods to meet the objectives, although we are strongly encouraging some type of below-grade disposal. As seen by scheme 4, a surface impoundment plan may be acceptable if it is shown to result in a reclaimed impoundment area that has comparable erosion characteristics as the surrounding environs.

The programs I have discussed are proposed for new milling opera-tions. We have also been extending the development of acceptable tailings management programs for currently operating mills within NRC jurisdiction. We asked each operator to propose programs meeting our performance objectives four through eight. Siting and impoundment area liners were no longer options for the piles already in place. We now have proposals addressing the objectives from all our mill operators which have been authorized or are in various phases of review. The proposals include (a) continued use of the existing tailings area with a firm reclamation program commitment, (b) discontinued use of the existing area with newly generated tailings impounded in mined out pits with a firm reclamation program for both areas, and (c) newly generated and existing tailings, that are now piled on the surface, impounded in mined out pits with a firm reclamation program. In all cases, surety arrangements covering the authorized program are required.

TABLE I. FAILURE MECHANISMS

A. ELEMENTAL

    1. CAP

        a) Differential settlement
        b) Gullying
        c) Water sheet erosion
        d) Wind erosion
        e) Flooding
        f) Chemical attack
        g) Shrinkage

    2. LINERS

        a) Differential settlement
        b) Subsidence of subsoil and rock
        c) Chemical attack
        d) Physical penetration

    3. EMBANKMENT

        a) Differential settlement
        b) Slope failure
        c) Gullying
        d) Water sheet erosion
        e) Wind erosion
        f) Flooding
        g) Weathering and chemical attack

    4. REVEGETATION

        a) Fire
        b) Climatic change

    5. WATER DIVERSION STRUCTURES

        a) Slope failure
        b) Obstruction

B. NATURAL PHENOMENA

    1. Earthquakes
    2. Floods
    3. Windstorms
    4. Tornadoes
    5. Glaciation
    6. Fire and Pestilence

The NRC review of proposed tailings management programs utilizes a report recently prepared for us by Colorado State University.[1] The central focus of the study was to identify and describe the potential failure modes which, over long time periods, could cause release of radioactive components of the tailings. The analysis of these potential failure mechanisms includes a description of the failure mechanism itself, a discussion of the natural or geotechnical processes that control it, an assessment of the magnitude of release that could result from a failure and the likelihood that the failure would occur within long time periods. The time periods considered range from a few hundred years up to 100,000 years.

An integral part of the NRC analysis of the proposed tailings management programs is the evaluation of site and design characteristics that could influence the magnitude and likelihood of failure for each mechanism. The failure mechanisms considered in the NRC evaluation are contained in Table I.

SUMMARY

The NRC feels that as a result of recent industry proposals and studies performed for the generic statement on uranium milling, below-grade tailings impoundment programs are economically viable and very attractive environmentally by providing greater assurance of containment over the long term. Therefore, in all cases the primary tailings management alternatives to be considered should be below-grade burial methods. If a specific site does not lend itself to an economically or environmentally sound below-grade program, a surface impoundment may be found acceptable. Its acceptability will depend on showing that location and design characteristics will assure long-term stability and that erosion of the reclaimed area will not take place at an accelerated rate when compared with the surrounding environs.

We do not feel that the four programs discussed here are the only acceptable ways to resolve the tailings issue. Some of the other methods that are currently being investigated are deep mine disposal, separating sands and slimes and "fixing" the slime portion with a solidifying agent such as concrete, the possibility of removal of the radium and thorium, and the burial of neutralized tailings between and in groundwater tables. We are looking for the industry to continue to take the lead in developing innovative tailings management programs that minimize the environmental and health effects attributed to uranium tailings.

REFERENCES

(1) Nelson, J. D. and T. A. Shepherd, Evaluation of Long-Term Stability of Uranium Mill Tailing Disposal Alternatives, Colorado State University, Fort Collins, Colorado, April 1978.

DISCUSSION

N. SAVIGNAC, United States

    I noticed that in the first two alternatives you presented, somewhere around 12 feet of material over the top of the tailings is mentioned, I am comparing that, I believe it was 6½ feet of overburden in the Bearcreek situation. Would you care to comment on that increased amount of material in relationship to the amount of radon emanation relative to your two times background radon emanation.

R.A. SCARANO, United States

    Well there are two concepts that we have to deal with when you talk about how much material to put over the top. There is really a relationship that we have to look at as far as dealing with the emanation rate problem and the long term problem. So that, it may even be shown that to deal with the emanation rate, you only need 12 inches of whatever kind of material that you are going to put over. We all know that for the idea of long term containment, 12 inches would not be enough. That is all I can say about that, because I do not know the details as far as the evaluation was concerned. You will find that in the draft environmental impact statement, by the way, there was an evaluation of the amount of material needed which would deal with those two items. How much do you need to assure a long term containment and how much you would need to deal with the specific emanation rate problem.

J. VELASQUEZ, United States

    I am looking at performance objective, which essentially says : locating the tailings such that disruption and dispersion by natural forces is eliminated or reduced to maximum extents possible. In light of the information that was given to us up to now, I was wondering if you would care to comment on what the NRC's position might be in regard to performance objectives and the concept that perhaps the ultimate disposal mechanism would be dispersion rather than containment.

R.A. SCARANO, United States

    Well, we feel that the proper way to handle this problem is to assure long term containment. Now there is no doubt when you talk about slurring tailings into a pit, that whatever water we are going to have in there will have a form of a sluck. There will be a certain amount that will be moving through regardless of what kind of liner that you are talking about. As, for that particular portion, that will move. But, the tailings themselves, we feel very strongly that they should be contained. Now that does not mean that through the normal process of learning, just as we are a little smarter now, than we were even one year ago, that there may be some slight flexibility in any of these ideas. However, we feel very strongly about securing long term containment.

D.C. McLEAN, United States

    Your specification for twice the rate on emanation of twice above background has a very powerful impact on the design and cost of the tailings pond installation. Will you tell me what the basis is for having selected that twice background figure ?

R.A. SCARANO, United States

        Well, as it really came about we really addressed this
problem for the first time when we were going through the Bear Creek
Project. And at that time, here again, we had to deal with how much
cover do you put on it, how much do you knock down the emanation
rate and look at the long term containment problem. When going
through the exercise, it was found out that for that particular
project and for the soils that were used and the amount of clays
that were available that the determination of an adequate amount
was somewhere around, and someone from Argonne might correct me,
I think it was about 8 feet. It so happens that the emanation
rate that came down to,was about twice what the background was. So
really what it came down to was that you can really get to the point
of twice the emanation rate, it is really achievable and it was in
our opinion within reason. As it turns out there have been some
studies following that, which have come to the conclusion that
the twice emanation rate was a good number. It seems to have
performed the goal of doing away with the radon problem. I would
like to take this opportunity to quote some   information
we got last week. In the opening presentation there was some word
about the results of a hearing about a week ago. This was a power
reactor hearing, and it was called upon because of the radon problem.
They were saying, we should not have this power reactor because of
the problems in the mines and mills. We have a population radon dose
that is too high. As they had the hearing and the conclusion was
that the board is of the opinion that what has happened with respect
to tailings piles has changed greatly within the past year. We are
no longer pressed with abandoned and unstabilized piles. The new
requirements will assure that they will no longer be a major source
of radioactivity. The NRC staff recognized the problem and they have
moved to handle it. The conclusion of this hearing is that this
board carefully considered available information concerning the
releases of radon-222, and concludes that these releases and impacts
are insignificant in striking the cost benefit balances for the
Perkins Nuclear Power Station. So, I would say that the choice of
twice the emanation rate was a good choice.

D.C. McLEAN, United States

        Well it may be fortunate or unfortunate, depending upon
your point of view,but in the case of the Bear Creek plant, if the
soil available for cover had a permeability such that the background
could be reduced to twice background with 8 feet that would not
necessarily be the case in another area. In the first place covering
soil might not be available and its permeability might be such that
you would need 20-30 feet. That difference in the amount of cover,
and the type of soil available makes a tremendous differencein the
cost of the tailings area of capital investment.

R.A. SCARANO, United States

        Well, I think according to the Oak Ridge study, that even
with soils that were quite permeable, that the maximum they were
looking at to get it down twice background was about 20 feet ;
so that might be the case if there is not any other method of
knocking down the radon except using the natural soils that you
might indeed have to look at the placement of 20 feet of soil over
the top. However, I would encourage you to take a look at using
other means of keeping the emanation rate down and maybe not using
as much as 20 feet of soil, not ever losing sight of the fact that
you still have to worry about long term containment, so that you
would still need enough soil on that pile to deal with the long
term wind and soil erosion.

D. MARKLEY, United States

Going back to Mr. Savignac's question and the discrepancy between the approximate 6 feet and the approximate 10 or 12 feet. I would just like to reflect back on the Bear Creek Statement, whose numbers were based in large part upon Sears et al's work, which did range from 6 to 20 feet ; these were the figures in that report. And of course completely depended upon the soil characteristics which were going to be used for cover. That was just to clarify it a little bit further. The second part of this actually goes back to what has just been raised in the last question. The discrepancy which comes up, where you find radon levels or radon emanation rates which differ in geographical locality. I think the comment is very valid that what may work in one situation and one geographical locality is going to be a completely different situation of course in another area. I guess the question that continues to come to my mind is whether or not some type of base level or safe level, if you prefer, or acceptable level of radon is ever going to be adopted or if we are ever going to stick to what I consider to be of an arbitrary two times type factor.

R.A. SCARANO, United States

Well as you know these are strictly guidelines that are used until we have the results of the GEIS, which you are going to hear about soon. If you did notice, I hope that in the performance objectives that we do use the words twice, what the natural rates would be. So, we do have some flexibility and we do look at the specific project. There is always a negotiation that goes on between what the environmental impacts are and what the gains are. I would not say we are exactly tied to two times background, maybe it is a 2.05 or 1.95. There is that kind of flexibility built in. Were you not glad to hear that Mr. Markley.

D. MARKLEY, United States

Just a bit further Mr. Scarano, in the sense that you brought up that 0.05 of a percent difference there.

R.A. SCARANO, United States

Let me make a comment before you go on. The kind of calculations that we are doing just do not lend themselves to that kind of accuracy. There is a lot that we have to know yet, even to the point of how do you calculate what kind of emanation rate you are going to have by using the 5½ feet or the 8 feet what we have done within our shop anyway, is to standardize on the method of calculating so that, recognizing as the industry is becoming involved, and the industry's consultants are becoming involved, we are going to learn how to go through calculations of that nature. Therefore you know I wish you would not be hung up on this two times because there is a wide band of air built into the ways of calculating that right now.

D. MARKLEY, United States

My only further comment is that I think we just need to continue to be cognizant that a metal might be situated in an area for instance where our radon emanation rate happens to be 1/3 the rate of another area, such as if it is going to be some type of objective we are shooting for we may have an approximate two times level in one area that does not even achieve the natural radon emanation rate in another geographical area.

R.A. SCARANO, United States

       That is a valid comment.

J. VELASQUEZ, United States

       I guess the germaine question is in regard to all of this controversey over twice background is, if a company could not demonstrate at the time of licensing that they could not meet twice background without being some economically feasible alternative, would the NRC consider giving them a license at the time they required it ?

R.A. SCARANO, United States

       I feel confident that there are many ways that anyone in the industry can meet that. So, I think that your case is really hypothetical and it would not happen. I have seen too many projects that have come across my office now, that show using the standard method of calculating that performance objective, can be met.

D.C. McLEAN, United States

       I do not argue the point that you could in some way reduce the emanation to twice background, but at what cost and that is the key to the whole thing here. We are saying that we want to do this within a reasonable cost and that is the crux of the whole thing and it gets back to the question that I asked in the first place ; where did this figure of two times background come from on a scientific basis as far as health impact is concerned ?

R.A. SCARANO, United States

       Well, you know I would like to handle the early part of that question and leave part two until later because NRC will make a presentation later on this point. But, let us talk about the economics involved. The whole idea of the NRC putting on these performance objectives is because of a cry from you people out in the industry. I have heard since I have been involved in this that we want to do what is right ; you told us what is right and we will do it. So, the NRC has indeed given you people in industry some goals to go by. Now, it is up to you in the industry to come up with a programme which meets these goals and to look at the economics involved. If the economics are not there, if it is not an economically viable project, do not do it. The whole idea in the tailings management issue is to assure that we are not going to have another problem five years from now ; that the industry knows what has to be done and plans to do it right at the beginning in the early planning phases of that project.

A.M.G. ROBERTSON, International Atomic Energy Agency

       I would like to ask two specific questions, if I may, related to the solutions you proposed being put to disposal. The first is related to the position of the water table and placing of the tailings 10 feet above the water table and I presume this is because you attach a lot of significance in keeping clear of long term solution seepage problems. The problem in deciding whether you are 10 feet above the water table appears again to be a very long term problem. What sort of data do you look at to establish the water table over periods of one or ten thousand years. That is question 1.

Question 2 is related to the clay liner, which is indicated. In the solution, which is figure 4, where you have shown the clay liner as going below the water table and under a layer of compacted fill. The reason for this I presume is because you want to raise the tailings above the water table, number one, and number two, you do not want to place the clay liner above this layer of compacted fill because of the differential settlements between the compacted fill and the rock material in which you are forming the whole. Now this eliminates the differential settlements to which the clay liner is subjected by pressing the interface between the rock and the compacted fill. However, you still have a differential settlement taking place in that compacted fill. This implies that you have the compacted fill which has to move downwards relative to every clay liner at the compacted fill clay liner interface. Alternatively, it could be that it does not move there, but it moves at the clay liner rock contact. This is a question of sheer strength. Now has this problem been addressed ? Which is the stronger I would suspect that if you have beautiful smooth wall, as we were shown earlier today, that in fact it would be the clay liner rock interface that would be the weakest. Now why I raise this question is because in conventional dams the contact between the cut off and a very steep rock abutment is a very serious weakness point, and this appears to me to be the main failing of such a solution. If we are talking about side slopes of 2 or 3 horizontal to one vertical, the problem does not arise because you do not have the sheet of information there. But, as soon as we start going to steep side slopes, 2 vertical to one horizontal, then this problem definitely exists. How is this problem being looked at in these particular assessments that you have done ?

R.A. SCARANO, United States

Let me just state that in the environmental impact statement we usually have a team of anywhere from 12-20 people. These are details that usually the specialist will handle. Let me tell you what I know about it, each of these, and then I would like to refer you to the statements that have gone out on each of these projects and you can probably get more information there. The water table indeed is a problem. To try and determine where you bring it up to assure that you not only have contact with that water table in a short time table, because that water table will be lowered while the mining is going on and it will take a while for it to get back up, but to try to determine where that water table is likely to be all I can do is to assure you that it has been looked at and the spacing above that has accounted for it. But, to explain to you what the details were as to coming up with that determination I am afraid I cannot do that for you.

Your question 2, about the clay on the bottom of the pit, and you know I put two on the board there, one had a clay liner on both the walls and the floor and the other only on the floor. The reason that the clay liner is not proposed for the top of the compacted overburden was exactly as you had stated. You know, there is just more likelihood than having a crack and a failure in that thing and it was put on the bottom of the pit specifically for that reason. The floor of the pit is not smooth, by no means, there are steps and everything else that go in with it. And, here again the specialists involved on the staff have looked into that and have concluded that it is indeed a stable way of putting it in.

A.M.G. ROBERTSON, International Atomic Energy Agency

The reason I brought the question up was in fact that I have looked at the assessments and I did not find anything that is published. I am sure there may be some things in your files that can be related to it, but nothing in the statements.

J.D. SHREVE Jr., United States

I keep getting more and more puzzled on why we keep talking about radon per se. First of all, I personally think it is a trivial problem. Secondly, I think we are making a basic mistake in that no radiobiologist that I know of, and I know many, would say that you could ever relate the health hazard to radon. As we know in uranium mining protection, the working level is the thing. And, if there is any weakness in the current monitoring programme, it is that we are not evaluating working levels. We are making some assumptions about an equilibrium co-efficient, and that is wrong. That can vary over wide, wide ranges and when we talk two times background with effect to radon, we can picture the situation as if I were standing a mile from the tailings site. Lesser amounts of radon farther up wind of me so that they had a longer travel time, could be more effective in doing me harm than a closer source with higher radon that did not allow the in-growth time of the daughters. So, I just register that as a suggestion. Measurement of a working level, even at these low environmental numbers should be part of the programme because now we are going to see a difference between a site that is in a valley, where you have mountain valley currents, such as if we fill up the bowl we lock that radon in, and it sits there cooking daughters, almost up to equilibrium, vs. out on the Wyoming plains where there is no way except for stagnation of the flow and that never happens in Wyoming. Therefore we are led to an over-simplified conclusion that just does not stand up against the facts.

M.C. O'RIORDAN, United Kingdom

Mr. Scarano, I would like to bring you back to your twice background. And, I would like to suggest to you that it is unfashionable to appeal to natural levels in setting limits from radiation protection.

R.A. SCARANO, United States

Well, all I want to say about it right now is that it is strictly a goal as part of the overall programme that the NRC is pursuing at this time, NRC is taking a hard look at what the standard ought to be, whether it ought to be in terms of a multiple of the background or in terms of a concentration, or in any other term. There will be a change taking place shortly. But, for the time being, and while we are moving towards that we need a goal. And that is what that is all about.

REGULATORY PRINCIPLES, CRITERIA AND GUIDELINES
FOR SITE SELECTION, DESIGN, CONSTRUCTION AND
OPERATION OF URANIUM TAILINGS RETENTION SYSTEMS

J. R. Coady, L. C. Henry
Atomic Energy Control Board
Canada

ABSTRACT

Principles, criteria and guidelines developed by the Atomic Energy Control Board
for the management of uranium mill tailings are discussed. The application of
these concepts is considered in relation to site selection, design and
construction, operation and decommissioning of tailings retention facilities.

PRINCIPES, CRITERES ET DIRECTIVES REGLEMENTAIRES
APPLICABLES AU CHOIX DES SITES, A LA CONCEPTION,
A LA CONSTRUCTION ET A L'EXPLOITATION DE SYSTEMES
DE STOCKAGE DES RESIDUS DE TRAITEMENT DE L'URANIUM

RESUME

Cette communication porte sur les principes, critères et directives
élaborés par la Commission de contrôle de l'énergie atomique en ce
qui concerne la gestion des résidus de traitement de l'uranium. On
étudie l'application de ces concepts eu égard au choix des sites,
à la conception et à la construction, ainsi qu'à l'exploitation et
au déclassement des installations de stockage des résidus de traitement.

1. INTRODUCTION

There are at present six operating uranium mine/mill facilities in
Canada with a combined throughput (1977) of approximately 5 million tonnes
of ore. There are expectations that the number of operations and the total
throughput will increase greatly in the next few years.

All such production plants are designated as "nuclear facilities" under
the Atomic Energy Control Act and as such require an operating licence issued
by the Atomic Energy Control Board. These licences are granted only after
the Board has been satisfied that operation of the facility will not
adversely affect the workers at the facility, the public, and the environment.
Prospective licensees must therefore indicate in their application documents
that there will be compliance with all the Regulations and criteria laid
down by the Board. Guidelines to assist the industry in meeting these
requirements have been and are being developed by the Board. The guidelines
have been designed to relate to the three normal stages of the licensing
process, site and construction approval, approval to operate, and
decommissioning approval.

That part of the facility designed for the management of the mill
tailings is assessed according to the general waste management principles
developed by the Atomic Energy Control Board in relation to all radioactive
wastes. One of the main objectives is that ultimately all radioactive wastes
should be disposed of in a manner which does not rely on the continued need
for human surveillance. It has not yet been demonstrated that present
methods of tailings management meet this objective and efforts are being made
to resolve this situation.

2.0 BASIS FOR LICENSING

The Atomic Energy Control Act authorizes the Atomic Energy Control
Board to make regulations governing the development, application and use of
atomic energy in Canada. Regulations, designated the Atomic Energy Control
Regulations have been formulated to control, among other things, nuclear
facilities and materials in the interest of national and international
security, and to ensure that the health and safety of workers and members of
the public are adequately protected. Both radiological and non-radiological
aspects of health and safety are included.

The Regulations define a nuclear facility as being -
"a nuclear reactor, a sub-critical nuclear reactor, a particle accelerator, a uranium or thorium mine or mill, a plant for the separation,
processing, reprocessing or fabrication of fissionable substances, a
plant for the production of deuterium or deuterium compounds, a facility
for the disposal of prescribed substances and includes all land,
buildings and equipment that are connected or associated with such
reactor, accelerator, plant or facility".

The definition of prescribed substances is contained in the Act and the
Regulations and it includes uranium mill tailings.

Section 25 of the Regulations relates to the abandonment or disposal
of prescribed substances and states:

"No person shall abandon or dispose of any prescribed substance except:

a)   in accordance with the conditions in any licence that is applicable
     to the prescribed substance and that is in force; or

b)   in accordance with the written instructions of the Board."

3.0 RADIOACTIVE WASTE MANAGEMENT PRINCIPLES APPLIED IN LICENSING

The approach adopted by the Atomic Energy Control Board in the licensing of facilities for the management of uranium mill tailings is based on certain principles which have been developed for the management of all radioactive wastes. The methods that may be used to meet these principles will vary with the type of waste. Factors such as the physical and chemical form of the wastes, the level of activity and the halflives of the isotopes involved will determine what particular form of management is required. However, the position of the AECB is that in performing this assessement there will be a consistent attempt to apply the following principles, regardless of the source of the wastes.

In all considerations related to waste management the AECB distinguishes between storage which is a method of containment with the intention and the provision for retrieval, and disposal which is a method of management in which there is no intent to retrieve but which, more importantly, does not rely for its integrity on the continued need for human intervention whether this be for treatment, monitoring or restriction of access. The objective, for all wastes, is that the ultimate form of management will be disposal.

Storage is essentially a temporary measure as, for example, in the case of wastes awaiting further treatment or awaiting the development of suitable methods of disposal. As a general rule storage also requires some form of surveillance. Disposal is intended to be a permanent step and, because of this, the concern for viability in the long term is paramount.

Where no recognized methods of disposal are available, then the system in use can only be regarded as storage. In such a situation the use of the method carries with it the normal obligation to provide all necessary surveillance but, in addition, carries a responsibility to search for acceptable means of disposal. The intention is that methods of disposal will be sought and implemented as soon as possible. The principle involved is that, to the extent possible - economic and social conditions being taken into account, continuing problems of management created by the producers of wastes will not knowingly be left to future generations.

One of the factors to be considered in discharging this responsibility to future generations is the period of time for which it is reasonable to require that institutional controls be in place to ensure the integrity of the disposal method. The goal that no disposal method should be dependent on the continued presence of humans is an indication of the low probability that institutional controls will survive for the very long periods of time usually associated with radioactive waste management - many thousands of years. On the other hand, surveillance systems are in operation now and will obviously continue to exist for many years to come. The question therefore arises as to where this changeover takes place.

At some time in the future surveillance and control can be assumed to have lapsed. There is, therefore, a necessity to determine the maximum period of time for which future generations should be committed, for their safety and the quality of their environment, to provide for the care of present wastes. The choice of any value will be to some extent arbitrary and sensitivity studies will need to be carried out to determine the cost-benefit of any particular number or range chosen. However, from the regulatory point of view, a period of several decades would appear to be much more acceptable than a period of several centuries. In this context, the AECB has selected 100 years as the value to be used in discussions, while further review is carried out.

4.0 MINE FACILITY OPERATING LICENCE

In situations where the waste management facility is considered to be an integral part of a larger nuclear facility, as in the case of a mine/mill/ tailings retention complex, the waste management facility may be regulated under the terms of a general facility licence. In this case, a Mine Facility Operating Licence (MFOL) is issued to include the tailings retention system. The normal period for a licence is one year. However, depending on the maturity of the operation and the compliance performance of the company, an MFOL may be issued for a longer period of time. Licences of shorter duration, usually six months, may be issued in the earlier stages of an operation or when improvements are required in the performance of the facility or its method of operations.

5.0 LICENSING STAGES

Licensing will normally progress through three approval phases:

1) site and construction approval,

2) licence to operate and

3) decommissioning approval.

The Regulations state that a facility operating licence shall not be issued unless authorization has been granted by the Board for the construction or acquisition of that facility. Subject to all requirements being met by the applicant, construction approval is granted concurrent with the granting of site approval.

The jurisdiction of the Atomic Energy Control Board extends over environmental considerations insofar as they relate to the health and safety of workers and the public. However, in licensing of radioactive waste management facilities such as tailings retention systems, the site selection and approval phase is normally coordinated with the appropriate review processes of the federal and/or provincial agencies concerned with the broader issues of environmental quality and conservation of natural resources. In the absence of public hearings or reviews, which may be conducted under the auspices of these agencies, the Board will require applicants to conduct a public information program to advise the local population of their intentions and to answer concerns expressed by the public related to the impact of the proposed facility.

Once site and construction approval has been granted, the Board will not normally require further public participation; however, it continues to stress the importance of the licensee/applicant keeping the public informed of developments.

6.0 LICENCE REQUIREMENTS AND CONDITIONS

6.1 GENERAL REQUIREMENTS

The following general statements summarize the requirements of the AECB in regard to the operation of all nuclear facilities and as such predetermine the position which is adopted in deciding on the acceptability of an application to operate a tailings management facility.

Maximum Permissible Doses

Nuclear facilities shall be operated in such a manner as to prevent any person from receiving a dose of ionizing radiation in excess of the maximum permissible doses specified in Section 19 of the Atomic Energy Control Regulations. Schedule II of the Regulations, Maximum Permissible Doses and Exposures, is appended.

ALARA Principle

Exposures of workers or the public to radiation or other toxic substances or releases of deleterious substances to the environment resulting from the operation of a nuclear facility shall be as low as reasonably achievable, economic and social considerations being taken into account.

Compliance With the Requirements of Other Agencies

In addition to AECB requirements there must also be compliance with the requirements of other appropriate agencies of the federal and provincial governments insofar as these requirements are not inconsistent with the Atomic Energy Control Act and Regulations.

In terms specifically related to tailings management, the following requirements must be observed when designing, siting, constructing and operating a tailings retention facility:

(1) the ground and surface water must be protected from contamination resulting from the movement of radioactive and chemical substances from the facility;

(2) the atmosphere must be protected from contamination due to radon gas emanation, dust release and direct gamma radiation from the tailings area; and

(3) the security measures must be adequate to restrict entry by members of the public and unauthorized use or removal of tailing materials.

6.2 STANDARD CONDITIONS OF LICENCE

In addition to the general requirements outlined above, the following conditions are normally included in licences for the operation of uranium tailings retention facilities.

Operation According to Writtern Submissions

The operation of the facility shall be governed by and shall be in accordance with the Atomic Energy Control Act and Regulations and with the document(s) submitted to and accepted by the Board.

Ownership and Control of Property

Ownership, control and use of the property described in the document(s) submitted to and accepted by the Board shall not be changed without prior notification to the Board.

Facility Available for Inspection

Access to the facility and to all plans, drawings, documents and records pertaining to the design, construction, testing and operation of the facility shall be available at all reasonable times for inspection by persons duly appointed by the Board.

Compliance with Board Requests

All tests, analyses, inventories, inspections, modifications or procedural changes as specified by persons duly appointed by the Board shall be carried out.

Reports of System Degradation

A preliminary report shall be made as soon as possible to persons duly appointed by the Board, on any significant deviations from or deficiencies in accepted documents, procedures, operating conditions or the integrity of components or systems when such deviations or deficiencies could result or could have resulted in an increased hazard to the health or safety of any person or to the environment. A complete, written report is to follow as soon as possible after the disclosure.

Reports of Incidents

A preliminary report shall be made to persons duly appointed by the Board within 24 hours of any incident(s) that results in or is likely to result in an increased hazard to the health or safety of any person or to the environment.

A complete, written report is to follow as soon as possible after the incident(s).

Reports of Excess Exposure to Toxic Substances

Excess exposure of any person to toxic substances stemming from the operation of the facility shall be reported within 24 hours to persons duly appointed by the Board. A complete, written report is to follow as soon as possible after the occurrence.

Annual Report

An Annual Report summarizing the operating experience, significant events, changes in procedures and modifications of equipment which occurred or were made in the preceding calendar year, and any information the Board may request shall be submitted to the persons duly appointed by the Board by 28 February of each year.

Physical Security

Security measures accepted by the Board for protecting the integrity of the facility and preventing theft, loss or any unauthorized use of prescribed substances shall be in force at all times.

Report of Breach of Security

A preliminary report shall be made to persons duly appointed by the Board within 24 hours of the occurrence of: any attempt at or actual breach of security, threats, and attempted or actual acts of sabotage of the facility, equipment or procedures. A complete, written report is to follow as soon as possible after the occurence.

Qualified Personnel

There shall be available at all times a sufficient number of qualified personnel to ensure the safe and secure operation of the facility.

Decommissioning Clause

The decommissioning of the facility shall be governed by and shall be inaccordance with the document(s) submitted to and accepted by the Board.

7. GUIDELINES

7.1 GENERAL COMMENTS

Licensing Guides are provided to publicize approaches and methods which are acceptable to the Atomic Energy Control Board for satisfying the requirements of the Atomic Energy Control Regulations.

Compliance with Licensing Guides, in whole or in part, is not mandatory. Where an applicant chooses to depart from the requirements specified in a Licensing Guide or its provisions, he must accept the responsibility of demonstrating, to the satisfaction of the Board staff and/or advisers, that the alternative approach or method adequately fulfills the intent and requirements of the Atomic Energy Control Regulations.

Comments and suggestions for new Licensing Documents and for improvement of existing Documents are encouraged. New Guides will be issued and existing Guides revised periodically to incorporate accepted suggestions and to reflect developing technology and practice.

7.2 GUIDELINES FOR TAILINGS RETENTION FACILITIES

7.2.1 Site Selection

Site Alternatives

The applicant should identify a number of potential sites for location of the tailings retention facility and conduct preliminary investigations on each. On the basis of the results applicants should choose the preferred site. Board staff, if requested, is willing to assist the applicant by commenting on the licensability of the particular sites.

Capacity

The tailings retention facility should ideally be located on one site and must provide sufficient capacity to store the total projected waste volumes.

Use of Rivers and Lakes

The use of rivers and other running waters for tailings management is not acceptable. The use of lakes, land-water, swamps and marshes will only be considered after it has been demonstrated that a suitable dry land area is not available and provided it can be established that the site is not a groundwater discharge area. Further, the area should not intercept runoff or other uncontaminated surface water sources or in the event that it does, appropriate actions must be taken to reroute the inflow away from the tailings area.

Use of Deep Lakes

The deposition of tailings in deep lakes, well below the water surface where chemical activity is expected to be limited by the absence of free oxygen, has not been ruled out; however, considerable supporting research will be required with respect to both short- and long-term performance of this type of management before a decision by the AECB could be made to licence such a facility.

Tailings Basin Permeability

The applicant must demonstrate, based on comprehensive field investigations, that seepage from the site will have an acceptably low impact on the downstream environment. Otherwise, appropriate sealing methods such as grouting or the use of liners or low-permeability blankets must be considered in the design or an alternate site chosen. A maximum permeability of 10^{-7} m/s has been suggested as a value which is likely to meet this objective.

Geologic and Hydrogeologic Studies

Site approval will normally require that the applicant demonstrate a comprehensive knowledge of the geology and hydrogeology beneath and in the vicinity of the tailings retention facility.

This will include a detailed knowledge of:

(1) the hydrostratigraphic units which contain the local and regional aquifers - by drilling, auguring and geophysical techniques to define the bedrock topography and the depth of unconsolidated deposits.

(2) the local and regional groundwater flow systems which may be affected by the waste management facility - by installing piezometers and observation wells in order to determine the hydraulic conductivity and groundwater velocity (in three dimensions) and the annual waterlevel fluctuations so that the direction, quantity and rate of groundwater flow is understood; and

(3) the baseline groundwater quality - by analyzing samples from piezometers and observation wells so that the naturally-occuring concentrations of potential contaminants that may be leached from the contained wastes may be determined.

Baseline Studies

Monitoring of ground and surface water should commence approximately three years prior to operations to establish natural backgrounds levels of all contaminants that may be released during future operations. These will normally include:

pH	chromium
hardness	iron
alkalinity	manganese
total solids	sodium
sulphate	potassium
total kjeldahl nitrogen	calcium
nitrate	magnesium
ammonia	copper
chloride	lead
phosphate	zinc
uranium	nickel
thorium - 227, 228, 230, 232	barium
radium - 226, 228	
lead - 210	
polonium - 210	

The program to be followed including types of samples, sampling methods, analytical methods and location and frequency of sampling should be as discussed and approved by the appropriate federal and provincial regulatory agencies and the AECB.

7.2.2   Design, Construction and Operation

Dam Permeability

Any impact on the environment which occurs as a result of seepage through tailings dams must be acceptably low. A maximum permeability of 10^{-8} m/s has been suggested as a value likely to achieve this objective.

Restrictions On Dam Construction Materials

The use of mill tailings for dam construction downstream of the impermeable core is unacceptable. Waste rock may be used downstream of the core provided it can be demonstrated that its impact on the general environment and on downstream waters will be acceptable small. Parameters that must be considered in demonstrating the acceptability of waste rock for dam construction include:

i)   leachability of radium-226 and other radioactive contaminants from the waste rock;

ii)   radiation fields associated with the rock; and

iii) acid generating potential (waste rock can be placed downstream of the core only if analysis shows it is a non-acid generating material).

The availability of other suitable materials for dam construction will be considered by Board staff when reviewing a proposal to use waste rock for downstream construction.

(The applicant is advised to refer to the Metal Mining Liquid Effluent Regulations and Guidelines, EPS 1-WP-77-1, and Prediction of Acid Generating Potential, A.R. Ballantyne, Technology Transfer Seminar Notes, November 1975, Mining Effluent Regulations/Guidelines and Effluent Treatment Technology, Environmental Protection Service, Fisheries and Environment Canada, regarding the use of waste rock.)

Dam Stability

Design and construction of embankment systems must conform to currently accepted practices and safety criteria. A dam stability analysis must be contained in the Safety Report.

(The applicant is advised to refer to the Pit Slope Manual, Chapter 9, Waste Embankments, CANMET 77091, Canada Centre for Mineral and Energy Technology, Energy, Mines and Resources Canada, regarding safety factors and design and construction practices.)

Flood Design

Sufficient free board shall be maintained at all times such that overtopping and/or damage of embankment crests or slopes will not occur in the event of flood conditions associated with the 100 year recurring storm. In certain regions, a historical storm may be used for design purposes. The incorporation of spillways in dam design to provide emergency protection for the crest and downstream face of the dam is encouraged.

Precipitation Facilities

Where chemical treatment (such as the addition of barium chloride or flocculant) is required to remove radium-226 and other contaminants, a precipitate removal system is required which has sufficient retention time and other features which assure that any process water being released to the environment will meet effluent requirements. Precipitation ponds if used must be engineered for ease of retrievability of the settled precipitate. The use of natural lakes, with no prior provision for retrievability and restoration, is no longer allowed.

Management of Precipitates

Precipitates, filtrates, etc., usually very high in radium-226 content, removed from settling ponds or other engineered systems may not be returned to the tailings area. Management will include immobilization and storage pending the establishment of a suitable means of disposal.

Recycle of Process Water

Recycle of decant water from the tailings pond to the mill circuit is strongly encouraged. Fresh water demand should be kept to a minimum.

Backfilling

The use of uranium tailings for backfilling of underground, mined areas is an acceptable practice provided it can be demonstrated that the health and safety of the underground worker is not being jeopardized and that the resulting impact on ground water quality is acceptably low. Backfilling is generally applicable in relatively "dry" mines.

Environmental Monitoring

Environmental monitoring will normally include discharge flows and qualities at the last point of control (e.g. at the weir discharge) and at a sufficient number of points downstream of this to determine the impact of the releases on ground and surface waters. All actual and potential seepage locations should be monitored for flow and quality and downstream impact.

Types of samples, sampling methods, analytical methods and location and frequency of sampling should be as recommended by the appropriate federal and provincial regulatory agencies and as approved by the AECB.

(The applicant is advised to refer to the Metal Mining Liquid Effluent Regulations, OSR/177-178, Department of Fisheries and the Environment Canada, for effluent quality criteria and monitoring guidance. Schedules I, II and III of the Effluent Regulations dealing with the objectives for substances and pH, frequency of sampling and analyses and analytical test methods are appended.)

Where Provincial standards on water quality exist they should be used to augment the federal standards. Where federal and provincial requirements differ, the more stringent of the two will normally apply.

(Tables PWS-1 and PWS-2 (in part), Province of Ontario's water quality criteria for public surface water and public groundwater supplies are appended. The applicant/licensee (Ontario) is advised to refer to Guidelines and Criteria for Water Quality Management in Ontario, Ministry of the Environment, Ontario, for details on water quality criteria and guidelines for the Province.)

(Tables 1 and 2, Province of Saskatchewan's surface water quality and municipal drinking water objectives are appended. Reference document: Water Quality Objectives, Environment Saskatchewan)

Analyses must include both total and dissolved radium-226. While the federal standard on mine effluent specifies dissolved radium-226 (10 pCi l on passing through a 3 micron filter), a standard based on total radium-226 is being investigated.

7.2.3. Decommissioning

The present objective is chemical and physical stabilization of the tailings and retention structures such that any drainage streams from the site will continue to meet acceptable regulatory levels of contamination and that the site is acceptable from the standpoints of aesthetics and safety. This will almost certainly require the recovery and disposal of radium precipitates from settling ponds.

Surface contouring and revegetation are recognized methods for promoting surface water run-off thus minimizing the percolation of water through the tailings.

Reliance on mechanical equipment, decant systems and treatment processes to control effluent releases for extended periods beyond close-down of the mine is not desirable. In the event that such procedures may have to be accepted in certain locations, consideration is being given to the formation of a fund to ensure their continued operation.

8.0 CONCLUDING COMMENTS

In the spectrum of radioactive wastes, the management of uranium mill tailings presents probably the most difficult example of trying to match acceptable practice with desirable criteria. This is mainly a consequence of the very large volumes of wastes which are produced.

The licensing procedures just described are those currently used to ensure that any impact of the Canadian uranium mining/milling industry on the public and the environment remains within acceptable limits. Because of these Regulations and guidelines, safe management during the operation of the mill is ensured. However, doubts are expressed in Canada, as in other countries, about the viability of present methods in the long term - after the mines have shut down.

As they are presently operated in Canada, tailings management areas are regarded by the Atomic Energy Control Board as storage facilities. The reason for this is simply that it would not be possible to walk away from a tailings pile, at the end of operations, with no further provision for supervision. In addition, is seems highly unlikely that it would ever be possible to do so without significant improvements to current procedures. The quantification of these apparent shortcomings and the means of overcoming them, thus arriving at an acceptable means of disposal, are the subject of greatly increased attention by the AECB, together with other government departments and industry in Canada.

The regulatory approach taken by the AECB is one which encourages technical innovation on the part the operators of nuclear facilities. This is considered to be particularly important if advancement in uranium tailings technology is to continue at a reasonable pace. Indeed, the Canadian uranium mining industry, is actively encouraged to participate in research and development programs intended to resolve present day and longer term problems associated with tailings and their management.

ATOMIC ENERGY CONTROL REGULATIONS CANADA

"SCHEDULE II

Maximum Permissible Doses and Exposures (1, 2)

TABLE 1

Maximum Permissible Doses (3)

Column 1	Column II		Column III		Column IV
Organ or Tissue	Atomic Radiation Workers		Female Atomic Radiation Workers of Reproductive Capacity		Any Other Person
	Rems per quarter of a year	Rems per year	Rems per quarter of a year	Rems per year	Rems per year
Whole body, gonads, bone marrow	3	5	1.3(4)	5(4)	0.5
Bone, skin, thyroid	15	30	15	30	3(5)
Any tissue of hands, forearms, feet and ankles	38	75	38	75	7.5
Lungs (6) and other single organs or tissues	8	15	8	15	1.5

TABLE 2

Maximum Permissible Exposures To Radon Daughters (6)

Column 1		Column II
Atomic Radiation Workers		Any Other Person
WLM per quarter of a year	WLM per year	WLM per year (7)
2	4	0.4

NOTES TO SCHEDULE 11

(1) The maximum permissible doses and exposures specified in this Table do not apply to ionizing radiation

(a) received by a patient in the course of medical diagnosis or treatment by a qualified medical practitioner; or

(b) received by a person carrying out emergency procedures undertaken to avert danger to human life.

(2) The Board may, under extraordinary circumstances, permit single or accumulated doses or exposures up to twice the annual maximum permissible doses or exposures for atomic radiation workers. Such variance will not be granted

(a) if appropriate alternatives are available;
(b) for irradiation of the whole body or abdomen of women of reproductive capacity; or

(c) for irradiation of the whole body, gonads or bone marrow if the average dose received from age 18 years up to and including the current year exceeds 5 rems per year.

(3) In determining the dose, the contribution from sources of ionizing radiation both inside and outside the body shall be included.

(4) The dose to the abdomen shall not exceed 0.2 rem per two weeks, and if the person is known to be pregnant, the dose to the abdomen shall not exceed 1 rem during the remaining period of pregnancy.

(5) The dose to the thyroid of a person under the age of 16 years shall not exceed 1.5 rems per year.

(6) For exposures to radon daughters, the maximum permissible exposures (in working level months) apply instead of the maximum permissible doses for the lungs (in rems).

(7) The WLM unit is not appropriate for exposures in the home or in other non-occupational situations. In such situations, the maximum permissible annual average concentration of radon daughters attributable to the operation of a nuclear facility shall be 0.02 WL."

Appendix

METAL MINING LIQUID EFFLUENT REGULATIONS
AND GUIDELINES – CANADA

SCHEDULE 1

PART I

Objectives for Substances

Item	Substance	Column I Maximum acceptable monthly arithmetic mean concentration	Column II Maximum acceptable concentration in a composite sample	Column III Maximum acceptable concentration in a grab sample
1.	Arsenic	0.5 mg/l	0.75 mg/l	1.0 mg/l
2.	Copper	0.3 mg/l	0.45 mg/l	0.6 mg/l
3.	Lead	0.2 mg/l	0.3 mg/l	0.4 mg/l
4.	Nickel	0.5 mg/l	0.75 mg/l	1.0 mg/l
5.	Zinc	0.5 mg/l	0.75 mg/l	1.0 mg/l
6.	Total Suspended Matter	25 mg/l	37.5 mg/l	50 mg/l
7.	Radium 226	10.0 pCi/l	20.0 pCi/l	30.0 pCi/l

All concentrations given are total values with the exception of Radium 226 which is dissolved value after filtering the sample through a 3 micron filter.

PART II

Objectives for pH

Parameter	Column I Minimum acceptable monthly arithmetic mean pH	Column II Minimum acceptable pH in a composite sample	Column III Minimum acceptable pH in a grab sample
pH	6.0	5.5	5.0

SCHEDULE 2

Determination of frequency with which undiluted effluents are to be sampled and analysed for particular substances

Item	Parameter	Column I At least weekly if concentration is equal to or greater than	Column II At least every two weeks if concentration is equal to or greater than	Column III At least monthly if concentration is equal to or greater than
1.	Arsenic	0.5 mg/l	0.2 mg/l	0.10 mg/l
2.	Copper	0.3 mg/l	0.1 mg/l	0.05 mg/l
3.	Lead	0.2 mg/l	0.1 mg/l	0.05 mg/l
4.	Nickel	0.5 mg/l	0.2 mg/l	0.10 mg/l
5.	Zinc	0.5 mg/l	0.2 mg/l	0.10 mg/l
6.	Total Suspended Matter	25 mg/l	20 mg/l	15 mg/l
7.	Radium 226	10.0 pCi/l	5.0 pCi/l	2.5 pCi/l

All concentrations given are total values with the exception of Radium 226 which is a dissolved value after filtering the sample through a 3 micron filter. Radium 226 need be measured in only those mines in which there is a radioactive ore.

SCHEDULE 3

Analytical test methods for determining concentrations of substances in liquid effluents

Item	Column I Substance	Column II Test method	Column III Procedure	Column IV Sample preservation	Column V References
1.	Arsenic	Colorimetric	HNO_3-H_2SO_4 digestion followed by AsH_3 reaction with silver diethyldithiocarbamate	to pH 1 with HNO_3	1
2.	Copper	Atomic absorption spectrophotometry	Sample is digested with HCl-HNO_3 before analysis	to pH 1 with HNO_3	2, 3, 4
3.	Lead	"	"	"	2, 3, 4
4.	Nickel	"			2, 3, 4
5.	Zinc			"	2, 3, 4
6.	Radium 226	Radon emanation	α Counting from Rn 222		5
7.	Total suspended matter	Gravimetric	Filter through Whatman GF/C or equivalent. Oven dry at 105°C to no further weight loss		1

PROVINCE OF ONTARIO GUIDELINES AND CRITERIA FOR WATER QUALITY MANAGEMENT

TABLE PWS-1

WATER QUALITY CRITERIA FOR PUBLIC SURFACE WATER SUPPLIES

(Unless otherwise indicated, units are mg/l)

Constituent or Characteristic	Permissible Criteria	Desirable Criteria
Physical		
Colour (platinum-cobalt)	75, units	< 5 units
Odour	Readily removable by defined treatment	Absent
Turbidity	— do —	Absent
Temperature	85°F	Pleasant tasting
Inorganic Chemicals		
Ammonia	0.5 (as N)	< 0.01
Arsenic*	0.05	Absent
Barium*	1.0	Absent
Boron*	1.0	Absent
Cadmium*	0.01	Absent
Chloride*	250	< 25
Chromium* (hexavalent)	0.05	Absent
Copper*	1.0	Virtually absent
Dissolved Oxygen	≥ 4 (monthly mean) ≥ 3 (individual sample)	Near saturation
Fluoride*	See footnote (1)	1.0
Hardness*	Acceptable levels will vary with local hydrogeologic conditions and consumer acceptance	
Iron (filterable)	0.3	Virtually absent
Lead*	0.05	Absent
Manganese* (filterable)	0.05	Absent
Nitrate plus Nitrite*	10 (as N)	Virtually absent
pH range	6.0 - 8.5 units	Least amount of inter-ference with treatment process

Phosphorus* (phosphates) | Not encourage growth of algae or interfere with treatment process

Selenium*	0.01	Absent
Silver*	0.05	Absent
Sulphate*	250	< 50
Total Dissolved Solids* (filterable residue)	500	< 200
Uranyl Ion*	5	Absent
Zinc*	5	Virtually absent

Organic Chemicals[2]:

Carbon chloroform extract* (CCE)	0.15	< 0.04
Cyanide*	0.20	Absent
Methylene blue active substances*	0.5	Virtually absent
Oil and grease*	Virtually absent	Absent

Radioactivity | (pc/l) | (pc/l)

Gross beta*	1,000	< 100
Radium-226*	3	< 1
Strontium-90*	10	< 2

Microbiological [3]

Coliform organisms (at 35°C)	5,000/100 ml	100/100 ml
Fecal coliforms (44.5°C)	500/100 ml	10/100 ml
Fecal streptococci (35°C)	50/100 ml	1/100 ml
Total Bacteria (20°C)	100,000/100 ml	< 1,000/100 ml
Clostridia (in water) (35°C)	50/100 ml	0/100 ml

* The defined treatment process has little effect on the constituents.

(1)

Annual Avg. of Max. Daily Air Temp. F.	Recommended Limit for Fluoride mg/l
50.0 to 53.7	1.7
53.8 to 58.3	1.5
58.4 to 63.8	1.3

(2) Organic chemicals should not be present in concentrations as to cause adverse tastes and odours which cannot be removed by the defined treatment and/or by chlorination only.

(3) A monthly geometric mean of the results of raw water samples collected on a weekly basis (minimum of one sample per week) should be less than the numbers given under the Permissible Criteria column. These figures do not imply a relationship between bacterial groups.

PWS-2 Criteria for Public Ground Water Supplies

With the exception of dissolved oxygen, fluorides and microbiological criteria, the water quality criteria for surface water apply to ground water supplies.

For fluorides, hydrogen sulphide and pollution indicator organisms, the following apply to ground water supplies:

	Permissible Criteria	Desirable Criteria
	(Unless otherwise indicated, units are mg/l)	
Fluoride	2.4	1.0
Hydrogen Sulphide	0.1	Absent
Pollution Indicator Organisms	Coliform and other pollution indicator organisms should be virtually absent from all ground water supplies.	

It is considered desirable to provide the maximum of treatment — chlorination — for all ground water supplies. This measure ensures that nuisance organisms which exist in virtually all waters do not get the opportunity to develop a foothold in a water distribution system and thereby create objectionable conditions.

PROVINCE OF SASKATCHEWAN WATER QUALITY OBJECTIVES

Table 1

SURFACE WATER QUALITY OBJECTIVES

These objectives have been prepared in co-operation with the Provinces of Alberta and Manitoba and represent water quality suitable for most uses either through direct use or prepared for use by an economically practical degree of treatment.

Parameter	Objectives
1. Bacteriology (Coliform Group)	(a) In waters to be withdrawn for treatment and distribution as a potable supply or used for outdoor recreation other than direct contact, at least 90 per cent of the samples (not less than five samples in any consecutive 30-day period) should have a total coliform density of less than 5,000 per 100 ml and a fecal coliform density of less than 1,000 per 100 ml. (The Maximum Permissible Limit of total coliform organisms in a single sample shall be determined by the Department based on the type and degree of pollution and other local conditions existing within the watershed.)
	(b) In waters used for direct contact recreation or vegetable crop irrigation the geometric mean of not less than five samples taken over not more than a 30-day period should not exceed 1,000 per 100 ml total coliforms, nor 200 per 100 ml fecal coliforms, nor exceed these numbers in more than 20 per cent of the samples examined during any month, nor exceed 2,400 per 100 ml total coliforms on any day.
2. Dissolved Oxygen	A minimum of five mg/1 at any time.
3. Biochemical Oxygen Demand	Dependent on the assimilative capacity of the receiving water. The BOD must not exceed a limit which would create a dissolved oxygen content of less than five mg/1.
4. Suspended Solids	Not to be increased by more than 10 mg/1 over background value.
5. pH	To be in the range of 6.5 to 8.5 pH units but not altered by more than 0.5 pH units from background value.
6. Temperature	Not to be increased by more than 3°C above ambient water temperature.
7. Odour	The cold (20°C) threshold odour number not to exceed eight.
8. Colour (Apparent)	Not to be increased more than 30 colour units above natural value.
9. Turbidity	Not to exceed more than 25 turbidity units over natural turbidity.

10. Organic Chemicals

Constituent	Maximum Concentration (mg/1)
Carbon Chloroform Extract (CCE) (includes Carbon Alcohol Extract)	0.2
Methyl Mercaptan	0.05
Methylene Blue Active Substances	0.5
Oil and Grease	— substantially absent no irridescent sheen
Phenolics	0.005
Resin Acids	0.1

Table 1 (continued)
SURFACE WATER QUALITY OBJECTIVES

Pesticides

To provide reasonably safe concentrations of these materials in receiving waters an application shall not exceed 1/100 of the 48-hour Tl_m. Consideration must also be given to sublethal effects such as taste and odour generation.

11 Inorganic Chemicals

Constituent	Maximum Concentration (mg/1)
Boron	0.5
Copper	0.02
Fluoride	1.5
Iron	0.3
Manganese	0.05
*Nitrogen (Total Inorganic and Organic)	1.0
*Phosphorus as PO_4 (Total Inorganic and Organic)	0.15 (0.05 as P)
Sodium (as per cent of cations)	between 30 and 75
Sulphide	0.05
Zinc	0.05

* These objectives are presently under study and may require adjusting according to naturally occurring concentrations or conditions.

NOTE: The predominant cations of sodium, calcium and magnesium and anions of sulphate, chloride and bicarbonate are too variable in the natural water quality state to attempt to define limits. Nevertheless, in order to prevent impairment of water quality, where effluents containing these ions are discharged to a water body the permissible concentration will be determined by the Department in accordance with existing quality and use.

12. Toxic Chemicals

Constituent	Maximum Concentration (mg/1)
Arsenic	0.01
Barium	1.0
Cadmium	0.01
Chromium	0.05
Cyanide	0.01
Lead	0.05
Mercury	0.0001
Selenium	0.01
Silver	0.05

13. Radioactivity

Gross Beta not to exceed 1,000 pCi/1.
Radium 226 not to exceed three pCi/1.
Strontium 90 not to exceed 10 pCi/1.

14. Unspecified Substances

Substances not specified herein should not exceed values which are considered to be deleterious for the most critical use as established by the Department.

NOTE: Unless specified, parameter concentrations refer to "total" measurements.

PROVINCE OF SASKATCHEWAN WATER QUALITY OBJECTIVES

Table 2

MUNICIPAL DRINKING WATER OBJECTIVES

1. Bacteriological Quality

(a) Total Coliforms

At least 90 per cent of the samples in any consecutive 30-day period should be negative for total coliform organisms and no one sample should have a Most Probable Number index of greater than 10 per 100 ml. or 2 per 100 ml. by membrane filter. Properly operated municipal waterworks should be free of coliform bacteria.

(b) Nuisance Biological Organisms

Biological organisms in concentrations which may produce objectionable colour, taste, odour and turbidity, or which may release toxic metabolites, or which may harbour pathogens are undesirable in drinking water and should be kept below such concentrations as to prevent any undesirable effects.

2. Radioactivity

The gross Alpha radioactivity should be less than 10 picocuries per litre (pCi/l). Certain maximum limits are as follows:

Strontium 90 — not to exceed 10 pCi/l
Radium 226 — not to exceed 3 pCi/l
Gross Beta Activity — not to exceed 1,000 pCi/l (in absence of Alpha)

For other specific radionuclides 0.1 of the International Commission on Radiological Protection Maximum Permissible Concentration in Water would apply.

3. Physical Quality

Water should not contain impurities that would be offensive to the sense of sight, taste or smell.

Parameter	Maximum Limit
Colour	15 units
Threshold odour number	3
Turbidity	5 units

4. Chemical Quality

Values marked with an asterisk (*) are modifications of the Canadian limits.

Chemical	Maximum Concentration in mg/1
Alkalinity (as $CaCO_3$)	500*
Chloride	250
Copper	1.0
Fluoride	1.5
Iron	0.3
Hardness (as $CaCO_3$)	800*
Magnesium	200*
(Magnesium & Sodium) and Sulphate	1,000*
Manganese	0.05
Methylene Blue Active Substances	0.5
Organics (carbon chloroform and carbon alcohol extractibles)	0.2
Phenolics	0.001*
Sodium	300*
Sulphate	500
Total Dissolved Solids	1,500*
Zinc	5.0

The pH range of the water should not fall outside the range of 7.0 to 9.5.

Toxic Chemicals	Maximum Concentration in mg/1
Arsenic	0.01
Barium	1.0
Boron	5.0
Cadmium	0.01
Chromium	0.05
Cyanide	0.01
Lead	0.05
Nitrates (including nitrites) as NO_3	40*
Selenium	0.01
Silver	0.05

Biocides In general biocides should be virtually absent or not detectable in any drinking waters. The presence of any biocide would require a special investigation and study.

DISCUSSION

R.E. WILLIAMS, United States

Is that three year background monitoring programme being implemented already and if so, how long has it been implemented for ?

L.C. HENRY, Canada

The three years is of course, a guideline period of time and our experience is showing that from the time we received the intent from a company to mine an area, meaning that they are probably going to have to proceed through the exploration period. We usually have no difficulty in allowing for three years. Consequently, we ask these people to begin their programme, particularly those parameters that vary seasonably. We ask them to commence their programme concurrent with their investigations expiration. At the very least we will require a four season evaluation of baseline parameters.

S.J.B. BAKER, United States

You talked throughout this paper about general decommissioning or shall I say waste management criteria for ultimate disposal vs. storage and I was wondering if you have actually formulated those criteria or if that is merely a statement ? If you are working on the policy or if you just generate them as you go along ?

L.C. HENRY, Canada

Well our current policy is that all radioactive wastes will be disposed of. Now, quite obviously we do not have the disposal facilities. On one hand we are working on regional waste management facilities. The facility gaining most attention these days is of course the deep geological disposal. However, one could visualize other waste management facilities for certain types of wastes. Now, concurrent with these developments, we are examining methods of rendering the tailings inocuous, if you wish from the standpoint of radiation. This would include the removal of radium and thorium and what not. Of course, we are also asking the mines to proceed with determining ways and means of removing sludges and precipitates from settling ponds, immobilizing them and storing them again. This would probably fit in with future developments for regional facilities, but we do not have a definite programme, it is very much in the research stage.

S.J.B. BAKER, United States

During the operations of a tailings retention facility, do you protect the atmosphere from contamination due to radon gas emanation, dust release and direct gamma radiation for the tailings area ? How does one match that criteria ?

L.C. HENRY, Canada

Well there are, I guess possibilities. One that comes to mind for me at least is the use of flooded tailings facilities. We have one facility that is under construction at the present time, that will, during the operating phase, maintain much of the tailings area submerged. Another possible approach is, of course, the removal of radium from the tailings. Quite obviously from that statement there is no intention on our part to overstep the bounds of technology, we know where it is at the moment.

W.E. KISIELESKI, United States

      In the joint US/Canadian Great Lakes Water Quality Board there was data reported of point source release values from radium that were higher than might be expected. In fact, it was suggested that some remedial action be taken. I just wondered if your board actually verified these numbers or indeed how your guidelines implemented any remedial action that was acted upon ?

L.C. HENRY, Canada

      I was not aware of this ; I know that the IJC in its recently published document indicated that there are a few radium levels in the Great Lakes waters in the vicinity of certain nuclear facilities that were somewhat elevated, certainly not beyond drinking water qualities, but the general quality of the lake was still very acceptable.

W.E. KISIELESKI, United States

      Well, what I was concerned about is that they indicated the Elliot Lake activities would impact Serpent Bay and Lake Huron specifically. I just wondered, is this verified ? And it looked like this would be an activity that your board would undertake.

L.C. HENRY, Canada

      I know that the Serpent River has elevated ratings, but by the time you get down to the north channel where it exits into Lake Huron, it is pretty much at the drinking water level at that time.

D. MARKLEY, United States

      Two questions. One is a general question on any contemplation you might have on specific requirements on capping or covering of tailings. The second question is also general, just in terms of your regulatory review period for applications.

L.C. HENRY, Canada

      I think I stated here that we could not see the tailings systems the way they are being constructed and used today that we could have a walk away situation in the future. And since we are not satisfied that we could have this walk away situation, I do not think we have addressed the matter of putting so many feet of topsoil, sealant or whatever. Since we feel that we do not have the ultimate solution there yet, I feel we have tended to shy away from expensive remedial work-topping, and rather have leaned toward revegetation simply to stabilize the tailings from wind blowing and to minimize the percolated water. There was a second question and I think we could probably say it would be typical of any waste management facility, assuming the technology has been developed, of course, twelve months would be a bare minimum from the time we received the letter of intent to when a construction license would be issued.

D.C. McLEAN, United States

      I may have misunderstood, but did you say that the barium sulphate sludge would have to be kept separate from the tailings ?

L.C. HENRY, Canada

        I was saying that the principal that we have is that no
concentrated radioactive material, that is concentrated in radioactivity, can be deliberately rediluted. This is the principal. On
that basis, experience is showing that sludges in barium radium
sulphate precipitate ponds is ranging upwards of 100,000 picocuries
per gramme of radium-226. Since tailings are typically in the order
of a few hundred picocuries per gramme, then it would be inconsistent
with that practice to allow that material once scraped from the
bottom of these ponds go be back into tailings. While it has not
happened frequently to date, there are settling ponds that are quite
shallow and it is quite conceivable that during the operating phase
of the mill, there will be a need to go into the pond and dredge it
out. What I said was that these dredgings would have to be immobi-
lized, presumably dried somehow, and contained in storage until
a suitable means of disposal is developed.

D.C. McLEAN, United States

        Could they be put back underground as fill material ?

L.C. HENRY, Canada

        I would not rule that out, although they would presumably
have to go back in a manner in which they would be retrievable, so
I do not think they could just be solidified - well, I know they
could not be solidified and sent back with other tailings. Conceivably, they could be stored underground. It has not been proposed, so
I say "conceivably".

GENERIC ENVIRONMENTAL IMPACT STATEMENT
ON U.S. URANIUM MILLING INDUSTRY

J. B. Martin and H. J. Miller

U.S. Nuclear Regulatory Commission

Washington, D.C.

ABSTRACT

The U.S. Nuclear Regulatory Commission is preparing a generic environmental impact statement on the U.S. uranium milling industry to the year 2000. This document will lead to regulations covering management and disposal of mill tailings and make recommendations on what institutional arrangements are necessary for long-term isolation of the tailings waste. The basis for these regulations will be an evaluation of alternative tailings disposal programs. Each alternative is being considered from several perspectives, such as risks to maximum exposed individuals, population doses and health effects, susceptibility of sites to natural weathering forces, potential for groundwater impacts and costs.

DECLARATION TYPE DES INCIDENCES SUR L'ENVIRONNEMENT CONCERNANT
L'INDUSTRIE DE TRAITEMENT DE L'URANIUM AUX ETATS-UNIS

RESUME

La Commission de la réglementation nucléaire des Etats-Unis établit actuellement une déclaration type des incidences sur l'environnement concernant l'industrie de traitement de l'uranium aux Etats-Unis jusqu'en l'an 2000. Ce document débouchera sur des réglementations applicables à la gestion et à l'évacuation des résidus de traitement, de même qu'il contiendra des recommandations sur les dispositions institutionnelles requises pour assurer l'isolement à long terme des résidus de traitement. Ces réglementations seront fondées sur une évaluation des différents programmes d'évacuation des résidus. Chaque solution possible est envisagée sous plusieurs angles, tels que les risques encourus par les individus les plus exposés, les doses d'irradiation délivrées à la population et les effets sur la santé, la vulnérabilité des sites aux effets naturels des intempéries, les incidences éventuelles sur la nappe phréatique et les coûts en cause.

INTRODUCTION

The U.S. Nuclear Regulatory Commission (NRC) is preparing a generic environmental impact statement on the U.S. uranium milling industry to the year 2000. Its purpose is to evaluate environmental impacts of milling operations on a local, regional, and national basis, and to propose regulation changes which are found to be needed in light of this evaluation. Although the scope of the study includes evaluation of potential nonradiological impacts from milling such as socio-economic effects, its primary focus from a decision-making point of view is on radiological hazards. More specifically, the study is (a) evaluating environmental controls which can be applied to reduce emissions from the mill and mill tailings during operation and (b) is evaluating alternative methods for final disposal of the mill tailings.

The study will not be complete for another three to four months so it is premature to talk about precise conclusions. However, we would like to discuss in general terms the approaches we are taking in developing the document and present some of our preliminary conclusions. We will first briefly discuss the matter of controlling emissions during operations. Then we will discuss the issue of tailings waste disposal, covering both technical and associated institutional aspects of the problem. With regard to technical requirements, we will discuss the alternatives we are evaluating and then describe the form we expect proposed regulations to take.

EMISSIONS DURING MILL OPERATION

With regard to impacts of milling operations, the purpose of our study is not to determine what level of emission control should be required; that is, we are not attempting to set new standards. Radioactive emissions during operations are effectively limited by standards recently developed by the U.S. Environmental Protection Agency (40 CFR 190) which limits annual dose commitments to off-site individuals to 25 mrem or less (doses to whole body or single organ excluding doses from radon and its daughter products). Because our agency has responsibility for implementing this standard which takes effect in 1980, our aim in this matter is to identify what must be done to meet this standard. Our emphasis here is on evaluation of technical alternatives for emission controls since we believe they must be the primary means of meeting the off-site dose limit as opposed to institutional methods, such as land use controls which could theoretically be appled. Relying on the latter would contribute to the problem of ultimate site decommissioning, since not controlling the emissions would lead to a buildup of ground contamination.

We cannot at this stage state conclusions, but a preliminary dose assessment carried out for the "model" mill and site being examined in our study strongly indicates that windblown tailings pose a threat to meeting the limit at residences near the tailings site. Further scrutiny of our existing data and assumptions remains to be done before our evaluation is complete, but one certainty here is the need for actual environmental measurements at active mills. The field studies described by others in this seminar being conducted in conjunction with our generic study are being relied upon to allow thorough assessment of the problems of implementing the 25 mrem limit.

TAILINGS DISPOSAL

Currently, there are no formal regulations covering disposal of mill tailings; although, as was mentioned by Mr. Scarano earlier, we have established interim performance objectives to guide continued licensing activities. Our study is being developed to support proposal of formal regulations. The regulations will determine the level of technical control that will be required in tailings disposal, as well as address institutional arrangements needed to carry out or supplement these technical requirements.

TECHNICAL REQUIREMENTS - EVALUATION OF ALTERNATIVES

Any tailings disposal program must address the following fundamental problems:

(a) Reduce or eliminate airborne radioactive emissions. Radon emissions are, of course, the main concern because of the mobility of this gas;

(b) Reduce or eliminate impacts on groundwater; and

(c) Assure long-term stability and isolation of the tailings.

Numerous specific methods and techniques have been suggested to address these problems. Table I lists the full range of alternatives initially considered in our study broken down into the following categories:

(a) Process methods which involve tailings treatment and handling during the period of active milling operations;

(b) Location of tailings disposal area;

(c) Treatment of the bottom of the disposal area; and

(d) Final stabilization and covering.

None of these alternative methods represent in themselves a complete tailings disposal program. Each offers potential for solving one or several, but not all, of the problems which must be addressed. They must be combined to form a complete tailings disposal program, and it is obvious that innumerable combinations exist.

However, we focused our attention on a limited range of alternative disposal programs for detailed evaluation. In addition to a base case, where no attempt is made to stabilize the tailings, the range of alternative programs considered can be broken down roughly into three classes. The three classes, with our preliminary conclusions about them, are characterized as follows:

o Class 1 - Tailings are disposed of in an above ground impoundment. Some steps are taken to eliminate or reduce airborne radioactive emissions, including radon, and to mitigate groundwater impacts. The distinguishing feature of this class is that no measures, however, are taken to assure long-term stability and isolation of the tailings. This class is illustrated schematically in Figure 1. By its location above grade, the tailings are susceptible to relatively rapid deterioration from exposure to natural forces such as wind and water erosion. Therefore, this class is characterized by a need for an ongoing level of active care, such as constant maintenance of vegetation and regrading, to preserve integrity of the pile.

For this reason, it is our conclusion that the level of control provided by this class, although it is an improvement on past practice, is unacceptable. It commits future generations to a significant, lingering obligation to care for wastes generated to produce benefits which they will only indirectly receive, if at all.

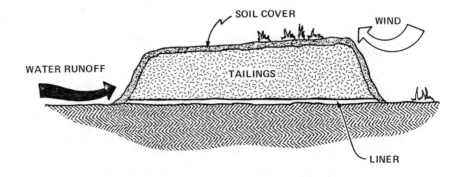

Figure 1. Class 1 Alternative — Above Grade Burial

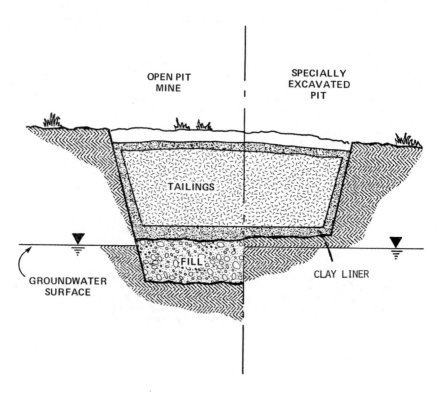

Figure 2. Class 2 Alternative — Below Grade Burial

o Class 2 - Tailings are disposed of below grade or in locations where they are sheltered from natural weathering and erosional forces. Steps are again taken to control airborne emissions and minimize impact on groundwater, but this class of alternatives is characterized by features which eliminate the need for ongoing, active care to maintain integrity of the pile.

Three situations are examined. The first is one where an open pit mine is available for disposal of the tailings. The second is one where a special pit is dug because an open pit mine is not available. These first two situations are depicted schematically in Figure 2. A third situation is difficult to depict graphically. It involves taking advantage of unique, natural characteristics augmented by design features, to provide protection from natural forces which is reasonably equivalent to that provided by the below-grade schemes. This may be necessary in some cases where groundwater formations are sufficiently close to the surface to foreclose below grade placement of tailings.

We will discuss more specifically how the matter of long-term stability is being addressed in our study in a moment, but the general conclusion is that beginning with below-grade burial as a prime option, this class represents a range of acceptable alternatives which can be developed to avoid an ongoing commitment to active care. This can be done with available technology at what we judge to be reasonable costs. We cannot at this stage talk in precise and absolute terms about costs, but they would generally represent only a few percent of total product price.

o Class 3 - This is a loose collection of alternatives which represent marked departure from current technology; to one degree or another, they provide added measure of isolation and protection above that provided by Class 2 alternatives. The leading alternatives falling into this category are: (1) disposal of tailings in very deep locations, (2) fixation of the tailings in asphalt or concrete, and (3) nitric acid leaching of ore.

Our preliminary conclusion is that because of any combination of the following--costs, negative environmental factors, and uncertainty about technological feasibility--we cannot require that these alternatives be adopted by the industry. This conclusion is, furthermore, based on uncertainty about the value of incremental benefits which these alternatives provide. Let us illustrate by discussing more specifically some of the problems with these alternatives.

Deep disposal offers potential for isolating the tailings so completely that no land use controls whatever would be required. (We are examining disposal in pits that are between 100 to 200 meters deep.) However, our preliminary evaluation is that this does not appear feasible without causing unacceptable groundwater problems. It is unlikely that a situation can be found near the uranium mines where groundwater formations are sufficiently deep to permit implementing this alternative. Such deep groundwater formations could probably be found, but they would most likely be far from the mines; and the costs and environmental problems associated with hauling the tailings to such a location appear to be unacceptable. Finally, we note that the question of how deep the tailings should be placed to eliminate the need for land use controls is arguable. Scenarios where deep wells are drilled into the tailings can always be postulated; and hence, the incremental benefit of such an alternative is questionable.

Fixation of tailings by cement or asphalt is not a commercially developed technology. There is also uncertainty as to the long-term stability of bonding between the tailings and the cement or asphalt. Furthermore, our preliminary cost figures indicate that this fixation alternative would cost anywhere between 2 to 15 times the upper range of costs for alternatives in Class 2. This alternative does provide an added measure of isolation for the tailings, but the costs for this incremental benefit do not appear warranted especially in light of the technological uncertainties involved.

Nitric acid leaching offers potential for reducing the radiological hazard of the tailings. Radium and thorium are removed from the ore during the same leaching process that removes the uranium. However, there are several problems. Laboratory studies to date indicate that residual radium concentrations in the tailings are still significantly above background soil concentrations. The values reported by Ryon et al[1] are between 17 and 60 pCi/gram which is certainly above what would be accepted as a "deminimus level" of activity. Therefore, isolation of the tailings in a manner similar to that provided conventional tailings would still be required. Nitrates formed from the nitric acid leach process also pose a more severe environmental problem than anion species formed from conventional sulfuric or alkaline leach processes. Costs are high. Our preliminary cost estimates indicate that the incremental lifetime costs of the nitric acid leach process (compared with conventional mill operating and tailings disposal costs of a Class 2 alternative) would be between about $50 to $200 million. Finally, there would still be a problem of disposing of the radium and thorium concentrates from this process.

FORM OF REGULATIONS

In view of what we have just presented, we expect the technical aspects of our proposed regulations to be very much like the interim performance objectives Mr. Scarano presented in the preceding speech which allow for flexibility with regard to specific methods of tailings disposal. The regulations will present a numerical limit on radon and gamma emissions, and state broad objectives with regard to groundwater protection, siting, and long-term stability. Working out detailed tailings disposal programs meeting the performance objectives of our regulations will remain a site specific task.

Radon Control

We expect the limit on radon exhalation rate to be an allowable increment above (expressed as $pCi/m^2/sec$), as opposed to a multiple of, background rates. We expect further that the increment specified will be essentially equivalent to the interim limit of twice background exhalation rate. The increment will be based upon selection of a background rate averaged over the western milling regions, leading to an incremental rate of about 1 $pCi/m^2/sec$ above background rates. This will mean that an equal level of radon control will be required at each site as opposed to effectively varying levels of control which would result if the limit is specified as a multiple of a variable background rate.

We should also note that the specific methods for determining compliance upon final stabilization remain to be worked out. This limit will, therefore, be applied in reviewing proposed designs of tailings disposal programs. Proposed plans will have to provide sufficient cover material on tailings areas to result, on a <u>calculated</u> basis, in the specified radon exhalation increment.

Let us elaborate for a moment on the "rationale" for the level of radon control we will be specifying. This is basically a matter of judgment, with ours being based on a number of different perspectives. We feel it is a reasonable value based on the following:

1. It would result in very small increments in population exposures and associated potential health effects above that occurring as a result of natural and technologically enhanced sources. Our preliminary calculations show that annual releases resulting from tailings accumulated by the entire U.S. industry to the year 2000 (about 750 million tons expected) will produce only a small fraction of that produced by natural sources, if emissions from these tailings are controlled at a level which is about two times an averaged background release rate. Specifically, it is estimated that the annual release rate from the total accumulation of mill tailings piles existing in the year 2000 will be less than 0.01 percent of natural releases from the continental United States.

2. The proposed release levels will result in acceptable exposures to residents that might be living nearby tailings disposal areas. These levels are comparable to those which have been established in other cases involving radon exhalation from radium-bearing materials (such as that previously established or proposed for remedial action at U.S. inactive mill tailings areas by the U.S. Surgeon General and for phosphate tailings by the U.S. Environmental Protection Agency).

3. The costs for providing this control will vary from site to site and the availability of some materials, but they typically will be less than a half of the overall lifetime cost of a tailings disposal program. Costs of cover material should amount to about one percent or less of the product price.

Some appear to expect a strict cost benefit balance, such as weighing cost of radon control against potential health effects monetized in some fashion. However, no precisely defined and commonly accepted formula for first quantifying, and then balancing, costs and benefits exists. Producing such a formula would require obtaining consensus on such questions as how much a life is worth in monetary terms, or how far into the future should potential health effects be considered? To date, no clearly accepted answers to these questions have emerged; they are likely to be the source of debate for many years to come. We will not engage in this kind of "medieval scholasticism" in our study.

Another point to state here is the interrelationship among objectives. One cannot reasonably separate the objective of radon control from that of providing long-term isolation of the tailings. It is unreasonable to weigh cost for stabilization cover just with potential health effects from radon exposure since burial of the tailings also contributes to the long-term isolation objective.

Therefore, what we will be doing in our study is proposing a level of radon exhalation which is very near background levels, and presenting the proposed level in terms of a full development of the various perspectives and considerations discussed above. With the benefit of this, the issue of how much radon control should be required can be debated openly by the public, industry, and other Governmental agencies based on a complete "record" of information.

Long-Term Stability

Although specific methods of tailings disposal will remain to be developed on a case-by-case basis, we expect the regulations will state that the primary option for disposal is below-grade burial. The burden will be on the industry to show how alternative above-grade schemes, where proposed, provide reasonably equivalent protection. It is not possible to state in simple terms how this can be done, but the basis for evaluating the long-term stability potential of alternative disposal programs is presented in a special investigation conducted in support of our generic study./2/

The purpose of the investigation was to identify potential failure mechanisms and associated natural processes which must be considered in developing a tailings disposal program. For each potential failure mechanism described, specific siting and design features which should be taken account of or incorporated in the planning process are systematically identified. Some of the major potential mechanisms and natural processes treated are wind erosion, water erosion and gullying, differential settlement of the tailings material or subsoils beneath it, earthquakes and floods.

Obviously, attempting to predict as far into the future as the tailings will remain hazardous (hundreds of thousands of years) is not possible. Over such long periods, major climatic and geological changes are certain; but their precise effects on tailings disposal schemes cannot be fully predicted. The study indicates that below-grade burial of tailings is an optimum mode of disposal in that it eliminates or reduces significantly exposure to the natural processes that can lead to disruption and dispersion of the tailings and remain stable for very long periods of time. It indicates, however, that with the right combination of siting and design measures, reasonably equivalent protection can be afforded the tailings placed above grade. As stated above, this factor must be addressed on a site specific basis.

INSTITUTIONAL REQUIREMENTS

As a practical matter, NRC has already adopted a policy of requiring financial sureties from mill operators to assure that the tailings disposal programs which they commit themselves to are carried out. In our study, we are evaluating a range of specific surety mechanisms and are concluding that there are a number of acceptable ways to accomplish the surety objective. We expect that the regulations will reflect this, allowing mill operators a measure of flexibility in this matter.

Other institutional issues of concern relate to long-term isolation of the tailings waste. Our evaluation of technical alternatives is being guided by the broad objective that tailings be disposed of in such a manner that no active maintenance or care be required. A conclusion we have formed, however, is that there should be some form of Government land ownership of disposal sites, and this has been incorporated into a legislative proposal which our agency is initiating.

The primary means of isolating wastes must be through physical barriers which do not require active care and maintenance. However, the potential for disruption or disturbance of tailings sites by human activities still exists. As discussed above, we are concluding that deep burial of tailings, with the intent of providing absolute protection from human intrusion and exposure, does not appear to be reasonable for high volume, diffuse radioactive waste such as mill tailings. Therefore, Government land ownership is not considered to be inconsistent with the goal of eliminating the need for active care and maintenance. It is instead viewed as a prudent, added measure of control of tailings sites.

Another question related to the issue of long-term control of disposal sites is whether or not there should be a tax or levy on milling operations in order to establish a "long-term fund" for costs which may be incurred following stabilization by the mill operator and termination of his license. In theory, all would agree that the waste generator should pay costs for waste disposal including any long-term care required. However, before making such a recommendation, a number of practical factors must be considered.

Because we are requiring that tailings disposal be carried out in such a way that no active care is required, we expect that a very low level of effort will be needed to control the tailings sites. For example, an annual or semi-annual visual inspection of each site may be all that is involved. Therefore, we expect costs of controlling sites to be small.

Another consideration is that the notion that a fixed, interest-bearing fund can be set up to perpetually provide a return to cover ongoing costs is unrealistic. Once milling operations terminate and revenues to the fund cease, its value would depend solely on interest and inflation rates which fluctuate markedly. Frankly, we believe it unlikely that such a fund administered by the Government could be sustained for any extended period of time under these conditions. This underscores the need to dispose of the tailings in a way that will not require active care.

It has also been suggested that funds be established to cover the costs of any unexpected remedial actions which may be required. Due to the very nature of this concern, setting the amount of such a fund would have to be completely arbitrary. Therefore, in addition to the question of whether or not there should be a fund, we are faced with the problem of establishing a proper total amount for such a long-term fund.

We are leaning towards the conclusion that the costs associated with the low level of surveillance effort we expect to be required at most sites would be small enough not to warrant hiring the army of bureaucrats that would likely be needed to set up and monitor a long-term fund and encountering the practical problems discussed above. Our hesitancy in stating this as a firm conclusion stems partly from the uncertainty about meeting the goal of eliminating the need for ongoing care at some existing sites. At existing sites, where in some cases very large volumes have been accumulated, disposal options are more limited than for new sites. How we should treat the matter of existing sites will have to be examined further before making a final conclusion about the long-term fund question.

SUMMARY AND CONCLUSION

We are evaluating a full range of alternatives for disposal of mill tailings in our generic study to support proposed regulations. We are concluding that there are a number of specific alternatives which can be implemented to solve the major problems of reducing radioactive emissions, protecting groundwater, and eliminating the need for active care of sites. Therefore, our regulations will provide flexibility; detailed disposal programs will continue to have to be worked out on a case-by-case basis within the limits prescribed by our regulations.

With regard to institutional requirements, we have concluded that it would be prudent to have Government land ownership of tailings disposal sites to avoid problems of disruption of disposal areas by human activity. An institutional question that we are still uncertain about is whether or not long-term funding arrangements should be established to cover any future cost of site control.

We will issue our study initially as a formal draft statement, proposing rules on mill tailings disposal at the same time. There will then be a period of public comment, most likely including public hearings, before a final version of our document and a final rule are issued. This process will provide full opportunity for involvement of the public, industry, and other Government agencies in the decisions we make.

REFERENCES

1. Ryon, A. D., et al, Nitric Acid Leaching of Radium and Significant Radionuclides from Uranium Ores and Tailings, Oak Ridge National Laboratory, Oak Ridge, Tennessee, U.S., ORNL/TM-5944, August 1977.

2. Nelson, J. D., and T. A. Shepherd, Evaluation of Long-Term Stability of Uranium Mill Tailing Disposal Alternatives, Colorado State University, Fort Collins, Colorado, April 1978.

TABLE I

ALTERNATIVE TAILINGS DISPOSAL
AND TREATMENT METHODS

1. PROCESS ALTERNATIVES

   o Nitric acid leaching to remove residual radioactivity from
      tailings

   o Segregation of slimes from sand portions of tailings for
      separate treatment

   o Lime neutralization

   o Barium chloride treatment

   o Removal of toxics by ion-exchange

   o Drying of tailings by thermal evaporators, by filtration,
      or by solar drying

   o Fixation of tailings to asphalt or cement

2. DISPOSAL LOCATION

   o Above grade

   o Below-grade near surface in available open mine pit or
      special excavations

   o Deep Disposal

3. TAILINGS SUBSURFACE TREATMENT

   o Do nothing

   o Compact soil

   o Clay in varying thicknesses

   o Synthetic liners

4. TAILINGS COVER AND STABILIZATION

   o Do nothing

   o Clay cover in varying thicknesses

   o Natural soil cover in varying thicknesses

   o Rip rap

   o Combinations of above

DISCUSSION

J. MONTGOMERY, United States

You discussed the excavation of a special pit designed to dispose of the tailings. Then a little bit later, you mentioned that we cannot require that as an alternative - we cannot require that be adopted by the industry. Could you give us some indication of just how strongly you are going to push for this particular method of a special pit designed strictly for disposal ?

H.J. MILLER, United States

I think the alternative you are speaking of is the deep burial scenario, I talked actually about two things. One is digging a specially excavated pit near the surface, a situation similar to what Mr. Scarano showed earlier with one of his alternatives. In regard to the deep burial, I think you used the words "are we going to push for that". We are openly considering it.

J. MONTGOMERY, United States

What I am really saying here is that I can see that this would be an extremely expensive excavation, if that were the only purpose for digging the special pit, and industry I am sure would say we cannot afford it - it is too expensive. My question really is, what would be your response ? Would you say, okay, we will try something else or would you try to put pressure on. If your policy is going to be to try to put pressure on this method ? Just how strongly are you going to push the industry ?

H.J. MILLER, United States

As I said, our preliminary conclusion is that in looking at it, the cost would be very high because of - not only the cost but the environmental impacts would be very high because you are talking about most likely having to isolate a region or a place that is far from the mines. So I do not know if I am begging the question, I just think that our preliminary conclusion that it is not something we want to do, that the negative cost in environmental benefits do not warrant doing. Now we are looking at it because, as you in the United States may know, the United States Environmental Protection Agency is developing broad waste management criteria that would apply to all kinds of waste, and one of the themes that runs through those criteria is the concept of no institutional controls. And about the only way that you can take that literally as you look at the scheme of things is putting it so deep that you do not have to rely upon institutional controls. So, as I say, we are looking at it.

J.B. MARTIN, United States

Let me interrupt here a minute. I think there seems to be some confusion. Mr. Montgomery, you are talking about the near surface, where if you do not have a mine pit available, you have to dig one and the question you answered was about finding one 300 or 400 meters deep. Mr. Scarano, could you shed some light on just what our position is on that and some current activity on that ?

R.A. SCARANO, United States

You made a conclusion, that it was going to be awfully
expensive and the industry would not go for it. I put a plan on the
board, I think it was plan three, and let me emphasize that that was
not my proposal. That is a proposal which we are evaluating which has
been proposed. You will find it in the draft statement. It is the
Sweetwater Project. The plan which is proposed in the draft statement
is not the authorized plan. When we come out with the final environmental statement in about a month, that plan will be in there, and
that is a proposal by the Sweetwater people, so it is not out of the
realm of what we would call a viable alternative.

P.C. REKEMEYER, United States

One comment I would like to make on that last statement
is the geology of the area is going to determine what those costs
are going to be. If you are doing hard rock mining, that is one
thing. If you are doing open pit mining like we have in Wyoming,
that is another situation.

The first question I had to ask is what mechanism do you
propose for implementing the 25 millirem standard in the agreement ?
Are you going to give the states the authority to go ahead and as
license conditions have the licensee make these measurements and
turn around and report to you ? Or are you going to have an independent agency, or an independent arm I guess you said, such as
you have in the non-agreement states, come in and do the inspections
and the checking ?

H.J. MILLER, United States

I cannot answer that question.

P.C. REKEMEYER, United States

The next question I have deals with the fact that we
presently have 22 inactive sites and I believe there is a bill
in the Senate to dispose of at this point in time. We also have a
number of sites, which Union Carbide operates which are now operating above surface, in the so-called the standard old mode of
operation. And then, of course, you have these mills that you are
licensing now in agreement or non-agreement states. What I would
like to ask is what, if any requirements, will there be, or will be
coming out of the GEIS relative to the movement of these existing
or operating sites which are still above ground ?

H.J. MILLER, United States

Well, I think that is a site by site problem, and that is
what we will say in the conclusions that this is what you do in the
future and we apply these principals to the maximum extent on each
case. Mr. Scarano talked about applying the interim objectives to
active sites where there already is an accumulation of tailings.
I expect that those same principals would apply and that same
kind of logic will be used in applying regulations to currently
active sites.

D. VAN AS, International Atomic Energy Agency

It is not clear to me why, in an environmental impact statement like this, that you address only the radioactive impact and not the impact of other things, especially when it has been shown in earlier papers that the things really found offside are selenium and arsenic and so forth. Certainly that must play a role in the control.

H.J. MILLER, United States

Yes, we are looking at that. I did say that we were excluding that. I did not mean to say that. The decision making aspect of our statement is based upon NRC's role to protect public health and safety from a radiological point of view. We do have a broader mandate, though, from our national environmental policy act, to look beyond that, and we are. We will be looking at it in assessing groundwater, for instance, arsenic, selenium. In some of the papers presented earlier, work is mentioned in connection with this. We are looking at those nonradiological contaminants.

J. HOWIESON, Canada

I have a couple of questions. I may just not be up to date, but this 25 millirem regulation caught me by surprise and I am rather wondering maybe its the wrong forum, but where did it come from and is it universal? Is it applied to bones, lungs, as well as whole body or what is it ?

H.J. MILLER, United States

It applies to the light water uranium fuel cycle. It applies to maximum exposure to any individual in the public from all fuel cycle facilities that he may be exposed to. But, typically, it is just one facility. The way I worry about it is in terms of mills. Now there are members of the Environmental Protection Agency here and I suggest that afterwards, if you are interested, I can point them out to you and they can give you information about how it was developed.

J. HOWIESON, Canada

The other question is related to your remark about deminimus levels of concentration. You mentioned the figure, I think, somewhere between 17 and 60 as being definitely higher than any deminimus level. I had the feeling that there was some consideration of a level of about 20 picocuries per gramme as being an acceptable level for building materials in Europe and so therefore, I am not quite sure of the relationship. Could you discuss that ?

H.J. MILLER, United States

Not very well. If you compare it to residual contamination levels that have been accepted in other cases, phosphate remedial actions and remedial actions at inactive sites, it is pretty much higher than those levels.

M.C. O'RIORDAN, United Kingdom

        I want to tell you that we did not decide in Europe to go
for 17 picocuries per gramme of radium. In fact, the OECD Nuclear
Energy Agency is working on a document which is thinking of suggesting about 3 or 4 picocuries per gramme. I think it is 500 bequerels
per kilogramme now, but I am not sure about this translation.

D. MARKLEY, United States

        Could you discuss some of the provisions in the proposed
legislation concerning federal ownership of land. I guess I have a
thought that comes to mind in this regard. I know it would be outside your jurisdiction, but are these similar plans in EPA or any
other agency to come up with this same scheme for the numerous
hazardous waste disposal sites in the country ?

H.J. MILLER, United States

        Well, I do not know if it is fair of me to involve the
chairman in answering a question, but Mr. Martin and myself and
Mr. Scarano have all worked on this.

J.B. MARTIN, United States

        Perhaps I can answer that legislative question. There is
a variety of bills presently being considered by Congress and a
variety of testimony being given on them, and of course, what the
final product looks like, we would not know for some time. But it
seems as if the general consensus is that there should be some sort
of government ownership. The NRC's position in testifying on the
bill was that it really did not make a whole lot of difference,
whether it be federal or state. The states testifying on the bill
thought it ought to be federal, and I expect that may be what will
happen. But that seems to be the general discussion - there ought
to be some government control and the states almost unanimously
think that it ought to be the federal government that winds up with
ultimate custodialship of it. But there has been no discussion in
that bill of trying to broaden it to all sorts of hazardous wastes
simply because you would never get a bill at all.

N. SAVIGNAC, United States

        In the generic environmental impact statement that is
coming out, are they going to be addressing the health effects
situation with selected radon levels ? Is this going to be one of
the topics addressed ?

H.J. MILLER, United States

        Yes. As I pointed out in my paper, we will be looking at
several points of view and the point of view of population dose
and population health effects as well as what might be received by
people near a pile.

N. SAVIGNAC, United States

        Will the economic impact of the various recommendations
likewise be addressed in the impact statement ?

H.J. MILLER, United States

       Yes.

S.J.B. BAKER, United States

       I do not see, from what has transpired today, where the
below grade system, the immediate below grade, not 200 meter depth
burial, but immediately below grade, would preclude continued monitoring, to verify that the groundwater table is not fluctuating
through a period of hundreds to thousands of years. My other point,
and to clarify here, I am questioning whether or not you are setting
the regulations before establishing a solid basis for the criteria.
Specifically, with respect to the fixed increment on the radon. The
fixed incremental level for radon - flux values - for your over-
burdened material - your cover material. Mr. Scarano stated earlier
that the design objectives attaining two times background was somewhat a value that could fluctuate depending upon cost benefit analysis
of the amount of radon flux or attenuation versus the cost of overburden and the percentage effectiveness in reduction thereby, and
that value does not necessarily extrapolate back to two times background, as I understand it. So, I am questioning whether or not you
are generating regulations ahead of the criteria.

H.J. MILLER, United States

       Well, I said that these were preliminary and I meant that.

Session 5

CONCLUSIONS OF THE SEMINAR

The Panel Discussion has been prepared by four Working Groups
chaired by the Chairmen of the previous Sessions.

The conclusions and recommendations of the four Working Groups as
well as the following general discussion are reproduced
in this part of the Proceedings.

Chairman — Président
Mr. R.E. CUNNINGHAM
United States

Séance 5

CONCLUSIONS DU SÉMINAIRE

La Table Ronde a été préparée par des Groupes de Travail, présidés
par les quatre Présidents des séances du Séminaire.

Les conclusions et les recommandations de ces quatre Groupes de
Travail ainsi que les discussions qui ont suivi sont reproduites
dans cette partie du Compte rendu.

Panel discussion

Table ronde

R.E. CUNNINGHAM, United States

     This is the concluding session of our Seminar on uranium mill tailings. I think perhaps this is probably the most important part of the meeting because it is a status report, and hopefully we will define, both for our individual governments and international agencies some direction we may go in pursuing our goals of minimizing public risk from uranium mill tailings. I am going to call on the chairmen from each of the sessions to read their summary report from this morning's working groups ; I think it might be best to hear all these reports and then discuss them in general.

D.M. LEVINS, Chairman of Session on Source Terms

     I would like to begin by making some general comments about the consensus of opinion that has emerged from the Seminar. Firstly, I think there is an acceptance that past practices of tailings management were, in many cases, inadequate. There is also a consensus that, if sound engineering practice is used in conjunction with present technology, there will be no gross environmental degradation. It is also clear, however, that we do not have all the answers at present and that further R & D is necessary. It is part of the task of this final session to make recommendations as to the most fruitful areas of research.

     Our working session concentrated on "Source Terms" and resulted in a very lively discussion. The most important source terms are :

- radon gas,
- airborne dusts containing radionuclides and toxic elements,
- liquid process wastes discharged with the tailings,
- contaminated rain-water that is released from the tailings dam or enters the environment by seepage.

     Radon : Doubts were expressed by the group as to whether the exhalation rates from a tailings pile could be calculated with any degree of accuracy. Some of the difficulties still unresolved are the dearth of knowledge about the emanation power and its variation with moisture content and other variables, the effect of convection near the surface, possible non-homogenities in tailings and earth cover, and a lack of knowledge of effective diffusivities of radon and the soil moisture profile. With the present state-of-the-art, measurement of radon flux is regarded as far more reliable than prediction.

     A number of methods are available for experimentally determining the radon source term including the inverted barrel technique, charcoal canisters and the use of continuous flow scintillation cells such as those developed at ANL and described at this Seminar. Another approach involves measurement of air velocities and radon dispersal patterns downwind of the tailings. Each method has its particular advantages and disadvantages and all should be used in conjunction. Calibration methods need to be developed, and national and international standards for interlaboratory comparison should be made available. This is one area for international cooperation through the OECD/NEA or the IAEA. The group considered that measurement of radon flux at the 1 pCi m^{-2}s^{-1} level required by the NRC may be difficult. Radon flux is often very dependent upon location on the pile due to sand-slime segregation. While it is recognized that radon exhalation is the appropriate source term, it must be appreciated that dose rate determination must be based on working levels which require appropriate measurements in buildings downwind from a radon source.

Airborne dusts : Theoretical calculations of this source term are very difficult due to the complex terrain and lack of information on a number of points such as the effect of moisture content, particle size and wind speed on the airborne dispersal of tailings. The papers presented at this Seminar represent a significant fraction of the total research which has been carried out to date. There is, at present, a lack of fundamental understanding despite the fact that it is a significant problem and may, under some conditions, contribute a greater dose than radon itself. Thorium-230 is the most significant radionuclide and toxic elements such as selenium and molybdenum also need to be considered. In the long term, even with stabilization, wind erosion is a potential source of dispersal of tailings.

Liquid sources : There appear to be wide variations in the concentration of radium-226 and thorium-230 in active tailings dams as reported at this Seminar. Radium is a very significant source term when process wastes must be discharged and its behavior in a tailings dam and the environment is clearly not fully understood at the moment. Leaching of heavy metals and toxic elements such as selenium and arsenic is another source term which is site specific. For ores or waste rocks containing high concentrations of sulphides, bacterial oxidation is likely to be the most important source of water pollution.

Recommendations for R & D : These are by no means complete and are confined mainly to Source Terms.

- Fundamental studies are required to determine the effects of particle size, intraparticle porosity, ore grade, mineralogy and moisture content on the radon emanation coefficient of ore and tailings.

- Simple methods need to be developed to estimate the effective diffusivity of radon in earth cover, clays and tailings under wet and dry conditions.

- Tracers should be used to supplement measurements of airborne tailings dispersal particularly in complex terrains. Sulphur hexafluoride was suggested as a possible tracer and radon exhalated from tailings may also be a convenient natural tracer.

- Studies of the dispersal of toxic elements (such as selenium and molybdenum) in windblown tailings should be made in addition to radionuclides such as thorium-230.

- Airborne tailings dust concentration should be measured at distances of up to 5 miles (8 km) from acid and alkaline leached tailings to test theoretical models of tailings dispersal.

- Concentration and disposal methods are required for barium/radium sulphate sludges. The leachability of these precipitates also needs to be studied.

- Research should be undertaken with the aim of eliminating or substantially reducing the bacteriological oxidation of tailings and waste rocks containing high sulphide concentrations.

- Consideration should be given to advanced methods of waste management which, although not feasible with today's technology, could become practicable with further research. These include possible low cost methods of radium and thorium removal, sand-slime separation followed by separate disposal, agglomeration or consolidation of the slime fraction, and methods of uranium ore processing that eliminate the need for a tailings dam.

The papers presented here have shown that countries have different emphases in both the technical aspects of waste management and in the regulatory controls. These arise primarily because the environmental impact of uranium milling is site specific and related to climate (eg. rainfall versus the evaporation), hydrology (eg. the possibility of surface versus ground water pollution) and ore characteristics such as the presence of sulphides or toxic elements in the ore and tailings. This has resulted in development of specific waste treatment practices to cope with particular problems. Examples are barium chloride treatment, neutralization, and covering and vegetation of tailings piles. There are varying emphases, particularly in regard to radium and radon. While local technical considerations are the main reason for different emphases, there has been a degree of unawareness about other countries' approaches. I think this Seminar has largely solved this problem for the moment but there will be a continuing need for cooperation and information exchange both at the national and international level. Waste management practices undertaken through necessity at some sites and in some countries are genuinely applicable at other locations but might represent advanced practices that could be implemented in the future.

M.C. O'RIORDAN, Chairman of Session on Environmental Aspects

We tried to stick to our terms of reference, which were to identify areas where work might usefully be done and where international effort might be most fruitful. It will not surprise you to hear that we give the physical aspects of the problem mainly in national terms and the other aspects in international terms.

One idea that appears to have gripped this conference is the inevitability of dispersion. As one of our participants put it, with the vigour of American English, you cannot build a box that will last forever. This led the group to suggest that studies of relevant earth processes should be undertaken so that one might model the behaviour of wastes in a geological time scale and perhaps even discover better ways of channeling them back into normal geological processes. This led us on to a specific suggestion that it might be profitable to study ancient waste tips and chart the dissemination of material from them. There was less than unanimity on this point, and I feel that it might be worthwhile exploring it here in the plenary session. Such sites are only a few thousand years old, but that is surely better than a few tens of years.

The other idea that emerged strongly was a need to give due emphasis to nonradiological pollutants, this topic might also be explored.

What dominated our discussion, however, was the question of risk, and the word that was used over and over again in this connection was "perspective". Some of us were more sanguine than others about the possibility that firmer risk factors for low exposure to natural radiation might emerge in the future, and that is why we suggest that the variations in natural levels should be charted more extensively to give a better sense of perspective to increments. We had some quite passionate outbursts on this point. There is no doubt that as a group we felt that minor perturbations in, for example, ambient radon concentrations had no significance, and that it was a waste of precious effort to engage in an assessment of them. The other way we saw of giving perspective to the problems posed by uranium tailings was by comparing any deleterious effects with those of other industries. Indeed, we felt that research and development on the effects of pollutants should be deliberately set up across industrial lines.

We saw a great need for effective public information, and we felt that there was a particular role for the Agency here. We would urge that the information be scrupulously honest, of course, setting out both the facts and the uncertainties.

We believe that we identified a need for standardization in analytical and field techniques with particular emphasis on the hydrogeological aspects of tailings disposal, and I am sure that some of my colleagues will wish to elaborate on that point.

One topic that underlines the geographical variability in the problems posed by waste disposal is the Canadian idea of deep lake disposal. We felt that this was virtually virgin territory deserving careful study.

That more or less completes my brief summary.

J.R. COADY, Chairman of Session on Management and Stabilisation

Our session was called "Management and Stabilisation", but except by implication the two words did not occur, I think, throughout the entire discussion. We accorded ourselves the freedom of discussing the whole of the subject at our disposal, but we did attempt to do this with each item in terms of the questions that Mr. Cunningham asked us to address. I hope that this was adequately achieved.

The first item we addressed was, this Seminar as a record of the present situation, or, the state of the progress which has been achieved in the management of tailings. The impression that we all had of the papers which were presented at this Seminar was that there is a very clear difference in emphasis between the two major contributors, the U.S. and Canada. There is definitely a greater emphasis in the U.S. on atmospheric flux rates, the airborne aspects, radon, and windblowing problems, whereas in Canada, and to some extent in Japan and Sweden, there is more concern with problems of solution transport. France demonstrated in their paper an aware-ness of the need to stabilize and revegetate tailings surfaces which is something which can be fitted into both of these areas, I think. Australia also has interests in both areas.

It was strongly argued, however, in our group, that the difference of opinion does not represent a difference in fundamental approach to tailings management, rather, it has occurred of necessity, due to certain, present, site specific differences. But there do appear to be the beginnings of concern, more concern, in the U.S. over the possibility of hydrogeological problems, and in Canada we are becoming more aware of the fact that we should be more concerned with radon emission. We are not unaware of it, of course, but we may have to look at it more seriously, as you are, to see where the problems are. This all indicated to us a very clear area for an exchange of information in these complimentary areas.

We then agreed that there was unanimity on the desirability of achieving passive, long term solutions for looking after tailings; that this was the goal of everybody in industry and regulatory agencies alike. We agreed that, to the extent possible, we should not leave problems for other people. There was also unanimity that it is difficult to predict what the problems might be in the long term and there was equal unanimity on the fact that we were not sure how the long term aspects could really be adequately addressed. This is an area which does have to receive much more attention. Some feeling was expressed that the long term aspects may be site specific, not so much with respect to the meteorology at those sites but with respect to the mineralogy of the tailings themselves.

In the papers which were presented, we perceived a lack of innovation and novelty in the approach to the subject of tailings management. We think this calls for more cross-fertilization, perhaps between countries, but certainly between the various disciplines whose expertise should be brought to bear on this problem. It is a multidisciplinary problem. To date, however, it seems that the area, and this is not to denigrate engineers, but it does seem that the search for solutions to tailings management problems has been dominated by the engineers. Just as an interesting aside, we passed a sheet around to obtain the technical background of the twenty-five people who were in the group at that point, and the results are interesting. Of the 25 people, there were 17 engineers, 4 chemists, 2 metallurgists, 1 physicist and 1 biologist, so it does rather bear out the point that was made. The lack of novelty was mentioned. Certainly, some new aspects were introduced in the papers, but the lack of novelty was perceived in there being little mention in the session of new head-end methods and in the lack of extraction processes being designed with a view to the tailings management problems. Insitu leaching, for example is being assessed and it represents a fairly new approach in uranium mining. If it could be assessed from the point of view of its being useful for waste management, there might have been more papers. Although it is recognized that it is a site specific subject.

It was expressed that little was mentioned during the meeting, the Symposium, on the cost effectiveness of measures which are being taken. Several people spoke to the fact that society only has so many resources and so much money and that should keep aspirations and solutions in perspective. This is not, perhaps, a job for us who are charged with providing the solutions, the technical solutions, but it is something we could bring to the attention of our governments and our international groups who do concentrate on problems related to the dividing up of the resource pie. The group also expressed the thought that there could have been more papers on the cost effectiveness of particular waste management methods. Perhaps as a follow up to that, and as a follow up to the previous comment about the engineers, again not to denigrate engineers, but it seemed that a point was being made that to date we have been mainly concerned with engineered solutions. Examples of these are, the use of plastic liners as being the state of the art for improving the impermeability of dams, or lowering the gross permeability, perhaps, as opposed to taking a more careful look at the natural geochemical processes which may help us in retarding the rate of transfer of undesirable species into the environment. The use of absolute, technical solutions like plastic liners, particularly if not used in conjunction with suitable drainage systems may make for more problems in the long-run than they cause alleviation in the short term.

The point was also made that we should not regard the tailings produced from the mill in isolation from the other solid wastes produced at mine sites. Consideration should be expanded to include management of waste rock and overburden, and it was noted that, in some cases, the tailings themselves, the radioactive part of them, may only represent 5 % of the total material excavated. The same comment, regarding the need for a general perspective on waste management problems at a tailings site, was made with respect to liquid effluents. In addition to the effluents from the tailings area itself, mine water and runoff should also be kept in mind.

Another aspect that Mr. Cunningham asked us to consider was, were there any points of contention. I do not think there were many points of contention. In fact, I do not think we actually perceived one good point of contention at the table. However, there is one thing that could lead to contention and which does lead to a lot of discussion. This is the lack of uniformity in the analytical methods being used in this industry. This applies to the methods being used for radium analyses, the question of the size of the filter, and whether we should or should not use filters. In many cases, different answers are being obtained at sites using the same method. The point was also made that many hydrogeological techniques are actually out of date and that a lot of hydrogeological information being produced in support of environmental impact analyses and licensing applications is incomplete when it is first submitted and, in many cases, it is obtained by methods which are out of date. It would seem that a lot of argument and perhaps differences of opinion might be avoided if there were more uniformity in these areas, and comparison of techniques would certainly be one way to begin this. Another point of discussion, not fully explored, was related to the allowance or the non-allowance of dilution to achieve release limits.

The next question that we addressed ourselves to was the question of R & D that should be undertaken, and in this area I am not about to be very specific in listing the programmes which should be launched. We simply did not have the time so this is a very general summary of R & D requirements. The first one is that there should be an investigation of the analytical aspects of sample collection. There are problems in the sampling of water, air and soil. Sample preservation & sample transport need comparison and evaluation ; so do the methods of analysis themselves. Also, within the analytical regime, site specific interferences must not be over-looked. The second point to make in regard to R & D is that, the information base is extremely inadequate and does not really allow us to answer the questions which are being asked. Questions such as the influence of neutralization on the final state of the tailings ; the stability coefficients of materials within tailings, depending upon how they are prepared ; the stability coefficients of precipi-tates resulting from effluent treatment. Distribution coefficients for radium in unconsolidated materials are practically unavailable. Information on the effect of head-end processes, present and pro-posed, on the tailings system are woefully inadequate and on that point we will leave R & D. There is a lot of R & D that can come out of the remarks just made and perhaps another session is needed to discuss R & D on its own.

With regard to international initiatives, we made the observation that first, the components of the international commu-nity should start at home with registers of their own basic and applied research because intranationally there seems to be a lack of awareness of what is going on. Once this has been achieved on a national basis, it should be expanded and there should be available from the NEA or the IAEA a compendium, on as current a basis as possible, of the R & D that is being done in various countries. This was thought to be most useful but does not yet exist in this area. There are one or two systems already in use which could be used as models by the agency prepared to undertake this task. These registers should contain not only R & D being conducted but, the point was made and agreed to, I believe, that perceived problems should also be listed. If programmes are not actually, currently under way in given countries on problems that they perceive that they have, the problem may be being worked on in another country but yet not have reached the R & D list in the country. Having an awareness of the problems of other countries, this leading to the sharing of information, may well be one way to commence their solu-tion. A great deal of information is not published and that is another conclusion that was made as a result of hearing the papers of this Seminar. This underlines the necessity for some kind of exchange. Another task for international groups would be a review of standards and codes, standards particularly, although we did get into semantic problems about what we meant by standards. I think that we all have the same goals. We readily admit that. But there still exists on a province to province, a state to state and a nation to nation basis, different ways of achieving this. And it appears that the differences are not always based on rationality. While the admission is always made that there are site specific aspects to be taken into account, an examination of the desirability of having more uniformity would be the first thing to accomplish. To achieve more uniformity in standards would be well worthwhile.

The cost effectiveness of some of the proposals that have been made for waste management appears, at this time, to be somewhat low, radium removal being one, perhaps. Because of this there does seem to be the need for what might be called a "regulatory cartel" in this world. That is, if we do agree that certain routes are desirable for waste management, and we do agree to that, then it may still be very difficult to implement them for financial reasons. Under such conditions, it would be difficult for any one nation to go unilaterally in that direction. To have prior international agreement that such directions were necessary would help smooth the path for each and every country faced with this decision.

That concludes the specific points we were asked to address. We do have some final notes, however. First, we think a seminar of this kind could well be commenced, in the future, with a review paper by someone who has taken the trouble to do some talking beforehand with people in the field, to bring the Seminar quickly up to date on the state of the art and to provide a perspective for the following papers. We feel that a review paper would be a useful addition to this kind of Seminar. Secondly, we feel that the working group in which we participated was extremely useful, but far too rushed. I share the comments of my colleagues in this respect. We make the other suggestion, therefore, that in addition to the review paper, it may be useful to have a short working session at the beginning of the Seminar to find common ground, then hear the papers and see what our reactions are afterwards. The final note that we make is that we encourage the secretariat with the help, of course, of the contributors to distribute the proceedings as quickly as possible.

J.B. MARTIN, Chairman of Session on Policies and Regulatory Aspects

Our report basically summarises major points of common view in the regulatory area and offers a few recommendations. First of all, as a general principle, it was agreed that it is desirable to work out and agree to a tailings disposal programme prior to generation of the tailings. This principle, of course, follows from the fact that options in dealing with tailings disposal become very limited after several million tons of them are generated. It is also based on the desirability to incorporate full costs of tailings disposal into the mill operating costs and ultimately to be reflected in the cost of power. The degree to which this principle is being implemented varies among countries, and it was the group's consensus that there should be a sense of urgency about establishing this practice as quickly as is reasonable everywhere.

Secondly, there was also general agreement that the ultimate tailings disposal programme should not be based upon on-going active care to assure long-term isolation of the waste. While the primary means of isolation should be physical barrier, continued land use control to avoid human activity, which could lead to disruption or breeching the confinement, is a desirable added measure which should be provided. It was also the feeling of the group that the responsibility for exercising this control over future activity should be that of some sort of a governmental body.

Thirdly, as has been discussed in some of the other reports, it was recognized that the problem of tailings disposal is highly dependent upon specific site characteristics. The methods developed to meet the objectives outlined should therefore be tailored to site-specific conditions. Regulations developed should not be method specific but should provide flexibility to accommodate variations which exist.

Fourthly, standards and regulations applicable to mill tailings should take account of the fact that tailings are basically a high volume, diffuse type waste which in contrast to other types of waste, particularly high-level waste, presents more of a chronic as opposed to an accute kind of problem. For example, additional flexibility may be warranted in regard to institutional control to isolate the tailings from human intrusion.

Fifthly, it is desirable to dispose of the tailings in such a way that, at a later time, residual uranium and other valuable materials could be extracted if that became attractive.

Lastly, there was consensus that it would be desirable to make sure that provisions had been made to finance disposal of the tailings prior to their creation to avoid very large downstream costs that are not anticipated at the time when the operations are started.

These are six general principles that permit a great deal of local variation, but I think they reflect the overall consensus of the papers given and the summary of group discussion.

Our recommendations are fairly limited since regulation is largely a national issue. I think it is hard enough to develop regulations on a national basis, much less internationally, but we did have three or four points that could be termed recommendations. First of all, NEA should urge uranium producing countries to develop their goals and criteria for tailings disposal on an expeditious basis. Secondly, standards and criteria should be broad so as to provide for flexibility and application of ingenuity on developing methods. This is necessary in terms of the site-specific nature of the tailings disposal problem. We felt that NEA could serve as a useful vehicle for exchanging regulatory information experience among countries. This could include information about standards and criteria being developed and their supporting data base. Particularly because of the variations in technical problems, a discussion forum such as this Seminar is a very useful means to exchange experiences gained in working through specific cases. It is often difficult to get the full value of information on specific problems encountered by reading an abstract. Face to face discussions of the problem could be very useful. We also felt that NEA might be useful in coordinating effort in some areas. Working groups might be usefully employed on certain regulatory issues, like quality control, effluent and environmental monitoring, dam construction, and tailings stabilisation methods. Most of these are dependent on the validity of analytical methods, and that lends itself to some international exchange and cooperation. It may be possible to develop some international standards as a result of such cooperative work. That concludes the work of our group.

R.E. CUNNINGHAM, United States

Again, I think each of the session chairmen did a fine job in summarizing the essence of what went on and what the conclusions were in their individual sessions. Before discussing some of the points that were brought up by the session chairmen, I might take this opportunity to invite anyone in the audience, the participants, observers, if they would like to add anything to these summaries.

I think that for purposes of providing information to the NEA and other organisations like IAEA, we have a rather broad range of specific areas where work might be undertaken. For example, each session chairman mentioned the need for standardisation of analytical methods, whether they are collection of field data, how you analyze that data, and so forth. That obviously is an area where it might be well worthwhile for an international agency to develop a report that can be given to interested countries on standardizing techniques. I think it goes beyond that. I heard some things like quality assurance on dam construction and so forth. I think we have enough information in the record and we can proceed from there. I think, more importantly, there were some points brought out that require some thought on the direction we are going. One of the points of interest that merits some discussion is the question brought up by Dr. Levins, and that is predicting what happens in the long term to mill tailings. We talk about stabilization, and containment of these materials for millenia, but that probably is not realistic and it certainly is not supported by the data we have now. The point is, the importance over the long term, that these tails be absolutely contained as opposed to eventually working their way into the various ecological systems. This is something that we need to pursue. First to identify the mechanisms by which tails might get out into the environment and, secondly, to assess how that might impact mankind. Could we have some discussion on that point ?

J.D. SHREVE Jr., United States

Some of this entered the discussion that occurred in the last group. The ideal, that seems to be the consensus, is to end up at the closing of a mill by packaging up the tailings with a situation wherein, either with or without government ownership of the surface rights, you can allow people access, but you are really obliged to limit activities in an unrestricted sense where deep digging, etc., would not breech the containment. I think, however, while this is not an ideal and we ought to try to do this, on the first several cases we should act as though we are doing it and allow all this human activity to occur within the bounds set. But, by the same token, monitor the first several cases to see that if, indeed, we have created a scene that was this trustworthy. And I am particularly concerned about the statement made in the last session that you just want to reduce it to a point where you drive over it in a jeep twice a year and say, "Hey, it looks good. Let's go to Joe's and have a beer". I am more worried about what happens underground. As I said in my paper, I think that is where the real drama is and that is where the long term implications may be more important because they are less visible. I think we must develop a programme in line with ever better monitoring systems, and ever better design of monitoring systems so that we can track this. In the first group's summary, they said, "Go out five miles and watch". I fully agree with this. So while we have the ideal and we will act as though we are turning it over with great aplomb and great confidence, we are to check a little on the inside so that we have a few check measurements backing us up.

R.E. CUNNINGHAM, United States

I might bring up a second point, that is source terms. The first summary identified a substantial amount of research that seems to be needed to further identify source terms and I wonder if I might ask you, Dr. Levins, if you really feel that this is cost effective. We know a fair amount about source terms, now. At least, I hope we do. The NRC has been spending a fair amount of money trying to get this information. Do we need greater detail on source terms today than we already have ? I have given the uncertainty of what the actual risk is even if we are able to calculate dose ? Would you care to comment on that ?

D.M. LEVINS, Australia

        I think this depends upon how seriously you take these
NRC agreements of one picocurie per square meter per second. If
these are regarded just as an indication only, then maybe these
very approximate indications that we have at the moment might be
satisfactory. If you want to improve your accuracy, and I think
ultimately you want, to be able to identify the sources of pollution
more and more accurately, I really could not comment about the cost
effectiveness of the whole thing because I do not really know the
cost. But I do not think this research is all that expensive.

R.E. CUNNINGHAM, United States

        How long do you think it would take to get the type of
information you feel you need ?

D.M. LEVINS, Australia

        Five years perhaps. But you know that is a fairly low
level involvement. We are perhaps talking about ten people internationally.

R.E. CUNNINGHAM, United States

        Another point that does not seem to be well covered.
Listening to the papers, it seems to me that there was a lot of
difference of opinion about what is really important in mill
tailings management and stabilization. There seemed to be some
controversy about it although I am not sure that it is a real
contentious point. I think the question is whether or not we are
prepared today to identify, perhaps through an international organization, those factors which must be considered, prior to starting
a milling operation which will bear on our ability to manage and
stabilize the mill tails. Have these parameters that should be
considered been adequately identified ? And, if they have, should
they be somehow clarified so that everybody can share the knowledge
of what is important and what might not be so important for evaluation. I might ask Mr. Martin to respond to that.

J.B. MARTIN, United States

        Yes, I think that bears a little bit on one of our
recommendations. It might be useful to work through some specific
cases periodically with an international group like this. There
certainly have been factors raised during this meeting that I was
not aware were causing people problems. I was quite surprised at
the different nature of a technical problem being examined in Canada,
for example, and that we have been grappling with. I think the other
thing to bear in mind, like all technical problems, the whole thing
does not get solved at one time. It is a continuing process, and I
am certain that two years from now, we are going to be a lot smarter
than we are now. We will probably be in the position of saying,
"If I had it to do over again, I would have done it differently". We
will probably be doing this for the next 5-10 years, or even longer.
But I sort of sense a reluctance among some people, to do anything
until all the returns are in and my own personal view is that what
we are doing today may not be the ultimate answer. But it is far
worse to do nothing, do a very little amount, because I think it
is very difficult to solve any of these problems in the abstract.
Ultimately, it seems as if actually doing some of this management
waste disposal work and seeing how it works out is a flexibility
that we should not use which we seem to have right now. But if the

problem gets to the point, like with some of the other forms of
waste where nothing can be done until all problems are answered
absolutely, you build in an almost impossible situation to solve.
So, it seems to me, and I thought it was the sense of our group,
that there are an awful lot of things that perhaps are not known.
But that should not serve as an excuse for not trying to deal with
the tailings as best you can today and recognize that tomorrow you
may be doing it somewhat differently.

R.E. CUNNINGHAM, United States

That brings up the next issue. You say that we should deal
with tailings as best we know how today, recognizing that we might
be doing something different tomorrow. Dr. Coady mentioned in his
summary a lack of innovation in the approach to mill tailings management and he cited the composition of the group in his session.
The background of this group - engineers, physicists, chemists -
is an indication of perhaps a one-sided viewpoint on how to approach
the problem. I wonder if you would care to expand on that, Dr. Coady ?
If you say we need to be more innovative, and I am inclined to share
that view, how do we go about getting that approaching the question
of innovation and doing something about it ?

J.R. COADY, Canada

I am glad you did not ask me if I am an engineer. Since
I am not I might not have got out of here unscathed. If I may merely
begin by following on what Mr. Martin talked about. In effect, what
he was saying was, let us not attempt to wait until we have all the
solutions before we begin to implement properly those solutions
which we have now. Our group would agree with that. I think I would
extend that to say that our group, had unanimity that there were
long term problems that we would like to passage solutions in the
long term. I think nobody in the groups would have said that there
are no long term problems. Some have a clear idea, from an intuitive
feeling perhaps, of what the long term problems may be. Some are
less reticent to be so dogmatic. What is absolutely clear, though,
is that as soon as you begin to try a quantitative assessment of
what is likely to be a long term problem, you lack the data, and the
knowledge of the interior of the tailings problem. One thing was
pointed out ; it may come down in some cases to a knowledge of source
terms. To be able to decide what we will need to do in the long run
to achieve our goal of facility in the long run, we need to have
the basis to do the analysis to find out what it is we need to
change and I am homing in on the innovation aspect. We need to scope
out the things that need to be done to see where they fit into our
present ability to decide what is wrong in the long term. To set
ourselves some fairly reasonable long term goals in terms of R & D.
Some of these were touched on at the Seminar. Even the representative
from Kerr-McGee went as far as to say that we could do slime sand
separation and, if necessary, wash the sand again. These are all
new things that have not really been done from the point of view
of waste management and it is only by beginning to scratch the
surface of that kind of approach that we will begin to learn what
the ins and outs of it are. So we must set ourselves some goals of
this kind. Get research going on a national-international basis on
some of these more reasonably achievable goals that we see cannot
obviously be achieved until a few years have passed. In the meantime,
it certainly applies now and we may have to live with the fact that some
of the tailings we have produced to this point will not be managed
by the methods which will be developed in the future. As each year
passes the millions of tons that are produced will presumably be
managed by some method which is just a little better than that
which went before. But I think we have to have a resolve to take a
close look at what we need to do to achieve our goal of passivity,

set ourselves some long term goals and make sure we do the work on them. Otherwise, they will never happen, and we will just coast along in never never style. I think some of those have been clearly identified. I think the work in the paper from Australia did identify clearly some of those areas where we can do further work. There were two papers from Canada that did not arrive, and one of them was on this innovative approach in terms of removal and washing of sands. I think we need to set these goals. In the meantime apply the methods we have and with each step, try to make them better. One day will not achieve what we want, but I think we have to get as soon as possible as clear an idea of what we want as possible. I think there is a bit of resolve needed there, otherwise we will just keep coasting. I do not know if that answers your question.

D. VAN AS, International Atomic Energy Agency

        I want to speak in my personal capacity. I think we are far too pessimistic about what we do not know. No industry in the world knows what its position is going to be a hundred years or a thousand years from now. I do not think they have even attempted to try and do this. I think we should concentrate on those things that we do know. We have learned a lot in this week here. There is a lot of information that has now become available, that can be applied and that would certainly improve the state of the art in the management of waste. If you think that over the last 20-30 years that uranium has been mined, the way this stuff has been managed in many cases and we have not had a catastrophe anywhere. In my country, South Africa, Johannesburg is built on mine dumps and it is hard to find anything. You go around looking for radioactivity, and you just do not find it. So the way we have managed it without much knowledge has not caused real serious problems. The way we are going to manage it with the knowledge we have now will probably again cause us problems in trying to look at the issues and trying to really find where it goes. I mean if you move more than a kilometer away from such a mine dump, you do not find anything, so I think we are really overly pessimistic and if we use this sort of data, we can certainly manage this very well. That is certainly not to say that we should not do anything any more, the research and development should go on.

V.I. LAKSHMANAN, Canada

        Dr. Coady just mentioned about sand and slime separation and I could make a comment here. We in our company are beginning to look into this particular area, separating the slime before it reaches the pond and we have started doing some work in this particular field.

J.B. MARTIN, United States

        I would just like to further emphasize something that Dr. Coady said. I think it is sometimes hard to quantify, but it is extremely important here, and that is the resolve. You used resolve. I generally use the word commitment. But one thing I have seen, in the past couple of years I have been involved in this problem, as some of the others, is that frequently on these kind of problems, the innovation goes into rationalization rather than actually applied to solving a technical problem. In every case I have seen where the particular company that is faced with the problem, develops a sense of commitment by the management to solve the problem, and the people start working on it, some really surprising things come up. So I think that the question of commitment to solve the problem, rather than a commitment to do better and protect is extremely important here. It lends a sense of credibility where the public and other people are very concerned about these kinds of problems, and will then provide the technical people the slack we need to work through these kinds of problems, many of which are ultimately judgments anyway. Without that sense of resolve, I think very quickly you lose the flexibility and the slack to innovate, in fact to do anything. I have often heard the most troubling form of human creative thought is rationalization and you know there is a lot of that that has gone on in the past few years. I think that those words innovation and commitment or application are extremely important here.

W.M. TURNER, United States

I reside here in Albuquerque. I am a geologist – hydrogeologist – and I would like to come back to point that I made in the working group discussion, and that is more or less in suppor\: of Dr. As's comments earlier. I sort of have the impression that we are trying to re-invent the wheel here. The point I made in the working group discussion was that throughout the world we have had over the centuries a great deal of mining taking place. We have tailings disposal areas covering half the face of the earth as far as I know. These have been weathering over geologic time. I myself happened to have mapped several slag piles overlying ground water systems where the water table is not more than 20 feet beneath the bottom of the slag pile. Water wells all over the place and people drinking that water happily today. I know of no cases of illness that were derived from water from beneath these slag piles. It seems to me that when we create a uranium tailings pond, all we are really doing is creating an ore deposit, we are changing the location of an ore deposit. For example, here in the Grants mineral belt we will extract uranium ore from 3,000 feet, and we will process it for the uranium and place the waste on the surface. All we really have done is create an ore deposit on the surface of materials that were some-what different than which we originally began with. We have added perhaps some sulphate to the water within the tailings disposal area, and I can take you 20 miles south of Albuquerque here and show you ground water that is very heavily contaminated with sulphate because the area for this particular arroyo which provides recharge is behind the Manzaon Mountains and the area is just full of gypsum. So, all we have got there is sulphate in the ground down there and what happens is that the pattern of settlement is somewhat different. It is not a very heavily inhabited area because, frankly, the water is not fit to drink. It seems to me that one of the problems we have in dealing with tailings disposal areas is perhaps one of shifting the location of people, a settlement problem, as well as in really trying to contain the tailings forever and ever which I think is a geological impossibility. I think it may be more realistic to recognize tailings disposal areas for what they are rather than trying to find ways in which to contain them forever and ever. And as one of the individuals said earlier, put it in a permanent black box. I do not think that is physically possible.

M.C. O'RIORDAN, United Kingdom

There were two small points I wanted to say. One was that I think the feeling in the working group was that relatively secure implacement of the tailings was not an insurmountable problem. The real problem was inadvertent use of these tailings in times to come, therefore, the problem is the question of the permanence of govern-ment and civilization which we thought was really the numb of the problem. The other thing that I felt came out of the group, was that we were more or less unanimous in suggesting that concentration of radium in any fraction was not desirable and we felt that it would impede dispersion and create, perhaps create more problems than are solved.

W.P. STAUB, United States

I have had a number of private discussions with engineers here at this conference and I believe I am about to go away with a gut feeling that some of the tailings management that is done here in the United States is done differently from that in other countries around the world, and that we need to have greater exchange of these tailings management alternatives with our friends from other countries. Perhaps they, in some respects, are doing things in better ways than we are, and we need to look into those things. Perhaps our friends from South Africa, who evidently do not have any problems, have some innovative ideas that we can use here in the United States that would help our situation immeasurably.

W.N. THOMASSON, United States

Over and over today and doing this week we have heard the term used, dispersion of effluents, of these materials. We have also heard about containment. I think we need to combine these two for the record because the industry has historically been in hot water because of the dilute and dispersion attitude that was used for so long without any attempt at control in various industries. Here what we are really talking about is controlled dilution and dispersion, and control conditions as well as how we can control them. I would like to think that is what we are talking about and not have the public get out and think, oh, we are just dumping it in the ground.

R.E. CUNNINGHAM, United States

I think that is a very good point and certainly it is consistent with U.S. policy on waste management. Our objective is to contain, recognizing that over very long time spans some of the material we are attempting to contain might be dispersed. We have talked about that problem in mill tailings management since the tails will be near the surface, but on close examination the same applies to deep geologic disposal of high level wastes. I think the only sure way to dispose of high level wastes if you are thinking in those terms is to use outer space. You can remove it from earth's environment, at least. But I quite agree with your comment and I think it is an important one.

D.M. LEVINS, Australia

I assume that most people here are not familiar with the situation in Australia. There has been approval for mining of the Ranger Deposit and I think some of the environmental safeguards that have been applied here are about the most stringent in the world. I think that what we have actually done is we have learned from past practice in the United States and Canada and applied the best features of both. For example, the wastes we intend to neutralize are those with lime to about pH8. We intend to maintain a water cover over the tailings during operation. We intend to line the tailings dam with clay. The basic philosophy is containment and I believe that can be met on a scale of the half lives of radium and thorium. I believe the amount of radium that leaches out is controlled by the sulphate level and under the actual assumptions we have made you get essentially containment of the radium. I believe that thorium is an even simpler problem to contain because it is basically under alkaline conditions, rather immobile. Finally the agreement is that the tailings will go back in the mines and actually restore the region to essentially the same condition that it was previously. I think, under today's conditions, if this is carried out, you have got a very safe operation. I do not think you have really got any problems, if you do the job properly.

M.H. MOMENI, United States

A study of the uranium mines in this country and looking
at relative equilibrium between the daughter products, in many areas
indicates almost equilibrium. The age of some of these mines exceeds
millions of years. Under those geological conditions, these materials
have not migrated and they have not moved. Why should they do it in
the future ? Unless the hydrogeology of the area drastically has
changed where they become more mobile and they revert to the conditions which forced them in to deposit. I feel our situation toward
tailings should be more optimistic than the pessimistic attitude
which has been portrayed before.

J.B. MARTIN, United States

We started all this by putting an exclamation point behind
innovative work in the future. I think we might go a step further
and define or describe what our hopes are and for those who could go
along with this, I would say that innovation in the future will take
the tack of more clever manipulation of the very process we now use
with an ultimate simplification of treatment and containment. I feel
that with the methods we have described, we have gone almost to brute
force to be safe. We are almost super safe. We are talking clay
liners, neutralization, containing and covering. My forecast will
be that with an injection of a little more science, a little more
attention to first principles to supplement the engineering and
balance it out, we are going to end up with simpler systems doing
better things. Were it otherwise, I would worry that a mill going
onstream ten years ago, now with changing regulations, will have a
different cost for the operation as new things are laid on it.
Every mine, say one today versus one five years from now would have
a different set of requirements to meet, and unless the scientific
input simplifies with the same or better control, then we will face an
economic problem. But if it does go as I predict then some of these
economic problems will level out and go away because it is going
to be cheaper to be smarter and contain with less of the brute force
intervention.

R.E. CUNNINGHAM, United States

I think that is a good point. I had expected in talking
about innovation that it might put a little different cast on the
problem. I think that all comments I have heard thus far this
afternoon about innovation refer to innovative technology. There was
mention made by one of the speakers here about setting some goals
and then trying to achieve them technologically. The problem I see
with that is that you set those goals on what you think today's
technology is capable of accomplishing. Given time, that is always
going to change, so I see no end to it if you base your goal on
technology alone. I had thought someone might bring in the idea, when
we talked about innovation, of having more involvement with people
who set public policy in this decision making process than we had
heretofore. The sooner we start involving those people and those
disciplines that bear on setting public policy, the sooner we will
have more firm goals, the sooner we will be able to put the mill
tailings management problem in perspective with other environmental
problems and the sooner we will be able to do something about the
problem. But as long as we look at this in isolation of other
problems, in the absence of agreement with those who set public
policy and depend solely on technology to provide our solutions,
I do not think we will have a solution because technology will
always change. We will be here ten years from now talking about
new technology and trying to achieve some higher level of isolation
based on the new technology.

I.I. ITZKOVITCH, Canada

I think the thing we are all forgetting is that in essence uranium processing was designed prior to our having a social conscience ; i.e., the environment. I think the innovation that Dr. Coady is talking about is the new technologies that are going to be developed to process uranium ores more cheaply, more quickly with a view to waste management. I think the waste management concept has to come first and a process designed with that view. I think this is what we must get across.

R.E. CUNNINGHAM, United States

I think that is a good point, except I do not know if I would agree with you on that, we have just developed a social conscience in protecting the environment. I think it has become a popular sort of issue, but I think it has been of concern to very many people in the past even before it became very popular within the last 10 or 15 years.

J.R. COADY, Canada

I do not want to convey the impression that I just want to have new technology for new technology's sake. The point I was making s that, as was just said, the goal for the introduction of technology would be for better waste management. I think it is fairly clear the direction you could go in an innovative sense to arrive at better waste management technology. The question that I think imposes itself on that area is, is that really necessary ? I am talking about radium removal for instance, as one example. I think if we took it out to the extent it was desirable, tailings management problems as we perceive them would largely disappear if we took the thorium as well as other things. But let us just concentrate on radium for the moment and thorium. If we did get them down to certain levels that we could live with, wastes with diminuous levels, that surely would solve the problem of need for restriction of access and the need for continuous surveillance. It is obvious, I think, that would happen. What is not clear, though, and we are finding this through our consultants who are doing this work for us, are the long term problems of tailings. In establishing that you find immediately that you do not have the information to make a concrete tight case that you do have to go to that extent. I think there are two things involved. One is having enough information to make the case iron clad for his country that you do or do not have to change the present process. Secondly, on the assumption that you do, introducing on a reasonable time scale, technology which will allow you to achieve that waste management goal. I hear the representative from Australia saying that the method he describes as being one that will meet all the goals we want. Yet some of the problems we have in deciding whether the methods we have will be good in the long term come as a result of examination of methods very very similar to that. For instance, you raised the point you need more substantial information but you seem to have concluded already that your method of management will be okay in the long term. There is a little dichotomy there it seems to me. So, I think that the solution is that the technology become waste management oriented rather than simply uranium production oriented. That would be the innovation.

R.E. CUNNINGHAM, United States

If I may comment on that. I do not know if I am prepared to agree with you. I think you get back to the social question. If you are, for example, to remove radium-thorium from the tails, we know that you are not going to remove 100 % of it. You might remove

90 %, I do not know what the number might be, but we know we would not get 100 % of it. We remain faced with the question of what amount of radiation in those tailings is going to be acceptable to leave out in that pile. You will have to deal with the social issue of what is acceptable and what is not acceptable. The second part of the issue of moving in the direction of waste management technology as opposed to looking at the tails as part of the milling complex and the problem I have in going to waste management technologies, is the simple fact that if you remove the radium, and radium has a toxicity roughly in the order of plutonium, it becomes quite clear that we do not have any way to dispose of plutonium today that is satisfactory. What are we going to do with plutonium ? The purpose of using the high level waste repositories is to isolate the plutonium for long time spans. We do not have that. We do not have it in the United States. We do not have it in Canada where a major amount of uranium is milled. And a country like Australia, that does not have a high level waste disposal problem to the best of my knowledge, will be faced with a new problem. So, I think it is an oversimplification just to say to move to waste management technology because that technology will remove the radium and thorium.

J.R. COADY, Canada

I will, in what I am saying now, ignore the possibility of radium removal, even though I do not share some of the concerns you have about the importance it makes really. If we do not take it out, what we are left with is what we have got now and we are not certain of the extent to which we are going to have problems. There is not unanimity that there is no long term problem and since some of us do feel that there potentially are some long term problems, how do you rationalize the present method in the light of our goal that we will attempt not to leave problems of a continuing nature for future generations ? If we do not look to something better, we have got what we have got now. I am not sure that is the best.

D.M. LEVINS, Australia

I think I should comment about this drawing a parallel between tailings disposal and high level wastes. I think this is an unfortunate comparison. We are talking about tailings and materials that still exist on the earth and have existed on the earth for a very long time. The level of activity is completely different. You are talking about megacuries in some cases ; you are talking about picocuries in other cases. I think this analogy is very unfortunate.

R.E. CUNNINGHAM, United States

May I comment on that ? First, I have not drawn an analogy between tailings and high level waste and I had no intention of doing so. The second point is, the high level waste problem in the long term is not the number of megacuries you have in the high level waste, it is the long term activity you have in the long-lived isotopes of plutonium that is not megacuries, it is a small amount of curies. And its roughly comparable to what you have in radium if you were to extract it from mill tailings. If you are just worried about all those megacuries that may go for 400 to 1,000 years, you could build surface storage facilities with reasonable confidence to take care of that problem.

I think we are approaching the time when I should try to sum up and draw this meeting to a close. Although I have been arguing a little bit with Dr. Coady, I do not think we really have a basic disagreement. We must be innovative. There is no question about that. We have to be innovative with technology and we have to be innovative in how we approach the problem of establishing with the public what is an acceptable risk.

We set three goals during the course of the meeting. One was to establish a record which presents the status of technology as we understand it today, and I think, in general, we have accomplished that purpose or it will be accomplished when we have the proceedings of this meeting published.

The second was to identify what further research, development and standards might be useful. I think during the work sessions and the summaries presented by the chairmen, there have been a number of very specific issues brought forth that would be desirable to explore further, both nationally and internationally. These too will become part of the record and will be identified. I think the summary statements identified those, and they should be considered by national governments and international agencies.

The third goal was to identify what international agencies, the NEA and perhaps the IAEA, might do that would be useful from this point on. There were a number of recommendations made. Just briefly to mention a few, I think it becomes quite clear that an international guide or technical report on uniform analytic methods would be very useful. Perhaps a number of other guides would also be useful, such as one on tailings dam construction. Another area that was mentioned, I suppose for both governments and international agencies, is to purge a lot of the standards we have floating around that are now antiquated. I think that would be a useful effort, both within our own countries, and within international agencies. The third issue that was brought to our attention was whether international agencies could be useful in serving as a means for the exchange information, and I view this seminar as such a means. Perhaps in the future there should be more opportunities to exchange information. I believe that consideration should be given to defining the topics to be discussed on a more narrow basis, so we can get more highly specialized people to work on specific issues through the international agencies. Perhaps NEA and IAEA might want to consider mechanisms for doing this. Finally, I think that we have identified areas where some international standards might be useful, or perhaps standards is not quite the word, recommendations, for example may be a better way to identify those things that are important to consider in siting the mills, siting the tailings pile and what must be done to stabilize the tailings pile as we understand the technology today. Finally, it is my personal opinion that international agencies could be of assistance in establishing an international consensus on radiation protection objectives for uranium mill tailings stabilization. The risk could be placed in context with other processes to produce energy. Our experience in this country is that sometimes, having international goals can be very helpful in establishing national goals. With these comments, I would propose to close the concluding session of this Seminar.

LIST OF PARTICIPANTS

LISTE DES PARTICIPANTS

AUSTRALIA - AUSTRALIE

LEVINS, D.M., Dr., Australian Atomic Energy Commission, Research
  Establishment, Private Mail Bag, Sutherland, NSW 2232

THOMPSON, K.E., Office of Protection Environment, P.O. Box 1890,
  Canberra City 2601

CANADA

BROWN, J.R., Dr., Geology Department, University Western Ontario,
  London, Ontario NGA 5B7

COADY, J.R., Dr., Atomic Energy Control Board, P.O. Box 1046,
  Ottawa, Ontario K1P 5S9

DODDS, R.B., Dr., Morton Dodds and Partners, 50 Galaxy Blvd,
  Unit 11, Rexdale, Ontario NW 4Y5

FLETCHER, R.H., Kilborn Limited, 36 Park Lawn Road, Toronto, M8Y 3H8

FOURNIER, P., c/o Amok Ltd., Box 9204, Saskatoon, Saskatchewan S7K 3X5

FROST, S.E., c/o Eldorado Nuclear Limited, Suite 400, 255 Albert
  Street, Ottawa, Ontario K1P 6A9

GILLHAM, R.W., Department of Earth Sciences, University of Waterloo,
  Waterloo, Ontario N2L 3GI

GORBER, D.M., c/o James F. MacLaren Limited, 435 McNicoll Avenue,
  Willowdale, Ontario M2H 2R8

GOODE, J., Kilborn Limited, 36 Park Lawn Road, Toronto M8Y 3H8

GRAHAM, R.G., c/o James F. MacLaren Limited, 435 McNicoll Avenue,
  Willowdale, Ontario M2H 2R8

GUIDOTTI, F., c/o Denison Mines Limited, P.O. Box B-2600, Elliot
  Lake, Ontario P5A 2K2

HENRY, L.C., Dr., Atomic Energy Control Board, P.O. Box 1046,
  Ottawa, Ontario K1P 5S9

HOWIESON, J., Energy, Mines and Resources Canada, 588 Booth Street,
  Ottawa, K1A OE4

ITZKOVITCH, I.J., Dr., Ontario Research Foundation, Sheridan Park,
  Mississauga, Ontario L5K 1B3

JANTZON, F.G.H., Dr., c/o Uranerz Exploration and Mining Limited,
  204-229 Fourth Avenue South, Saskatoon, Saskatchewan S7K 4K3

LaHAYE, G.J.L., 445 Albert Street East, Sault Ste. Marie, Ontario

LAKSHMANAN, V.I., Dr., c/o Eldorado Nuclear Limited, Research and
  Development Division, Suite 400, 255 Albert Street, Ottawa,
  Ontario KIP 6A9

LaROCQUE, J.E., c/o Denison Mines Limited, P.O. Box B-2600,
  Elliot Lake, Ontario P5A 2K2

LUSH, D.L., Dr., Peal Consultants Ltd., 6870 Goreway Dr., Malton,
  Ontario I4V 1L9

POTTER, C.L., Department of the Environment, Water Pollution Control
    Board, 1855 Victoria Avenue, Regina, Saskatchewan S4P 3V5

SCHMIDT, J.W., Dr., Wastewater Technology Centre, Environmental
    Protection Service, Department of Fisheries and the Environment,
    P.O. Box 5050, Burlington, Ontario L7R 4A6

SIMMONS, G.R., Atomic Energy of Canada Limited, Whiteshell Nuclear
    Research Establishment, Pinawa, Manitoba, ROE 1LO

SMITHSON, G.L., Dr., Saskatchewan Research Council, 30 Campus Drive,
    Saskatoon, Saskatchewan S7N 0X1

TARRY, B., Eldorado Nuclear Ltd., Box 7010, Eldorado,
    Saskatchewan SOJ 0TO

VIVYURKA, A.J., P.O. Box 1500, Elliot Lake, Ontario P5A 2K1

WILES, D.R., Dr., Chemistry Department, Carleton University,
    Ottawa, Ontario I1S 5B6

DENMARK - DANEMARK

SKYTTE JENSEN, B., Dr., Chemistry Department, Risø National
    Laboratory, DK-4000 Roskilde

FRANCE

BAUER, J., Elf Aquitaine, Aquitaine Mining Corporation, First
    Denver Plaza Suite 1700, 633 17th Street, Denver,
    Colorado 80202, United States

CHAUVEAU, J.C., Urania Exploration, 1536 Cole Blvd., Building IV
    Suite 310, Golden, Colorado 80401, United States

FOURCADE, N., Mme, Commissariat à l'Energie Atomique, STEP,
    B.P. n° 6, 92260 Fontenay-aux-Roses

PERSONNET, M.P., Minatome, 69-73 rue Dutot, 75538 Paris Cedex 15

FEDERAL REPUBLIC OF GERMANY - REPUBLIQUE FEDERALE D'ALLEMAGNE

BARTHEL, F.H., Dr., Bundesanstalt für Geowissenschaften und
    Rohstoffe, P.O. Box 510153, D-3000 Hannover 51

ITALY - ITALIE

BASSIGNANI, A., AGIP Nucleare, Corso di Porta Romana 68, Milano

VEGGIANI, A., Dr., AGIP - Attivita Minerarie, 20097 San Donato
    Milanese (Milano)

JAPAN - JAPON

IWATA, I., Power Reactor and Nuclear Fuel Development Corporation,
    9-13, 1-chome, Akasaka, Minato-ku, Tokyo

KITAHARA, Y., Power Reactor and Nuclear Fuel Development Corporation,
    9-13, 1-chome, Akasaka, Minato-ku, Tokyo

THE NETHERLANDS - PAYS-BAS

GOEDKOOP, J.A., Prof., Netherlands Energy Research Foundation ECN,
    Westerduinweg 3, 1755 ZG Petten (N.H.)

SPAIN - ESPAGNE

IRANZO, E., Dr., Head of Protection and Medicine Services, Junta
    de Energia Nuclear, Avenida Complutense, Madrid 3

PERARNAU PERRAMON, M., Dr., Fabrica de Uranio G.H.V., Apartado 93,
    Andujar (Jaén)

SWEDEN - SUEDE

AGNEDAL, P.O., AB Atomenergi, Studsvik Energiteknikab, Fack,
    S-611 01 Niköping 1

BAECKSTRÖM, L.Å.I., L.K.A.B., Randstadsverket, S-520 50 Stenstorp

SNIHS, J-O., Dr., National Institute of Radiation Protection, Fack,
    S-104 01 Stockholm

SÖDERMARK, B.Å., The National Swedish Environment Protection Board,
    Fack, S-171 20 Solna

UNITED KINGDOM - ROYAUME-UNI

O'RIORDAN, M.C., National Radiological Protection Board, Harwell,
    Didcot, Oxfordshire OX11 ORQ

UNITED STATES - ETATS-UNIS

BAKER, S.J.B., Ms., c/o Western Nuclear Inc., 134 Union Blvd.,
    Suite 640, Lakewood, Colorado 80228

BAMBERG, S.A., Dr., Urangesellschaft USA Inc., 6880 E. Evans Ave 3-200,
    Denver, Colorado 80222

BEACH, G., Division of Health and Medical Services, New State Office
    Building, Cheyenne, Wyoming 82001

BLUBAUGH, R.E., Chevron Resources Co., P.O. Box 1000, Hobson,
    Texas 78117

BOHM, R.E., Gulf Mineral Resources Co., 1720 South Bellaire St., Denver, Colorado 80222

BOKICH, J.C., Ecology Audits Inc., 3428 Stanford Dr. NE, Albuquerque, New Mexico 87107

BROUGH, T., Dr., New Mexico Environmental Improvement Division, 708 Uranium Avenue, Milan, New Mexico 87020

BUHL, T.E., Dr., New Mexico Environmental Improvement Division, P.O. Box 968, Santa Fe, New Mexico 87503

BURGER, L., Ms., University of New Mexico, 328-A Jefferson NE, Albuquerque, New Mexico 87108

CHARLIE, R.H., San Juan Basin Regional Study, P.O. Box 1590, Albuquerque, New Mexico 87103

CLEVELAND, J.L., Kerr-McGee Nuclear Corporation, Grants, New Mexico

CROSS, F.T., Dr., Battelle Pacific Northwest Laboratories, P.O. Box 999, Richland, Washington 99352

CUDDIHY, R.G., Dr., Lovelace Inhalation Toxicology Research Institute, P.O. Box 5890, Albuquerque, New Mexico 87115

CUNNINGHAM, R.E., US Nuclear Regulatory Commission, Mail Stop 396-SS, Washington, D.C. 20555

CURTIS, D.B., Dr., Los Alamos Scientific Laboratory, P.O. Box 1663, MS-490, Los Alamos, New Mexico 87545

DANIELS, R.S., NUS Corporation, 4 Research PL, Rockville, MD 20850

DONAHOE, P., Environmental Improvement Division, State of New Mexico, P.O. Box 968, Crown Building, Santa Fe, New Mexico 87503

DREESEN, D.R., Los Alamos Scientific Laboratory, P.O. Box 1663, MS-522, Los Alamos, New Mexico 87545

DUDLEY, J., New Mexico Environmental Improvement Division, Box 968, Santa Fe, New Mexico 87503

EHRICH, H., MtN States Engineers, P.O. Box 17960, Tucson, AZ 85731

FILIPY, R.E., Dr., Battelle Pacific Northwest Laboratories, P.O. Box 999, Richland, Washington 99352

FRANKFORT, J.H., Burns & Roe/Pechwieg, 550 Knedenkanach Rd., Oradelle, New Jersey 07649

FRIEDLAND, S.S., 13900 Panay Way, R 305, Merino Del Rey, California 90291

GALLAHER, B., New Mexico Environmental Improvement Division, P.O. Box 968, Santa Fe, New Mexico 87503

GANCARZ, A., Los Alamos Scientific Laboratory, P.O. Box 1663, MS-514, Los Alamos, New Mexico 87545

GOAD, M., New Mexico Environmental Improvement Division, P.O. Box 968, Santa Fe, New Mexico 87503

GRAY, W.E., Anaconda Company, P.O. Box 638, Grants, New Mexico 87020

GRIFFITH, W.C., Lovelace Inhalation Toxicology Research Institute, P.O. Box 5890, Albuquerque, New Mexico 87115

GUREGHIAN, A.B., Dr., Argonne National Laboratory, E.I.S., 9700 South Cass Avenue, Argonne, Illinois 60439

HARSHAW, F., Harshaw Medical Physics Consultants, P.A., P.O. Box 1617, Socorro, New Mexico 87801

HAYDEN, J.A., Department of Energy, Rockwell International, Rocky Flats Plant, P.O. Box 464, Golden, Colorado 80401

HEID, K.R., Battelle-Northwest, P.O. Box 999, Richland, Washington 99352

HENDRICKS, D., ORP Las Vegas Facility, Environmental Protection Agency, P.O. Box 15027, Las Vegas, Nevada 89114

HOOVER, M.D., Lovelace Inhalation Toxicology Research Institute, P.O. Box 5890, Albuquerque, New Mexico 87115

JACKSON, B., TRW Systems Group Inc., One Space Park, Bldg 01/2151, Redondo Beach, California 90278

JACOBS, D., Oak Ridge National Laboratory, P.O. Box X, Oak Ridge, Tennessee 37830

JOLY, M.G., Minatome Corporation, 2040 S. Oneida, Denver, Colorado 80224

KABELE, T.J., Dr., Battelle Northwest Laboratories, P.O. Box 999, Richland, Washington 99352

KELLEY, N.E., Dr., Department of Biology, University of New Mexico, Albuquerque, New Mexico 87020

KENNEDY, E.E., United Nuclear Homestake Partners, P.O. Box 98, Grants, New Mexico 87020

KIRBY, R., Environmental Protection Agency, Effluent Guide Lines Division, 401 M. Street S.W. WH552, Washington, D.C. 20460

KISIELESKI, W.E., Dr., Argonne National Laboratory, 9700 South Cass Avenue, Argonne, Illinois 60439

KNIGHT, M., Ms., US-DOI, San Juan Basin Regional Uranium Study, P.O. Box 1590, Albuquerque, New Mexico 87103

KUES, G.E., American Ground Water Consultants, Inc., 2300 Canadelaria Road N.E., Albuquerque, New Mexico 87107

LANDA, E.R., Dr., US Geological Survey, 432 National Center, Reston, Virginia 22092

LAURIANO, L., Ms., Americans for India Opportunity, 20600 2nd NW Suite 403, Plaza Del Sol Building, Albuquerque, New Mexico

LUBINA, R.H., W.A. Wahler & Associates, 1023 Corporation Way, Palo Alto, California

MANO, T., Chem & Nuclear Eng. Dep., University of New Mexico, Albuquerque, New Mexico 87131

MARKLEY, D., Energy Fuels Nuclear, 3 Park Central, 1515 Arapahoe, Denver, Colorado 80202

MARPLE, M.L., Ms., Los Alamos Scientific Laboratory, P.O. Box 1665, MS-522, Los Alamos, New Mexico 87545

MARTIN, J.B., Nuclear Regulatory Commission, Mail Stop 396-SS, Washington, D.C. 20555

McLEAN, D.C., Western Knapp Engineering, Arthur G. McKee & Co.,
    2700 Campus Drive, San Mateo, California 94403

MEWHINNEY, J.A., Dr., Lovelace Inhalation Toxicology Research
    Institute, P.O. Box 5890, Albuquerque, New Mexico 87115

MILLER, H.J., Nuclear Regulatory Commission, Mail Stop 396-SS,
    Washington, D.C. 20555

MITCHELL, D.L., Dr., Department of Energy, Rockwell International,
    Rocky Flats Plant, P.O. Box 664, Golden, Colorado 80401

MOMENI, M.H., Dr., Argonne National Laboratory, 9700 South Cass
    Avenue, Argonne, Illinois 60439

MONTET, G.L., Dr., Argonne National Laboratory, Building 10,
    9700 South Cass Avenue, Argonne, Illinois 60439

MONTGOMERY, J., Colorado Department of Health, 4210 E. 11th Ave,
    Denver, Colorado 80220

NISHIKAWA, G., Kilborn NUS Inc., The Galleria Bldg Ste 930,
    720 South Colorado Blvd., Denver, Colorado 80222

NYLANDER, C., New Mexico Environmental Improvement Division,
    P.O. Box 968, Santa Fe, New Mexico 87503

OVERMYER, R.F., Ford, Bacon & Davis, P.O. Box 8009, Salt Lake City,
    Utah 84108

PEACOCK, R., Phillips Uranium Company, Box J, Crownpoint,
    New Mexico 87313

POWERS, H., Homestake Mining Company, 7625 West 5th Ave, Lakewood,
    Colorado 80226

RAHN, P.H., Dr., Argonne National Laboratory, Division of Environ-
    mental Impact Studies, 9700 South Cass Avenue, Argonne,
    Illinois 60439

REDMAN, H.C., Dr., Inhalation Toxicology Research Institute,
    P.O. Box 5890, Albuquerque, New Mexico 87115

REKEMEYER, P.C., P.O. Box 1029, Grand Junction, Colorado 81501

RITTS, J.D., Sohio Natural Resources, P.O. Box 25201, Albuquerque,
    New Mexico 87125

ROBERTS, C.J., Dr., Division of Environmental Impact Studies,
    Argonne National Laboratory, 9700 South Cass Avenue, Argonne,
    Illinois 60439

ROBINSON, B., Battelle-Northwest, P.O. Box 999, Richland,
    Washington 99352

ROBINSON, P., Southwest Research & Information Center, 1822 Lomas NE,
    Albuquerque, New Mexico

RODGERS, J.C., Los Alamos Scientific Laboratory, P.O. Box 1663,
MS-490, Los Alamos, New Mexico 87545

ROSWELL, R.L., Ms., Battelle-Northwest, P.O. Box 999, Richland,
    Washington 99352

RYAN, R.G., US Nuclear Regulatory Commission, Mail Stop MNBB-7211,
Washington, D.C. 20555

SAVIGNAC, N., United Nuclear Corporation, P.O. Box 3951, Albuquerque, New Mexico 87190

SAYALA, D., Dr., Triple S. Development, 140 Washington Dr. SE, Albuquerque, New Mexico 87107

SCARANO, R.A., Nuclear Regulatory Commission, Mail Stop 396-SS, Washington, D.C. 20555

SCHOMISCH, J., The Gallup Independent, P.O. Box 1210, Gallup, New Mexico 87301

SEHMEL, G.A., Dr., Pacific Northwest Laboratory, Battelle, P.O. Box 999, Richland, Washington 99352

SHELLEY, W.J., Kerr-McGee Corporation, 3808 NW 69th Street, Oklahoma City, Oklahoma 73125

SHEPERD, T., Civil Engineering Department, Colorado State University, Fort Collins, Colorado 80523

SHREVE Jr., J.D., Dr., Kerr-McGee Corporation, P.O. Box 25861, Oklahoma City, Oklahoma 73125

SILHANEK, J., Environmental Protection Agency, Office of Radiation Programs (AW 461), 401 M Street SW, Washington, D.C.

SMITH, H., US Department of Energy, P.O. Box 26500, Lakewood, Colorado

SORENSON, J.B., New Mexico Water Quality Control Commission, Box 968 Crown Building, Santa Fe, New Mexico 87503

STANLEY, J.A., Lovelace Inhalation Toxicology Research Institute, Box 5890, Albuquerque, New Mexico 87115

STAUB, W.P., Oak Ridge National Laboratory, P.O. Box X, Oak Ridge, Tennessee 37830

THATCHER, J., Homestake Mining Company, 7625 West 5th Ave., Lakewood, Colorado 80226

THOMASSON, W.N., Dr., Technology Assessment, US Department of Energy, Washington, D.C. 20545

TOPP, A., State of New Mexico, P.O. Box 968, Crown Building, Santa Fe, New Mexico 87503

TORMA, A., Dr., New Mexico Institute of Mining and Technology, Socorro, New Mexico 87801

TREVATHAN, S., New Mexico EID, P.O. Box 2536, Milan, New Mexico 87201

TURNER, W.M., Dr., American Ground Water Consultants, 2300 Candelaria NE, Albuquerque, New Mexico 87107

VELASQUEZ, J., Phillips Uranium Corporation, P.O. Box 26236, Albuquerque, New Mexico 87125

WAKABAYASHI, H., Department of Chemical and Nuclear Engineering, University of New Mexico, Albuquerque, New Mexico 87131

WIENKE, C.L., Mrs., Los Alamos Scientific Laboratory, P.O. Box 1663, MS-490, Los Alamos, New Mexico 87545

WILLIAMS, R.E., Prof. of Hydrogeology, College of Mines, University of Idaho, Idaho

WILSON, L., Envirologic Systems Inc., 155 S. Madison, Denver,
    Colorado 80209

WOLFF, T., Dr., Environmental Improvement Division, P.O. Box 968,
    Santa Fe, New Mexico 87503

INTERNATIONAL ATOMIC ENERGY AGENCY
AGENCE INTERNATIONALE DE L'ENERGIE ATOMIQUE

ROBERTSON, A.M.G., 405 Europa House, 32 Plein Str., Johannesburg 2000,
    South Africa

VAN AS, D., Dr., International Atomic Energy Agency, Division of
    Nuclear Safety and Environmental Protection, P.O. Box 590,
    A-1011 Vienna, Austria

OECD NUCLEAR ENERGY AGENCY
AGENCE DE L'OCDE POUR L'ENERGIE NUCLEAIRE

WALLAUSCHEK, E., Dr., OECD Nuclear Energy Agency, 38 boulevard
    Suchet, F-75016 Paris, France

SCIENTIFIC SECRETARIAT - SECRETARIAT SCIENTIFIQUE

DUBE, J., OECD Nuclear Energy Agency, 38 boulevard Suchet,
    F-75016 Paris, France

HEUMANN, E., Ms., US Nuclear Regulatory Commission, Mail Stop,
    SS-396, Washington, D.C. 20555, United States

LOCAL ARRANGEMENTS (CONFERENCE AND ADMINISTRATION)
ORGANISATION LOCALE (CONFERENCE ET ADMINISTRATION)

NORRIS, D.E., Dr., University of Denver, Center for Public Issues,
    University Park, Denver, Colorado 80208, United States

SOME
NEW PUBLICATIONS
OF NEA

QUELQUES
NOUVELLES PUBLICATIONS
DE L'AEN

ACTIVITY REPORTS

RAPPORTS D'ACTIVITÉ

Activity Reports of the OECD
Nuclear Energy Agency (NEA)

Rapports d'activité de l'Agence de
l'OCDE pour l'Energie Nucléaire (AEN)

- 6th Activity Report (1977)

- 6ème Rapport d'Activité (1977)

Free on request - Gratuits sur demande

Annual Reports of the OECD High
Temperature Reactor Project
(DRAGON)
- 17th Report (1975-1976)

Rapports annuels du Projet OCDE de
réacteur à haute température
(DRAGON)
-17ème Rapport (1975-1976)

The Project was terminated on 31st March 1976 and
the 17th Annual Report is the last of this series

Le Projet a pris fin le 31 mars 1976 et le
17ème Rapport Annuel est le dernier de cette série

Free on request - Gratuit sur demande

Annual Reports of the OECD HALDEN
Reactor Project

Rapports annuels du Projet OCDE
de réacteur de HALDEN

- 17th Report (1976)

- 17ème Rapport (1976)

Free on request - Gratuits sur demande

■ ■ ■

Twentieth Anniversary of the
OECD Nuclear Energy Agency
- Proceedings on the NEA
Symposium on International
Co-operation in the Nuclear
Field : Perspectives and
Prospects (in preparation)

Vingtième Anniversaire de l'Agence
de l'OCDE pour l'Energie Nucléaire
- Compte rendu du Symposium de
l'AEN sur la coopération internationale dans le domaine nucléaire : bilan et perspectives
(en préparation)

Free on request - Gratuit sur demande

NEA at a Glance

Coup d'oeil sur l'AEN

Free on request - Gratuit sur demande

SCIENTIFIC AND TECHNICAL PUBLICATIONS

PUBLICATIONS SCIENTIFIQUES ET TECHNIQUES

NUCLEAR FUEL CYCLE	LE CYCLE DU COMBUSTIBLE NUCLEAIRE

Uranium - Resources, Production and Demand	Uranium - Ressources, Production et Demande

1977
£ 4.40, $ 9.00, F 36.00

Reprocessing of Spent Nuclear Fuels in OECD Countries	Retraitement du combustible nucléaire dans les pays de l'OCDE

1977
£ 2.50, $ 5.00, F 20.00

Nuclear Fuel Cycle Requirements and supply considerations, through the long-term	Besoins liés au cycle du combustible nucléaire et considérations sur l'approvisionnement à long terme

1978
£ 4.30, $ 8.75, F 35.00

World Uranium Potential - An International Evaluation (in preparation)	Potentiel mondial en uranium Une évaluation internationale (en préparation)

1978

■ ■ ■

RADIATION PROTECTION	RADIOPROTECTION

Estimated Population Exposure from Nuclear Power Production and Other Radiation Sources	Estimation de l'exposition de la population aux rayonnements résultant de la production d'énergie nucléaire et provenant d'autres sources

1976
£ 1.60, $ 3.50, F 14.00

Personal Dosimetry and Area Monitoring Suitable for Radon and Daughter Products (Proceedings of the NEA Specialist Meeting, Elliot Lake, Canada)	Dosimétrie individuelle et surveillance de l'atmosphère en ce qui concerne le radon et ses produits de filiation (Compte rendu d'une réunion de spécialistes de l'AEN, Elliot Lake, Canada)

1976
£ 6.80, $ 14.00, F 56.00

Iodine 129   Iode-129
(Proceedings of an NEA Specialist   (Compte rendu d'une réunion de
Meeting, Paris)   spécialistes de l'AEN, Paris)

1977
£ 3.40, $ 7.00, F 28.00

Recommendations for Ionization   Recommandations relatives aux
Chamber Smoke Detectors in   détecteurs de fumée à chambre
Implementation of Radiation   d'ionisation en application des
Protection Standards   normes de radioprotection

1977
Free on request - Gratuit sur demande

■ ■ ■

RADIOACTIVE WASTE MANAGEMENT   GESTION DES DECHETS RADIOACTIFS

Monitoring of Radioactive   Contrôle des effluents radioactifs
Effluents   (Compte rendu du Séminaire de
(Proceedings of the Karlsruhe   Karlsruhe)
Seminar)

1974
£ 4.40, $ 11.00, F 44.00

Management of Plutonium-   Gestion des déchets solides
Contaminated Solid Wastes   contaminés par du plutonium
(Proceedings of the Marcoule   (Compte rendu du Séminaire de
Seminar)   Marcoule)

1974
£ 3.80, $ 9.50, F 38.00

Bituminization of Low and Medium   Conditionnement dans le bitume des
Level Radioactive Wastes   déchets radioactifs de faible et
(Proceedings of the Antwerp   de moyenne activités
Seminar)   (Compte rendu du Séminaire d'Anvers)

1976
£ 4.70, $ 10.00, F 42.00

Objectives, Concepts and   Objectifs, concepts et stratégies
Strategies for the Management   en matière de gestion des déchets
of Radioactive Waste Arising   radioactifs résultant des program-
from Nuclear Power Programmes   mes nucléaires de puissance
(Report by an NEA Group of   (Rapport établi par un Groupe
Experts)   d'experts de l'AEN)

1977
£ 8.50, $ 17.50, F 70.00

Treatment, Conditioning and   Traitement, conditionnement et
Storage of Solid Alpha-Bearing   stockage des déchets solides alpha
Waste and Cladding Hulls   et des coques de dégainage
(Proceedings of the NEA/IAEA   (Compte rendu du Séminaire
Technical Seminar, Paris)   technique AEN/AIEA, Paris)

1977
£ 7.30, $ 15.00, F 60.00

Storage of Spent Fuel Elements
(Proceedings of the Madrid
Seminar)

Stockage des éléments combustibles
irradiés
(Compte rendu du Séminaire de
Madrid)

1978
£ 7.30, $ 15.00, F 60.00

Management, Stabilisation and
Environmental Impact of Uranium
Mill Tailings
(Proceedings of the Albuquerque
Seminar, United States)

Gestion, stabilisation et incidence
sur l'environnement des résidus de
traitement de l'uranium
(Compte rendu du Séminaire
d'Albuquerque, Etats-Unis)

1978
£ 9.80, $ 20.00, F 80.00

In Situ Heating Experiments in
Geological Formations
(Proceedings of the Ludvika
Seminar, Sweden)

Expériences de dégagement de
chaleur in situ dans les formations
géologiques
(Compte rendu du Séminaire de
Ludvika, Suède)

1978
in preparation - en préparation

■ ■ ■

SAFETY

SURETE

Safety of Nuclear Ships
(Proceedings of the Hamburg
Symposium)

Sûreté des navires nucléaires
(Compte rendu du Symposium de
Hambourg)

1978
£ 17.00, $ 35.00, F 140.00

■ ■ ■

SCIENTIFIC INFORMATION

INFORMATION SCIENTIFIQUE

Generalized Data Management
Systems and Scientific
Information
(Report of the Specialist Study)

Systèmes de gestion de bases de
données et information scientifique
(Rapport d'étude de spécialistes)

1978
Free on request - Gratuit sur demande

Neutron Physics and Nuclear Data
for Reactors and other Applied
Purposes
(Proceedings of the Harwell
International Conference)
(in preparation)

La physique neutronique et les
données nucléaires pour les réac-
teurs et autres applications
(Compte rendu de la Conférence
Internationale de Harwell)
(en préparation)

1978
approximative price - prix approximatif F 210

LEGAL PUBLICATIONS

PUBLICATIONS JURIDIQUES

Convention on Third Party Liability in the Field of Nuclear Energy - incorporating provisions of Additional Protocole of January 1964	Convention sur la responsabilité civile dans le domaine de l'énergie nucléaire - Texte incluant les dispositions du Protocole addition- nel de janvier 1964

1960

Free on request - Gratuit sur demande

Nuclear Legislation, Analytical Study : "Nuclear Third Party Liability" (revised version)	Législations nucléaires, étude analytique : "Responsabilité civile nucléaire" (version révisée)

1977

£ 6.00, $ 12.50, F 50.00

Nuclear Law Bulletin (Annual Subscription - two issues and supplements)	Bulletin de Droit Nucléaire (Abonnement annuel - deux numéros et suppléments)

£ 4.40, $ 9.00, F 36.00

Index of the first twenty issues of the Nuclear Law Bulletin	Index des vingt premiers numéros du Bulletin de Droit Nucléaire

Free on request - Gratuit sur demande

Licensing Systems and Inspection of Nuclear Installations in NEA Member Countries (two volumes)	Régime d'autorisation et d'inspec- tion des installations nucléaires dans les pays de l'AEN (deux volumes)

Free on request - Gratuit sur demande

NEA Statute	Statuts AEN

Free on request - Gratuit sur demande

■ ■ ■

OECD SALES AGENTS
DÉPOSITAIRES DES PUBLICATIONS DE L'OCDE

ARGENTINA – ARGENTINE
Carlos Hirsch S.R.L., Florida 165,
BUENOS-AIRES, Tel. 33-1787-2391 Y 30-7122

AUSTRALIA – AUSTRALIE
International B.C.N. Library Suppliers Pty Ltd.,
161 Sturt St., South MELBOURNE, Vic. 3205. Tel. 699-6388
P.O.Box 202, COLLAROY, NSW 2097. Tel. 982 4515

AUSTRIA – AUTRICHE
Gerold and Co., Graben 31, WIEN 1. Tel. 52.22.35

BELGIUM – BELGIQUE
Librairie des Sciences,
Coudenberg 76-78, B 1000 BRUXELLES 1. Tel. 512-05-60

BRAZIL – BRÉSIL
Mestre Jou S.A., Rua Guaipá 518,
Caixa Postal 24090, 05089 SAO PAULO 10. Tel. 261-1920
Rua Senador Dantas 19 s/205-6, RIO DE JANEIRO GB.
Tel. 232-07. 32

CANADA
Renouf Publishing Company Limited,
2182 St. Catherine Street West,
MONTREAL, Quebec H3H 1M7 Tel. (514) 937-3519

DENMARK – DANEMARK
Munksgaards Boghandel,
Nørregade 6, 1165 KØBENHAVN K. Tel. (01) 12 69 70

FINLAND – FINLANDE
Akateeminen Kirjakauppa
Keskuskatu 1, 00100 HELSINKI 10. Tel. 625.901

FRANCE
Bureau des Publications de l'OCDE,
2 rue André-Pascal, 75775 PARIS CEDEX 16. Tel. 524.81.67
Principal correspondant :
13602 AIX-EN-PROVENCE : Librairie de l'Université.
Tel. 26.18.08

GERMANY – ALLEMAGNE
Verlag Weltarchiv G.m.b.H.
D 2000 HAMBURG-36, Neuer Jungfernstieg 21.
Tel. 040-35-62-500

GREECE – GRÈCE
Librairie Kauffmann, 28 rue du Stade,
ATHÈNES 132. Tel. 322.21.60

HONG-KONG
Government Information Services,
Sales and Publications Office, Beaconsfield House, 1st floor,
Queen's Road, Central.   Tel. H-233191

ICELAND – ISLANDE
Snaebjörn Jónsson and Co., h.f.,
Hafnarstraeti 4 and 9, P.O.B. 1131, REYKJAVIK.
Tel. 13133/14281/11936

INDIA – INDE
Oxford Book and Stationery Co.:
NEW DELHI, Scindia House. Tel. 45896
CALCUTTA, 17 Park Street. Tel. 240832

IRELAND - IRLANDE
Eason and Son, 40 Lower O'Connell Street,
P.O.B. 42, DUBLIN 1. Tel. 74 39 35

ISRAËL
Emanuel Brown: 35 Allenby Road, TEL AVIV. Tel. 51049/54082
also at:
9, Shlomzion Hamalka Street, JERUSALEM. Tel. 234807
48, Nahlath Benjamin Street, TEL AVIV. Tel. 53276

ITALY – ITALIE
Libreria Commissionaria Sansoni:
Via Lamarmora 45, 50121 FIRENZE. Tel. 579751
Via Bartolini 29, 20155 MILANO. Tel. 365083
Sub-depositari:
Editrice e Libreria Herder,
Piazza Montecitorio 120, 00 186 ROMA. Tel. 674628
Libreria Hoepli, Via Hoepli 5, 20121 MILANO. Tel. 865446
Libreria Lattes, Via Garibaldi 3, 10122 TORINO. Tel. 519274
La diffusione delle edizioni OCSE è inoltre assicurata dalle migliori
librerie nelle città più importanti.

JAPAN – JAPON
OECD Publications Center,
Akasaka Park Building, 2-3-4 Akasaka, Minato-ku,
TOKYO 107. Tel. 586-2016

KOREA - CORÉE
Pan Korea Book Corporation,
P.O.Box nº101 Kwangwhamun, SÉOUL. Tel. 72-7369

LEBANON – LIBAN
Documenta Scientifica/Redico,
Edison Building, Bliss Street, P.O.Box 5641, BEIRUT.
Tel. 354429–344425

MEXICO & CENTRAL AMERICA
Centro de Publicaciones de Organismos Internacionales S.A.,
Av. Chapultepec 345, Apartado Postal 6-981
MEXICO 6, D.F. Tel. 533-45-09

THE NETHERLANDS – PAYS-BAS
Staatsuitgeverij
Chr. Plantijnstraat
'S-GRAVENHAGE. Tel. 070-814511
Voor bestillingen: Tel. 070-624551

NEW ZEALAND – NOUVELLE-ZÉLANDE
The Publications Manager,
Government Printing Office,
WELLINGTON: Mulgrave Street (Private Bag),
World Trade Centre, Cubacade, Cuba Street,
Rutherford House, Lambton Quay, Tel. 737-320
AUCKLAND: Rutland Street (P.O.Box 5344), Tel. 32.919
CHRISTCHURCH: 130 Oxford Tce (Private Bag), Tel. 50.331
HAMILTON: Barton Street (P.O.Box 857), Tel. 80.103
DUNEDIN: T & G Building, Princes Street (P.O.Box 1104),
Tel. 78.294

NORWAY – NORVÈGE
Johan Grundt Tanums Bokhandel,
Karl Johansgate 41/43, OSLO 1. Tel. 02-332980

PAKISTAN
Mirza Book Agency, 65 Shahrah Quaid-E-Azam, LAHORE 3.
Tel. 66839

PHILIPPINES
R.M. Garcia Publishing House, 903 Quezon Blvd. Ext.,
QUEZON CITY, P.O.Box 1860 – MANILA. Tel. 99.98.47

PORTUGAL
Livraria Portugal, Rua do Carmo 70-74, LISBOA 2. Tel. 360582/3

SPAIN – ESPAGNE
Mundi-Prensa Libros, S.A.
Castelló 37, Apartado 1223, MADRID-1. Tel. 275.46.55
Libreria Bastinos, Pelayo, 52, BARCELONA 1. Tel. 222.06.00

SWEDEN – SUÈDE
AB CE Fritzes Kungl Hovbokhandel,
Box 16 356, S 103 27 STH, Regeringsgatan 12,
DS STOCKHOLM. Tel. 08/23 89 00

SWITZERLAND – SUISSE
Librairie Payot, 6 rue Grenus, 1211 GENÈVE 11. Tel. 022-31.89.50

TAIWAN – FORMOSE
National Book Company,
84-5 Sing Sung Rd., Sec. 3, TAIPEI 107. Tel. 321.0698

UNITED KINGDOM – ROYAUME-UNI
H.M. Stationery Office, P.O.B. 569,
LONDON SEI 9 NH. Tel. 01-928-6977, Ext. 410
or
49 High Holborn, LONDON WC1V 6 HB (personal callers)
Branches at: EDINBURGH, BIRMINGHAM, BRISTOL,
MANCHESTER, CARDIFF, BELFAST.

UNITED STATES OF AMERICA
OECD Publications Center, Suite 1207, 1750 Pennsylvania Ave.,
N.W. WASHINGTON, D.C.20006. Tel.   (202)724-1857

VENEZUELA
Libreria del Este, Avda. F. Miranda 52, Edificio Galipán,
CARACAS 106. Tel. 32 23 01/33 26 04/33 24 73

YUGOSLAVIA – YOUGOSLAVIE
Jugoslovenska Knjiga, Terazije 27, P.O.B. 36, BEOGRAD.
Tel. 621-992

Les commandes provenant de pays où l'OCDE n'a pas encore désigné de dépositaire peuvent être adressées à :
OCDE, Bureau des Publications, 2 rue André-Pascal, 75775 PARIS CEDEX 16.
Orders and inquiries from countries where sales agents have not yet been appointed may be sent to:
OECD, Publications Office, 2 rue André-Pascal, 75775 PARIS CEDEX 16.

PUBLICATIONS DE L'OCDE, 2, rue André-Pascal, 75775 Paris Cedex 16 - No.40.937 1978

IMPRIMÉ EN FRANCE

Proceedings of the

SEMINAR ON MANAGEMENT, STABILISATION AND ENVIRONMENTAL IMPACT OF URANIUM MILL TAILINGS

Compte rendu du

SÉMINAIRE SUR LA GESTION, LA STABILISATION ET L'INCIDENCE SUR L'ENVIRONNEMENT DES RÉSIDUS DES TRAITEMENTS DE L'URANIUM

organised by
THE OECD NUCLEAR ENERGY AGENCY
organisé par
L'AGENCE DE L'OCDE POUR L'ÉNERGIE NUCLÉAIRE
in/à ALBUQUERQUE, UNITED STATES
at the invitation of/à l'invitation de
THE NUCLEAR REGULATORY COMMISSION
THE ENVIRONMENTAL PROTECTION AGENCY
THE DEPARTMENT OF ENERGY

The Organisation for Economic Co-operation and Development (OECD) was set up under a Convention signed in Paris on 14th December, 1960, which provides that the OECD shall promote policies designed:

— to achieve the highest sustainable economic growth and employment and a rising standard of living in Member countries, while maintaining financial stability, and thus to contribute to the development of the world economy;
— to contribute to sound economic expansion in Member as well as non-member countries in the process of economic development;
— to contribute to the expansion of world trade on a multilateral, non-discriminatory basis in accordance with international obligations.

The Members of OECD are Australia, Austria, Belgium, Canada, Denmark, Finland, France, the Federal Republic of Germany, Greece, Iceland, Ireland, Italy, Japan, Luxembourg, the Netherlands, New Zealand, Norway, Portugal, Spain, Sweden, Switzerland, Turkey, the United Kingdom and the United States.

The OECD Nuclear Energy Agency (NEA) was established on 20th April 1972, replacing OECD's European Nuclear Energy Agency (ENEA) on the adhesion of Japan as a full Member.

NEA now groups all the European Member countries of OECD and Australia, Canada, Japan, and the United States. The Commission of the European Communities takes part in the work of the Agency.

The primary objectives of NEA are to promote co-operation between its Member governments on the safety and regulatory aspects of nuclear development, and on assessing the future role of nuclear energy as a contributor to economic progress.

This is achieved by:

— *encouraging harmonisation of governments' regulatory policies and practices in the nuclear field, with particular reference to the safety of nuclear installations, protection of man against ionising radiation and preservation of the environment, radioactive waste management, and nuclear third party liability and insurance;*
— *keeping under review the technical and economic characteristics of nuclear power growth and of the nuclear fuel cycle, and assessing demand and supply for the different phases of the nuclear fuel cycle and the potential future contribution of nuclear power to overall energy demand;*
— *developing exchanges of scientific and technical information on nuclear energy, particularly through participation in common services;*
— *setting up international research and development programmes and undertakings jointly organised and operated by OECD countries.*

In these and related tasks, NEA works in close collaboration with the International Atomic Energy Agency in Vienna, with which it has concluded a Co-operation Agreement, as well as with other international organisations in the nuclear field.

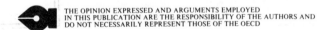